Pvt. Ltd.

Affectionately dedicated to

Simran

Mahendra

Advances in Integrated Pest and Disease Management in Horticultural Crops

Vol. 3: Ornamental, Medicinal, Aromatic & Tuber Crops

P. Parvatha Reddy

Former Director and Former Head,
Division of Entomology & Nematology,
Indian Institute of Horticultural Research
Bangalore – 560 089

2009

Studium Press (India) Pvt. Ltd.

Advances in Integrated Pest and Disease Management in Horticultural Crops

Vol. 3: Ornamental, Medicinal, Aromatic & Tuber Crops

ISBN: 978-93-80012-18-6

Published by:
Studium Press (India) Pvt. Ltd.
4735/22, 2nd Floor, Prakash Deep Building
(Near Delhi Medical Association),
Ansari Road, Darya Ganj, New Delhi-110 002
Tel.: 23240257, 65150447; Fax: 91-11-23240273;
jngovil@gmail.com; jngovil@hotmail.com

Printed at:
Salasar Imaging Systems
C-7/5, Lawrence Road Indl. Area,
Delhi-110035

Dr. M.V. Rao, MLC, Govt. of A.P. President, Agri Biotech Foundation Special Director-General, ICAR (Retd) Special Secretary, DARE, GOI (Retd) Vice-Chancellor, ANGRAU (Retd) Ex-President, Andhra Pradesh Academy of Sciences

Plot No. 4, BHEL Enclave Akbar Road, Bowenpally, Secunderabad 500 009, Phone: (R) 040-27954748/27955795, (O) 040 24002415 (Cell) 98494 77088; Fax: 040-24002417

MVR/2008/ **10 September 2008**

FOREWORD

Floriculture is estimated to cover an area of 126,000 ha with production of 694,000 MT of loose flowers and 2762 million numbers of cut flowers during 2006. Many corporate houses (more than 100) have set up 100% export oriented units with an investment of more than Rs. 1000 crores. With this, India is emerging as one of the powerful forces in world flower trade.

Medicinal crops form one of the important groups due to their demand in various pharmaceutical industries and also to earn foreign exchange by way of export. The export of these plants and their products has a tremendous potential in advanced countries like Europe, USA and Japan. Destination of foreign tourists is now aimed at India for health treatment. Kerala and Gujarat have become their favourite destinations since Ayurvedic treatment is very popular in these two states.

Aromatic crops occupy an area of 0.262 million ha with production of 0.202 million tonnes. In recent years, there has been an increased interest in the cultivation of aromatic plants to meet the requirements of cosmetic, flavouring and perfumery industries. The essential oils, are now becoming one of the more important export items from many developing countries of Asia. The annual exports of the derivatives of aromatic plants are to the tune of Rs. 600-700 million.

The annual tropical root crop production (potato, sweet potato and cassava) is in the range of 11.835 million MT grown in 1.909 million ha with productivity of 19.7 MT/ha. Tuber crops form an important part of the diet in many areas of the tropics and subtropics. They are mostly grown for the carbohydrates stored in underground storage organs viz., roots or rhizomes.

The horticultural sector in 2005-2006 has contributed around 28% of GDP in agriculture. Vegetables, fruits, plantation crops and spices contributed 59.8, 30.9, 6.5 and 2.1%, respectively of total horticulture production. About 500 ha of climatic control greenhouse are available for growing quality flowers for export. Many corporate houses (more than 100) have set up 100% export oriented units with an investment of more than Rs. 1000 crores. The cut flowers followed by live plants, cut foliage, seeds and bulbs have got an immense potential for export to Western countries where the cost of production is comparatively high. India exports roses, gladioli, carnations, chrysanthemums, lilies and jasmines to Gulf, Germany, France, England, Netherlands and Singapore. Our share in the export market in floriculture has registered a sharp increase from Rs.144.5 million in 1991-92 to Rs.2994.1 million in 2005-06. For this, focus should be more on the quality and phytosanitation aspects than on quantity.

Of the several factors responsible for lower productivity of ornamental, medicinal, aromatic and tuber crops, pests (insect pests, diseases and nematode parasites) are considered as important limiting factors. Some pest problems are of national importance which causes significant losses to ornamental, medicinal, aromatic and tuber crops. The annual losses due to pests to all the crops in India was estimated at Rs.60,000 million in 1983, which at today's prices could exceed Rs.200,000 million. Use of pest avoidance tactics, enhancement of biological pest suppression and adoption of other non-chemical methods of pest management would help to over come much of the pest problems. Use of pesticides when the pest density reaches ETL only needs to be advocated to reduce expenses to the grower.

The information on insect pests, diseases and nematode pests on ornamental, medicinal, aromatic and tuber crops is very much scattered. There is no book at present which comprehensively and exclusively deals with the above aspects on these crops. The present book deals with integrated pest and disease management in ornamental, medicinal, aromatic and tuber crops in detail using regulatory, physical, cultural, chemical, biological, host plant resistance and integrated methods. The book is extensively illustrated with excellent quality photographs enhancing the quality of publication. The utility of the book is further enhanced by chapters like pesticide residues, allied topics in pest management and sources of critical inputs for IPM. The book is written in lucid style, easy to understand language along with adoptable recommendations involving eco-friendly components of IPM. This book is a practical guide to growers of ornamental, medicinal, aromatic and tuber crops. Further, it is a useful reference to students, researchers and extension workers. The book can be used for teaching undergraduate and postgraduate courses as well.

In view of greater emphasis being given by the Government of India to horticulture by establishing horticulture mission and exclusive State Horticultural Universities, there is an urgent need for good text books to teach courses on different disciplines of horticulture. In this context this book written by Dr P Parvatha Reddy is of immense help to all those concerned with ornamental, medicinal, aromatic and tuber crops. With vast experience of more than three decades working with pests of horticultural crops, Dr Reddy has taken up this noble task single handedly and brought out an excellent publication which I am sure will be useful to all those who are interested in these crops – the farmers, scientists, extension workers, students and policy makers. He deserves all appreciation and commendation.

(M.V. Rao)

PREFACE

Floriculture is estimated to cover an area of 126,000 ha with production of 694,000 MT of loose flowers and 2762 million numbers of cut flowers during 2006. Another 500 ha of climatic control greenhouse are available for growing quality flowers for export. The area under cut flowers has increased significantly in recent years. Many corporate houses (more than 100) have set up 100% export oriented units with an investment of more than Rs. 1000 crores. India is bestowed with natural advantages like favourable weather conditions, soil, cheap labour which are likely to make a real mark in international trade. With this, India is emerging as one of the powerful forces in world flower trade.

Medicinal crops form one of the important groups due to their demand in various pharmaceutical industries and also to earn foreign exchange by way of export. The potential for earning foreign exchange by India from the exports of medicinal and aromatic plants is estimated over US$ 3,000 million per annum. The export of these plants and their products has a tremendous potential in advanced countries like Europe, USA and Japan. The international market for medicinal plants related trade is estimated at US$ 60 billion/year having a growth rate of 7% per annum. Destination of foreign tourists is now aimed at India for health treatment. Kerala and Gujarat have become their favourite destinations since Ayurvedic treatment is very popular in these two states.

In recent years, there has been an increased interest in the cultivation of aromatic plants to meet the requirements of cosmetic, flavouring and perfumery industries. Aromatic plants and their products, particularly the essential oils, are now becoming one of the more important export items from many developing countries of Asia. India has enjoyed a pre-eminent position in the manufacture of superior perfumes and aromatics by using essential oils. In the world market, India stands in second position. Most of the essential oils produced though marketed within the country, but a sizeable amount has also been exported. The annual exports of the derivatives of aromatic plants are to the tune of Rs. 600-700 million.

Tuber crops form an important part of the diet in many areas of the tropics and subtropics. They are most important after cereals and grain legumes. They are mostly grown for the carbohydrates stored in underground storage organs *viz.*, roots or rhizomes. Tuber crops are a major source of food and calories in many tropical countries. The annual tropical root crop production (potato, sweet potato and cassava) is in the range of 11.835 million MT in 1.909 million ha with productivity of 19.7 MT/ha.

Of the several factors responsible for lower productivity of ornamental, medicinal, aromatic and tuber crops, pests (insect pests, diseases and nematode parasites) are considered as important limiting factors. The annual losses due to pests to all the crops in India was estimated at Rs.60,000 million in 1983, which at today's prices could exceed Rs.200,000 million. Use of pest avoidance tactics, enhancement of biological pest suppression and adoption of other non-chemical methods of pest management would help to over come much of the pest problems. Use of pesticides when the pest density reaches ETL only needs to be advocated to reduce expenses to the grower.

The information on pests (insects, mites and vertebrate pests), diseases (fungal, bacterial, viral/mycoplasma/phytoplasma diseases) and nematode pests on ornamental, medicinal, aromatic and tuber crops is very much scattered. There is no book at present which comprehensively and exclusively deals with the above aspects on these crops. The present book deals with integrated pest, disease and nematode management in ornamental, medicinal, aromatic and tuber crops in detail using regulatory, physical, cultural, chemical, biological, host plant resistance and integrated methods. The book is extensively illustrated with excellent quality photographs enhancing the quality of publication. The utility of the book is further enhanced by chapters like pesticide residues, allied topics in pest management and sources of critical inputs for IPM. The book is written in lucid style, easy to understand language along with adoptable recommendations involving eco-friendly components of IPM. This book is a practical guide to growers of ornamental, medicinal, aromatic and tuber crops. Further, it is a useful reference to students, researchers, extension workers and policy makers. The book can be used for teaching undergraduate and postgraduate courses as well. Suggestions to improve the contents of the book are most welcome (E-mail: reddy_parvatha@yahoo.com).

In view of greater emphasis being given by the Government of India to horticulture by establishing horticulture mission and exclusive State Horticultural Universities, there is an urgent need for good text books to teach courses on different disciplines of horticulture. In this context the present book is of immense help to teach courses on plant protection in ornamental, medicinal, aromatic and tuber crops.

I profusely thank Dr. K. Srinivasan, Chief Agricultural Economist, TAFE, Chennai and Dr. B. Jhansi Rani, Division of Entomology and Nematology, Indian Institute of Horticultural Research, Bangalore who have kindly made available many of the photographs included in this book. The publisher, Studium Press LLC, USA deserves commendation for their professional contribution.

Bangalore
February 20, 2009

P. Parvatha Reddy
34, UAS Layout, Ist Main, 7th Cross
Sanjay Nagar, Bangalore-560 094

CONTENTS

1 ORNAMENTAL CROPS

India is known for growing of traditional flowers such as jasmine, marigold, chrysanthemum, tuberose, crossandra and aster. Commercial cultivation of cut flowers—rose, orchids, gladiolus, carnation, gerbera, anthurium and lilies—has also become popular. The important flower-growing states are Tamil Nadu, Karnataka, Andhra Pradesh, Maharashtra, West Bengal, Gujarat, Sikkim, Haryana, Orissa, Uttar Pradesh and Delhi.

With the liberalization of Indian economy, the domestic floricultural trade, which hitherto remained an unorganized sector, became an organized sector with the active participation of corporate giants. The establishment of state-of-art hi-tech floricultural units in early 90's for intensive cultivation of cut flowers contributes about Rs. 300 crores of valuable foreign exchange today. Various government agencies are actively involved in the promotion of this sector and are highly committed to resolve the teething problems.

The world demand for floricultural products is approximately US$ 40 billions and cut flowers contribute nearly 60% (US$ 25 billions) to the global trade. India's share in the world flower market is about 0.7% which is contributed mainly from open cultivation.

Floriculture is estimated to cover an area of 126,000 ha with production of 694,000 MT of loose flowers and 2762 million numbers of cut flowers during 2006. The area under cut flowers has increased significantly in recent years and the production has reached 2762 million numbers. Another 500 ha of climatic control greenhouse are available for growing quality flowers for export. Many corporate houses (more than 100) have set up 100% export oriented units with an investment of more than Rs. 1000 crores. India is bestowed with natural advantages like favourable weather conditions, soil, cheap labour which are likely to make a real mark in international trade. With this, India is emerging as one of the powerful forces in world flower trade.

Floriculture is an intensive type of agriculture wherein the income per unit area is much higher than any other agricultural product if it is done in a scientific way. The area used for production of ornamental plants is low, but the cash returns are high because of the intensive cropping. The flowers grown in large areas include jasmine, rose, marigold, chrysanthemum, crossandra, champak, tuberose, gladiolus and aster. New crops like lilies, tulips, anthurium, carnation, gerbera, etc., are also finding potential place in the foreign market. With the rapid development in urban planning and hotel industry, there has been an increasing demand for cut flowers and ornamentals. The cut flowers followed by live plants, cut foliage, seeds and bulbs have got an immense potential for export to Western countries where the cost of production is comparatively high. India exports roses, gladioli, carnations, chrysanthemums, lilies and jasmines to Gulf, Germany, France, England, Netherlands and Singapore. Our share in the export market has registered a sharp increase from Rs. 14.45 crores in 1991–92 to Rs. 299.41 crores in 2005–06.

Table 1.1. State-Wise Area and Production of Flowers (1999–2000)

State/UT's	Area (in ha)	Production (in MT)	
		Loose	Cut
Andhra Pradesh	18087	126609	—
Delhi	3450	21007	—
Haryana	2250	40500	760
Himachal Pradesh	133.3	976	181
Jammu & Kashmir	42	7	17.4
Karnataka	20801	124440	—
Madhya Pradesh	1387	832	—
Maharashtra	6600	28336	—
Orissa	175	868.5	9.3
Tamil Nadu	18120	142250	—
U.P.Plain	160	—	160
West Bengal	13227	14644	5678
Others	4174.7	8723.5	1.0
Total	**88607**	**509193**	**6806**

Pests and their Impact on Ornamental Plants: Large scale cultivation of ornamental crops face many pest problems. Attack by insects, mites and other pests is one of the key limitations in production of these crops. Pest scenario on ornamental crops varies with the growing conditions. Closed structures and uniform growing conditions favour multiplication of pests round the year on protected ornamentals like rose, carnation, gerbera, chrysanthemum, orchids and anthurium. Red spider mites, thrips, whiteflies, aphids and leaf miners are serious pests on these crops. Open cultivated ornamentals are mainly damaged by bud borers, foliage caterpillars, beetles and weevils, besides thrips and aphids. Damage by insects causes severe yield loss in terms of quantity by making holes and eating on leaves, buds, flowers and fruits. Curling, distortion, discolouration, browning and drying of plant parts as a result of damage cause qualitative loss.

Strategies adopted to control pest problems in ornamental crops mainly include use of insecticides specially imported and costly ones in case of protected ornamentals. Cultural methods like pruning, removal and destruction of infested parts and raking of soil are followed in some crops. Plant products such as neem and pongamia oils and kernel extracts are also used to some extent.

Understanding of the pest incidence, biology, distribution and symptoms of damage caused and the present control methods followed is important to develop ideal integrated management methods in these crops.

Nematodes and their Impact on Ornamental Plants: Monoculture, the common feature of the ornamental crops, provides congenial environment for one or the other nematode, the population of which if unchecked may touch an alarmingly high level and may even wipe out the whole crop. Apart from this, even a low nematode population may render the planting stock like tuberose and gladioli bulbs, unmarketable. The infested planting material may also act as a carrier of some nematode genera/species of quarantine significance to newer areas. In USA, losses in ornamentals due to nematodes were estimated to the tune of $60 million annually (Hague, 1972). Throughout the world in ornamental crops, plant nematodes are responsible for 11.1% losses (Sasser & Freckman, 1987). Information about the crop losses in ornamental crops in India are presented in Table 1.2.

Table 1.2. Avoidable yield losses in ornamental, medicinal and aromatic crops due to plant parasitic nematodes

Crop	Nematode(s)	Yield loss (%)	Reference(s)
Crossandra	*M. incognita*	21.64	Khan & Parvatha Reddy, 1994
Tuberose	*M. incognita*	13.78	Khan & Parvatha Reddy, 1994
Patchouli	*M. incognita*	47.00	Prasad & Reddy, 1984
Davana	*M. incognita*	50	Haseeb & Pandey, 1989a
Cymbopogon	*M. incognita*	20	Pandey, 1994a
Carnation	*M. incognita*	27	Nagesh & Parvatha Reddy, 2000
Gerbera	*M. incognita*	31	Nagesh & Parvatha Reddy, 2000
Coleus forskohlii	*M. incognita*	70.2	Senthamarai *et al.*, 2006c
Chethikoduveli (*Plumbago rosea*)	*M. incognita*	29.00–43.96	Santhosh Kumar & Sheela, 2004
Kacholam (*Kaempferia galanga*)	*M. incognita*	18–64	Sheela & Rajani, 1998
Colocasia	*M. incognita*	24	Anon, 1990b
Menthol mint	*M. incognita*	30	Pandey, 2003

Ornamentals like other crops, harbour a multitude of nematode fauna of which some nematodes based on their intrinsic potentials have caused tremendous economic losses. Nematodes like *Ditylenchus dipsaci* has been reported to be responsible for completely wiping out the narcissus industry of Britain (Hague, 1972). *Meloidogyne* spp. on gladioli, *Aphelenchoides ritzemabosi, Belonolaimus longicaudatus* and *Trichodorus* sp. on chrysanthemum in USA, *Xiphinema americanum* on roses in USA and *Xiphinema diversicaudatum* in Western Europe have been listed as the major pests resulting in considerable growth reduction (Hague, 1972).

A perusal of literature indicates that little information is available from India on these noxious pests. Hence, keeping in view the importance of these pests, a review at this stage would help to identify the progress made as also to pin-point our major deficiencies and gaps from the view point of practical approach to management of these pests.

1.1. ROSE

Rose (*Rosa* spp.) is the most ancient and popular flower grown the world over. In India, it is commercially grown for cut flowers and rose oil. Rose flowers without stems and loose flower petals are used in traditional markets for making garlands, for offering in temples, while the florist shops sell cut roses with stems mainly for bouquets and floral arrangements. In recent times, about 60 units have been established under joint ventures around Bangalore, Pune, Nasik, Mumbai, Hyderabad, Gurgaon (Haryana), Chandigarh and Saharanpur (Uttar Pradesh) for growing roses in greenhouses for export of cut flowers to Japan, The Netherlands, Germany and other European countries. Besides, the Damask rose (*Rosa damascena*) and Edward rose (*R. bourboniana*) are cultivated for rose *attar* and other products—*gulkand, gulabjal*, and *pankkurj*. It is cultivated in about 6,000 ha area in the states of West Bengal, Uttar Pradesh, Gujarat, Haryana, Jammu and Kashmir, Tamil Nadu, Karnataka, Rajasthan, Madhya Pradesh, Andhra Pradesh and Punjab.

1.1.1. Pests

1.1.1.1. California Red Scale, *Aonidiella aurantii*

(i) Damage: Scales appear in severe form before and after the monsoon season. Both adults and young scales suck the sap from mature shoots and deplete vigour of the plant. Infested shoots bear reddish brown encrustations (Fig 1.1), bear few small flowers and dry in case of severe damage.

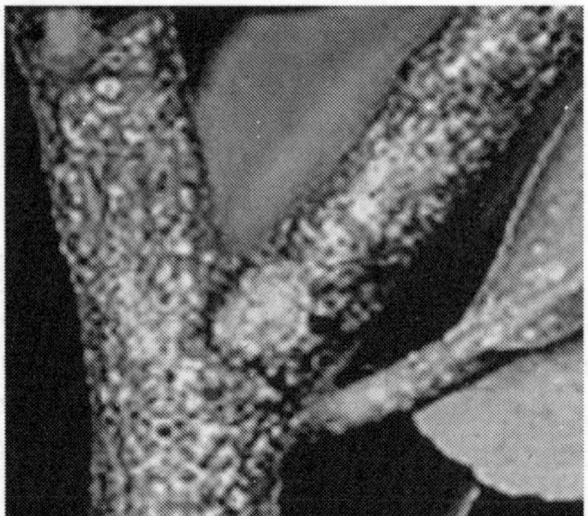

Figure 1.1. Red scale infestation on stems of rose

Besides infesting roses, these scale insects are serious pests of citrus fruit and they infest a wide range of ornamentals. They inject a toxic saliva as they feed and cause leaf-yellowing, leaf-fall, and die-back of twigs and stems.

Female red scales are circular, about 2.5 mm in diameter and have a slightly raised centre. The scale is reddish brown, and the insect under the scale is yellow. Nymphs are light-yellow, oval, and less than 0.5 mm long. They emerge from under the female scale and crawl about for a few hours before settling permanently. Scale numbers build up quickly in dry-dusty areas or dry seasons.

(ii) Management

- Selection of planting material free from scales is essential.
- Cutting and burning or burying of infested parts is also effective.
- Spraying of malathion 50 EC at 1 ml/l or chloropyriphos 20 EC at 2.5 ml/l + dimethoate 30 EC at 2 ml/l or ethion 50 EC at 1 ml/l causes significant reduction of the scale.
- Soil application of carbofuran at 1.0 kg. *a.i.*/ha (33 kg/ha of Furadan 3G) followed by application of 10% pongamia oil to infested shoots, immediately after pruning provides complete control of red scale within 3 weeks (Jhansi Rani & Jagan Mohan, 1997).

1.1.1.2. Thrips, *Rhipiphorothrips cruentatus, Scirtothrips dorsalis*

(i) Damage: Thrips are slim-winged insects 1mm in length, resembling fine black slivers of wood. Preferring light-coloured blooms and often appearing in large numbers, flowers are often left looking bruised and lusterless.

Thrips are sap-sucking insects that can cause deformities in flowers, leaves (Fig 1.2), stems, and shoots. The insects are difficult to see because not only are they small but also they hide deep in the flower or on the undersides of the leaves, which become silvery on the upper surface. Thrips are most active during spring and autumn.

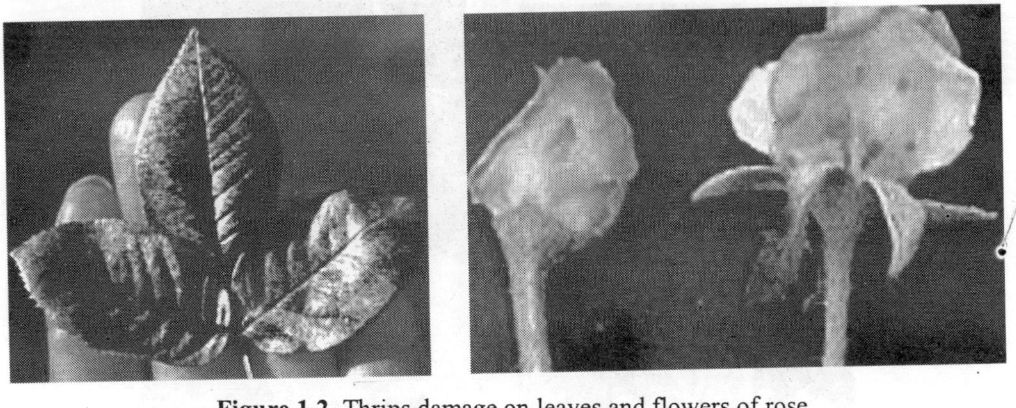

Figure 1.2. Thrips damage on leaves and flowers of rose

Female thrips lay eggs within leaf tissue which hatch out and cause damage to foliage by rasping the tissue and sucking the cell sap that oozes out. Matured nymphs pupate in the soil and remain inactive during winter.

(ii) Management

- Spraying of 0.1% acephate (75 SP at 1.5 g/l) or 0.05% dimethoate (30 EC at 2 ml/l) or 0.05% oxydemeton methyl along with 0.5% pongamia oil 2–3 times at 15 days interval with onset of new flush control thrips on open cultivated roses (Jhansi Rani & Jagan Mohan, 1997).

- In case of severe infestations, applying fipronil 5 SC at 1.5 ml/l is effective.
- Raking up of the soil during winter and drenching the soil with chlorpyrifos 20 EC at 5 ml/l kills pupae in the soil.
- Spraying pongamia oil 0.5% when early symptoms are observed control thrips in open cultivated roses.
- Spraying Vertimac (biopesticide of *Streptomyces avermitilis*), 4% NSKE, 2% neem oil, 1% jatropha oil or annona oil was found effective.
- Nair *et al.* (1991) reported effective control of thrips by soil application of carbofuran at 0.18 g *a.i.*/plant and spraying of quinalphos or phosphamidon both at 0.05%.

1.1.1.3. Aphid, *Macrosiphum rosae*

(i) Damage: Likely to be found on new shoots and buds, aphids are soft bodied insects 1–2 mm long. Often green but occasionally light-brown, and sometimes with wings, they may cover (in a colony) the complete growing tip of the plant (Fig 1.3). Aphids are most active in spring and summer and multiply at a prodigious rate feeding on the sap of the plant by piercing the plant cells via a proboscis. In large quantities they may seriously retard the growth of the plant and ruin buds. They are particularly damaging to the new shoots with subsequent damage to the emerging leaves which become malformed with much the same appearance as leaf-curl in peaches.

Figure 1.3. Aphid infestation on buds and stems of rose

Aphids feed on roses by puncturing and sucking plant fluids from the veins. Aphids cause damage by secreting large amounts of honeydew which collects on leaves and buds.

Honeydew encourages the growth of sooty mould which is unsightly and inhibits photosynthesis by reducing the amount of light reaching the leaves.

Only pregnant females survive after winter. In spring, females reproduce without mating and produce nymphs which form social broods and feed near their mothers. Winged females develop and disperse and aphid populations build up rapidly with spring growth.

(ii) Management

Small populations can be removed by hosing at high pressure, but this needs to be repeated every two or three days. For large infestations, spray with imidacloprid (such as Confidor), maldison, pyrethrums, or white oil. Home made recipes include a brew made from 100 g of fresh garlic and/or chilli, crushed and fermented for two days in two tablespoons of mineral oil. Add 500 ml of water and two table spoons of soap powder. Strain into a garden sprayer and dilute one part to 50 parts of water. The garlic acts as a repellent to further aphid attack.

- Spray *Verticillium lecanii* at 3 g/l.
- When infestation begin, spray neem or pongamia oil at 10 ml/l or dimethoate 30 EC at 2 ml/l to control aphids in open cultivated roses (Jhansi Rani & Jagan Mohan, 1997).
- If the incidence is severe, spray imidacloprid 200 SL at 0.4 ml/l or cartap hydrochloride 50 SP at 1 g/l.

Rajagopal and Kadire Gowda (1994) reported effective control of aphids with 0.1% liminool or 0.25% methyl parathion.

1.1.1.4. Bud Borer, *Helicoverpa armigera*

(i) Damage: Infestation of the bud borer is seen on open cultivated roses in severe form from January to March. Female moths lay cream coloured eggs singly on young buds. Hatched larvae bore into buds by making holes and feed on growing petals (Fig 1.4). Caterpillars also damage flowers by eating flowers and leaving excreta.

Figure 1.4. Bud borer damage on flowers of rose

(ii) Management

- Collection and killing of grown up larvae reduces further damage and population.
- NSKE 4% or neem oil 1% sprays offer protection to plants from borer damage (Jhansi Rani & Jagan Mohan, 1997).
- Spraying methyl parathion or methomyl each at 0.05% or endosulfan at 0.07% or ethofenprox at 0.01% controls the pest effectively.
- Application of *Ha* NPV at 250 LE/ha causes considerable mortality of early instar larvae.

1.1.1.5. Leaf Caterpillars, *Euproctis fraternae, Spodoptera litura, Achaea janata*

(i) Damage: Pest incidence is severe from September to October. Female moths of *Euproctis fraternae* and *Spodoptera litura* lay eggs in clusters on lower side of leaves. Early instars are gregarious in habit and cause skeletonization and drying of leaves. *Achaea janata* adults lay eggs in groups on tender parts of plants. There are four larval instars with a total developmental period of 33–41 days (Vyas, 1994). Matured larvae of *A. janata, E. fraternae* and *S. litura* caterpillars feed voraciously on leaves, buds and flowers resulting in severe defoliation and flower loss.

(ii) Management

- Deep ploughing in summer exposes pupae to predators.
- Collection and destruction of leaves with egg masses and early instars reduces further infestation.
- Setting up of light traps attract adult moths.
- Spraying of neem oil 1% or NSKE 4% checks damage by caterpillars (Jhansi Rani & Jagan Mohan, 1997).
- Complete control of leaf caterpillars, *E. fraternae* and *A. janata* was achieved with 'Thuricide' (*Bacillus thuringiensis*) dust and spray ($3x10^6$ spores/g) (Sundarababu, 1969).
- Application of *Sl* NPV at 250 LE/ha gives effective control of *S. litura*.

1.1.1.6. Chafer Beetles, *Adoretus* spp., *Apogamia* spp.

(i) Damage: Black, brown or metallic green coloured adult beetles emerge from soil with the onset of monsoon showers (May-June) after dusk. They feed on leaves, buds, flowers and tender shoots by making irregular holes and cuts (Rai, 1977) resulting in severe defoliation. Eggs are laid in the soil and grubs feed on roots.

(ii) Management

- Deep ploughing after pruning exposes eggs, grubs and pupae to natural enemies.
- NSKE 2.5% spray inhibits feeding by the beetles (Doharey & Singh, 1981).
- Fortnightly spraying of quinalphos or chloropyriphos or carbaryl each at 0.15% during evening protects foliage from beetle damage (Jhansi Rani & Jagan Mohan, 1997).
- Application of phorate or carbofuran each at 1.0 kg *a.i.*/ha kills grubs in the soil.

1.1.1.7. Weevils, *Myllocerus* spp.

(i) Damage: Weevils attack on rose starts in April and continue till October. Ash coloured adult weevils damage leaves by feeding from margins in a characteristic fashion. Eggs are deposited in the soil and grubs feed on developing roots and pupate there. Heavy infestation of the weevil results in severe foliage loss.

(ii) Management

- Deep digging after pruning exposes grubs and pupae to natural enemies.
- Application of 0.1% carbaryl or malathion dust or phorate at 1.0 kg *a.i.*/ha control soil inhabiting stages.
- Spraying of quinalphos or chloropyriphos or malathion each at 0.05% protects foliage from beetle damage (Jhansi Rani & Jagan Mohan, 1997).

1.1.1.8. Whitefly

(i) Damage: Whiteflies resemble a tiny, white 1.5–4 mm long moth. Most commonly, these whiteflies live in the upper regions of the plant. Whiteflies pierce leaf tissues (Fig 1.5) causing noticeable wilting, chlorosis, loss of leaves and/or stunted growth. In addition, they produce "honeydew" which attracts other pests and produces a medium for "sooty mold" fungus.

Figure 1.5. Whitefly infestation on rose leaves

(ii) Biology: The female whitefly will deposit up to 400 eggs in circular clusters on the undersides of leaves. Within 5–10 days, the eggs hatch into scale-like larvae which pass through several stages during the next four or so weeks, after which they pupate for about a week and emerge as a new adult to renew the cycle. Depending on temperature and other environmental conditions, however, the entire life cycle can span as little as 18 days. Optimum temperatures for infestation are 18°C to 24°C.

(iii) Management: Controlling these pests includes:

- Traps, which include sticky yellow traps, non-baited traps, light traps or pheromone traps.

- Biological controls like *Beauveria bassiana*, *Paecilomyces fumosoroseus*, or *Verticillium lecanii* fungi (the latter for greenhouse use); parasitoids like *Encarsia formosa* and *Eretmocerus californicus*; and a host of other predators like lacewings, lady beetles, and predatory mites.
- Extracts of Nicotiana plants and a variety of other organic and synthetic pesticides including neem oil, horticultural oils, pyrethrums and garlic water gives effective control. Research has shown that whiteflies can quickly mutate to pesticide-resistant strains. If using synthetic pesticides, rotate between two or three different chemical types.

1.1.1.9. Mealybugs

(i) Description: Mealybugs are soft bodied critters covered with a waxy material that resembles powder or cotton. The adult males have wings and the females look like a barnacle but with multiple legs and a tail. Size varies, but is generally 1.5–3.0 mm long. Because they do not have a hard outer covering, they are very sensitive to temperature and humidity.

(ii) Damage: Mealybugs pierce the plant, sucking its juices to the point of rapid defoliation, deformation and eventual death if not treated. In greenhouse conditions, where natural predators are scarce, control must be achieved quickly. Fortunately, their cotton-like exudates makes detection reasonably easy. Unfortunately, they tend to congregate on plant surfaces hard to reach with insecticidal sprays.

(iii) Biology: Mealybug is very prolific breeder. One species can produce up to 1000 eggs at a time, but half of that is more common. The mealybug can have as many as eight life cycles per year in southern climates which means one single female mealybug can be the progenitor of as many as 4000 offspring. It is no wonder many commercial growers throw out their infested plants rather than try to rescue them with treatment.

The female mealybug will lay eggs within a cottony mass that surrounds its body. In about a week, the eggs hatch into nymphs and begin crawling to a suitable location on the plant nearby. After four weeks of sucking on your prized specimen, those nymphs that will become males form a mass of cotton-like filaments where they change into winged adults about two weeks later. The females change into their scale-like bodies about two weeks after the males; they mate; and the process renews.

(iv) Management

- Hand removal with fingers or cotton swabs if populations are small.
- Biological control using parasitoid such as the *Leptomastix* sp. mini-wasp; and natural predators like lady bird beetles (specifically, *Cryptolaemus montrouzieri*, the "Mealybug Destroyer") and lacewings.
- Various synthetic and organic pesticides like diatomaceous earth (DE), horticultural oils, insecticidal soaps, and neem oil gives effective control.

1.1.1.10. Red Spider Mite, *Tetranychus urticae*

(i) Damage: Red spider mite, also known as two-spotted mite, is the most serious pest of rose. These arachnids prefer the underside of leaves and are difficult to see with an unaided eye. Evidence of their presence is silvering of leaves where the mites have destroyed individual leaf cells. Fine webbing and eggs on the undersides of leaves is further evidence of the presence of *Tetranychus urticae.*

Mites are sap-sucking arachnids that can cause yellowing or bronzing of leaves (Fig 1.6). If the population is large, leaf drop occurs and the health of the plant is affected.

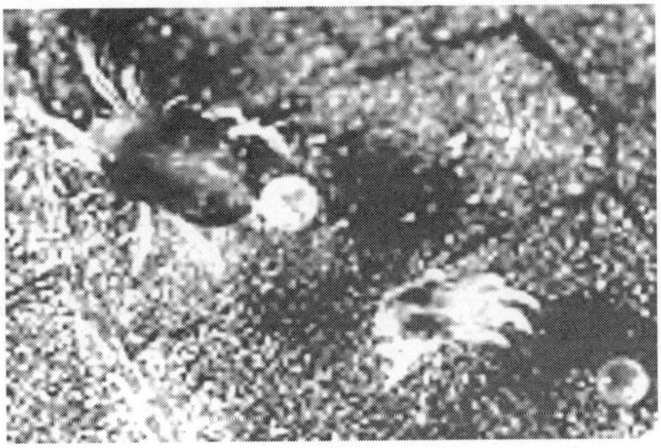

Figure 1.6. Red spider mite infestation on leaves of rose

Roses are damaged when nymphs and adults pierce the leaves and suck the contents of cells causing them to collapse and die. Heavily-infested leaves become lifeless and may become bronzed and shriveled. Plants may lose their leaves.

Females lay up to six eggs a day, totaling 70 or more. Eggs are minute, globular and almost transparent pale yellow-white. Males develop from unfertilized eggs and females develop from fertilized eggs. Egg hatch takes from three to 10 days depending on the surrounding temperature.

Newly-hatched larvae have six legs and are pale yellow-white, minute and oval. When they moult, larvae become eight-legged nymphs which mature after another two moults. Young mites take four to 12 days to mature and live for about three weeks.

In hot weather, the complete life cycle may be as short as two weeks and overlapping generations may occur.

Adult females are about 0.5 mm long and adult males are about 0.3 mm long. When feeding in summer, adult females are a yellowish-green with two pronounced dark spots, one on each side of the body. In males, these spots are less obvious and the body is smaller and tapered towards the rear. Both sexes also have two reddish eye-spots on the body.

Red spider mite is a profuse web-spinner in its active stages. Under artificial long-day length or in warm situations, such as glasshouses, red spider mite may continue to feed and reproduce through winter. Outdoors, when temperatures and day length decrease in autumn, males die and adult females stop feeding and change colour to orange-red. They usually migrate to sheltered sites where they assemble to overwinter. When temperatures increase in spring, females leave their shelters, begin to feed, and lose their red colour.

(ii) Management

- Cutting and burning of severely infested plant parts should be undertaken.
- Proper ventilation, clean cultivation and frequent irrigation to bring down temperature and to keep the pest under check.
- Spraying dicofol 18.5 EC at 2.5 ml/l or wettable sulphur 80 WP at 3 g/l (Pirone, 1978) or 0.05% oxydemeton methyl or 0.012% fluvalinate (Pokharkar *et al.*, 1986) or profenofos 50 EC at 1 ml/l or ethion 50 EC at 1 ml/l followed by pongamia or neem or jatropha oil at 5 ml/l controls mite problem.
- New miticides like vertimec at 0.0025%, flufenoxuron at 0.01%, diafenthiuron at 0.025%, profenphos at 0.05% and amitraz at 0.05% offer very good solution for the mite problem (Jhansi Rani & Jagan Mohan, 1998).
- Neem or pongamia oil 0.5 to 1% sprays should be given in evening hours and before onset of summer to avoid mite build up and better control under polyhouse conditions.
- Release of predatory mite, *Phytoseilus persimilis* at 26/m^2 successfully contains mite population.

Table 1.3. Pest management in rose

Pest	Chemicals
Aphids (*Macrosiphum rosae*)	Monocrotophos at 2 ml/l of water
Red spider mite (*Tetranychus urticae*)	Abamectin (at 0.5 ml/l of water), dicofol at 2 ml/l of water
Thrips (*Rhipiphorothrips cruentatus*)	Acephate at 2 g/l of water
Bud borer (*Helicoverpa armigera*)	Endosulfan (at 2 g/l of water), *Ha* NPV (at 400 LE/ha)
Whitefly (*Trialeurodes* spp.)	4% NSKE, Commercial neem formulations
Rose midge (*Dasineura rhodophaga*)	Monocrotophos at 2 ml/l of water
Red scale (*Aonidiella aurantii*)	In case of severe infestation, remove shallow layer of bark of stem above soil surface. Tie a cotton swab dipped in monocrotophos (10 ml/l of water) in the exposed portion. The scales will get killed. Afterwards, the exposed portion is to be swabbed with Bordeaux paste.

1.1.2. Diseases

As is the case with other plants, maintaining the health of roses makes them less susceptible to diseases and better able to withstand disease attack. Cultivars that are more resistant to certain diseases should be used. Humidity encourages fungal disease, as does poor air circulation. Avoid planting roses too closely; avoid overhead watering; and prune to maintain a fairly open structure.

1.1.2.1. Blight or Grey Mold, *Botrytis cineria*

(i) Symptoms: The disease is severe in roses during winter months. The fungus *Botrytis cinerea* affects most aboveground plant parts (flowers and buds). Typical symptoms on leaves and flowers appear as water-soaked lesions resulting in immature fading and drying of petals, which drop ultimately (Fig 1.7). The infection is generally seen on inner whorls and petals and leads to complete blight of flowers. Flowers often become spotted or may fail to open, eventually becoming covered in a grey or brown mould. Purple lesions occurring on canes are often caused by *Botrytis*. Infected buds rot on the stem and infection may progress down the stem. On petals, *B. cineria* produces pink rings. Spores are wind-borne and also can be spread on garden tools such as secateurs.

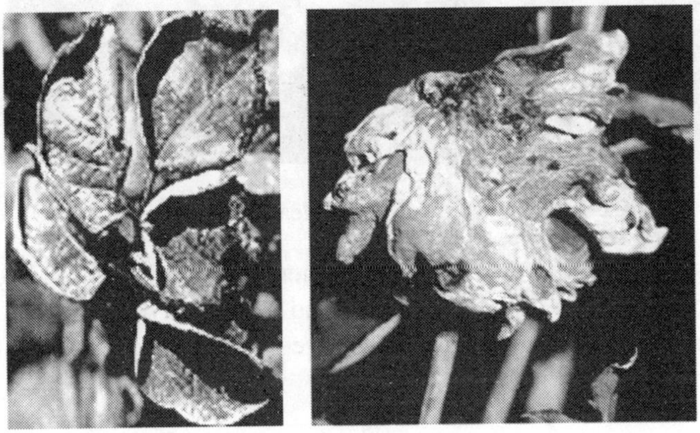

Figure 1.7. Grey mold on leaves and flowers of rose

(ii) Management

Since *Botrytis* sporulates readily on wounded and senescent tissue, these should be thoroughly removed. Infected plant debris should be removed. Bioagents like *Gliocladium roseum, Myrothecium verucaria, Trichoderma harzianum* and *Pencillium* spp. were found to be effective against rose blight. Spray carbendazim, thiophanate, triadimefan or iprodione to control the disease.

To treat *Botrytis*, spray bushes in late summer and autumn with chlorothalonil, iprodione (such as Rovral), or mancozeb. When pruning, wipe secateurs clean before progressing to the next bush, and apply fungicide immediately to protect wounds from infection.

1.1.2.2. Powdery Mildew, *Sphaerotheca pannosa* var. *rosae*

(i) Symptoms: This is a serious disease of rose in openly grown and polyhouse as well. The disease may affect all aerial parts of the plant but symptoms are more conspicuous on leaves (Fig 1.8). The disease manifests as grayish-white, powdery patches on tender leaves and flower buds. Mature leaves get malformed and flower buds fail to open and plants present a wilted appearance. The infection occurs on abaxal side of the leaves. Young canes may be infected downward from the tip and badly diseased canes are killed. Petals become discoloured, dwarfed and die.

Figure 1.8. Powdery mildew on leaves and stem of rose

The disease is serious in warm humid weather with cool nights. The fungal spores overwinter on the stems and in infected buds and fallen leaves. The disease spreads by wind-borne spores. Continuous wetting of the leaves with dew or rain do not favour disease development.

(ii) Management

To control powdery mildew, remove and destroy diseased material. Foliar application of 0.2% wettable sulphur fungicides like Sulfex, Wet-sulf, Wettasul or 0.1% systemic fungicides like benomyl or carbendazim or triademefon or fenarimol or penconazole or triforine offer good control.

Two antagonistic fungi, *Ampelomyces quisqualis* and *Pseudozyma flocculosa* are reported to control powdery mildew when their spore suspensions were sprayed on the foliage.

During the growing season, spray bushes with mancozeb, lime sulphur, or triforine.

1.1.2.3. Black Spot, *Marssonina rosae* (syn. *Diplocarpon roseae*)

(i) Symptoms: This disease is more of a problem in open-field cultivation. The characteristic symptom of the disease is the appearance of leaf spots which are coal-black, circular or irregular with black margins and yellow halos (Fig 1.9). The spots develop on either side of the leaf. Symptoms first appear on older leaves on the lower portion of plant causing severe defoliation, reducing yield and the size of flowers. On close examination with a hand lens, the spots show small, black blister-like fruiting bodies of the fungus. Infected leaves soon drop off from the plant. Continual defoliation will cause weakness, dieback or death of the plant. Some very susceptible species may have stems affected with a considerable reduction in plant vigour.

Figure 1.9. Black spots on rose leaves

The pathogen can survive in the canes. Spores are spread by wind and water-splash.

(ii) Management

Cultural practices like wider spacing, pruning the dense central portion of the canopy to allow good air circulation contribute to disease control. The debris from infected plants should be removed and burnt to minimize the inoculum potential.

It is desirable to give a dormant spray of lime-sulphur when the leaves begin to emerge. Spraying with 0.2% captan/chlorothalanil/dithionon/propineb at 15 days interval can control the disease. Foliar application of chlorothalanil/benomyl/carbendazim/propiconazole/ triforine is also effective. Lower surfaces of the leaves should be particularly covered with the fungicides.

1.1.2.4. Die-back, *Diplodia rosarum, Colletotrichum* sp.

(i) Symptoms: Die-back generally occurs after pruning. The infection occurs at pruned ends of stem (Fig 1.10) or through wounds caused by flower harvest and other operations. In severe cases, the infection may spread to the main stem and subsequently to roots thus killing the entire plant. The infected stem when split open show brownish discolouration of the internal tissues. The disease is accentuated by Digger Wasp which burrows into the pith through the cut ends and makes an empty tunnel.

(ii) Management

Pruning of infected stems 5–10 cm below the margin of infection and painting the cut ends with Bordeaux paste or rose paint (Copper carbonate, red lead and linseed oil in 4 : 4 : 5 ratio) or 0.2% carbendazim or benomyl or copper oxychloride is recommended.

Addition of 0.1% DDT or BHC to these paints helps in preventing the damage by Digger Wasp.

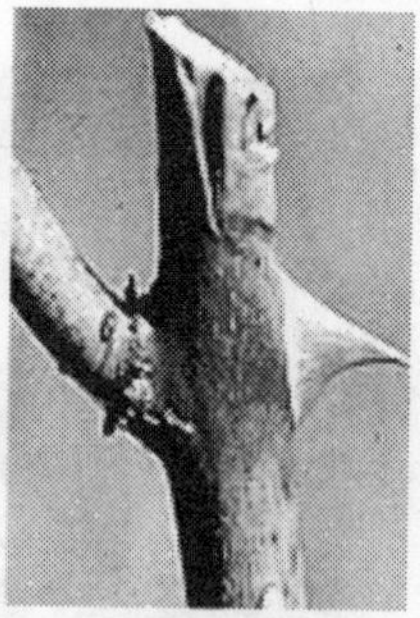

Figure 1.10. Die-back on stem of rose

Cvs. White Sign, Red Gold, Bhim, Summer Queen, Quebec, Resolute, Samba and Blue Moon are reported to be resistant.

1.1.2.5. Downy Mildew, *Peronospora sparsa*

(i) Symptoms: The fungus *Peronospora sparsa* causes downy mildew. New growth affected will be deformed. Purplish-red or dark brown, irregular, and often angular, blotches develop on leaf surfaces (Fig 1.11), and during humid weather a blue-grey, downy growth of fungus may appear on the undersides. The leaves droop, turn yellow, and may drop. Stems and calyces develop purple or blackish spots, streaks, or blotches. The infection on flower buds leads to total drying of flowers. Petals have brown, dead areas. Infected buds may produce deformed flowers. Spores are wind-borne but need free water on the leaf surface to germinate.

Figure 1.11. Defoliation due to downy mildew of rose

(ii) Management

To control downy mildew, remove and destroy diseased material and spray bushes with zineb. The disease can be controlled by spraying Al-Fosetyl or metalaxyl mancozeb or cymoxanil mancozeb or propineb all at 0.2%.

1.1.2.6. Rust, *Phragmidium mucronatum*

(i) Symptoms: Most rusts are host-specific. The fungus *Phragmidium mucronatum* that causes rust on roses does not infect any other plants. In spring and early summer, orange powdery pustules develop on the undersides of leaves (Fig 1.12), while upper surfaces become speckled yellow. These leaves drop prematurely. In late summer, black or brown spore masses appear with the orange pustules; these are the teliospores that can overwinter. The fungus spreads rapidly and can cause severe defoliation. The spores are spread by wind.

Figure 1.12. Rust pustules on rose leaves

(ii) Management

To control rose rust, remove and destroy diseased material. When new growth occurs, spray bushes with bitertanol (such as Baycor), mancozeb, lime sulphur, triforine, or zineb.

1.1.2.7. Anthracnose, *Sphaceloma rosarum*

(i) Symptoms: The fungus *Sphaceloma rosarum* causes anthracnose, which is sometimes confused with black spot on rose leaves. The differences are that anthracnose spots are smaller and initially have sharply defined margins (Fig 1.13). As the spots enlarge, the centres become ash grey and drop away, leaving the black margins intact. The leaves become red or yellow, and may drop prematurely, but not to the same degree as with black spot. As the spots enlarge, the centre becomes gray and may fall out resulting in a shot-hole appearance.

(ii) Management

To control anthracnose, remove and destroy diseased material. Spray bushes with chlorothalonil, mancozeb, triforine, or zineb.

Figure 1.13. Anthracnose leaf spots on Rose

1.1.2.8. Cane Canker, *Coniothyrium* spp., *Botryosphaeria* spp.

(i) Symptoms: Apart from *Botrytis* (see above), two of the commonest fungi to cause cane canker are *Coniothyrium* and *Botryosphaeria* species. These diseases are more commonly seen on plants stressed through poor management practices such as inadequate nutrition or irrigation. Water-borne spores enter mainly through pruning wounds or other injuries. The disease can affect any part of the stem, initially producing yellow or red spots on the bark. These enlarge into brown cankers with dark margins, and the bark cracks and becomes sunken. Tiny black fruiting structures are sometimes seen. Canker can encircle the stem or travel down it, killing the plant.

(ii) Management

To treat canker, prune stems back to healthy buds and destroy diseased material. After pruning, apply mancozeb to protect wounds from infection.

1.1.2.9. Root Rots, *Phytophthora cinnamomi, Pythium* spp.

(i) Symptoms: Roses affected by root rots wilt and may eventually die. The root ball may be water soaked and brown and none or few new white roots may be seen. *Phytophthora* root rot may be caused by more than one species of the pathogen. Common sources of *Phytophthora cinnamomi* (jarrah dieback) are infected soil or irrigation water that has been in contact with the soil, for example, dam water. Other species of *Phytophthora* may be imported in potting mixes.

Pythium affects plants in a similar manner but is more often a sign that the plants are being stressed in some other way, for example, by high salinity, low pH, or water logging.

These root diseases will often show up first after a few warm days when plants wilt because their water conducting mechanisms are disrupted. Sometimes the disease affects only one side of the plant.

(ii) Management

The *Phytophthora* disease may be suppressed with the use of phosphorous acid. The best treatment for *Pythium* is to eliminate the causes of stress.

1.1.2.10. Crown Gall, *Agrobacterium tumefaciens*

(i) Symptoms: The bacterium *Agrobacterium tumefaciens* causes crown gall. The galls that develop on stems or just under the soil surface are round and rough textured. Round growths about 5 cm in diameter appear at the base of the plant (Fig 1.14). Young galls are soft and white or pale green; older ones are dark and woody. Affected plants lose vigour and may become stunted, with abnormally small leaves and fewer flowers. The bacterium is spread on garden tools.

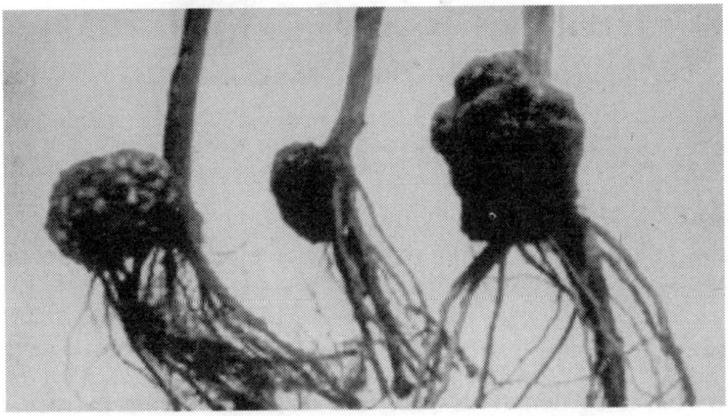

Figure 1.14. Crown galls on stems of rose

Crown gall, a disease caused by bacteria that live in the soil. The bacteria enter a plant through the roots or through wounds at the root area. The bacteria cause abnormal cell growth, which produces the galls. If the disease affects the plant whilst it is young the plant may be affected to the degree where it will not produce blooms. All affected plants wilt readily and grow poorly.

(ii) Management

Commercial formulation of non-tumerogenic strain of *Agrobacterium rhizogenes* (K-84 strain) is being used successfully to control crown gall.

1.1.2.11. Rose Mosaic Virus

(i) Symptoms: This disease is caused by a complex or viruses and is characterized by yellow patterns on the leaves. The patterns vary considerably, ranging between all-over fine blotches to patterns of lines in waves (Fig 1.15). The patterns may appear on a few or many leaves. Plants are infected by this virus at propagation using infected plant material.

Figure 1.15. Mosaic virus on Rose leaves

(ii) Management

Clean nursery practices, production of disease-free planting material and removing and destroying all infected plants helps in checking the disease.

1.1.3. Nematodes

The lesion and root-knot nematodes are the major problems in rose cultivation. The unproductive flowers in roses were attributed to the attack of the ectoparasites– *Hemicycliophora labiata* and *Xiphinema basiri* (Muthukrishnan *et al.*, 1975).

1.1.3.1. Lesion Nematodes, *Pratylenchus* spp.

The lesion nematodes are the major problems in rose cultivation. *P. zeae* was associated with poor growth of rose plants.

(i) Economic Importance and Losses: Sundarababu and Vadivelu (1988a) observed the pathogenic potential of *P. zeae* on Edward rose. Prasad and Dasgupta (1964) found *P. pratensis* in rose plants exhibiting stunted growth.

(ii) Symptoms: The affected plants showed chlorotic symptoms, poor and stunted growth with necrotic lesions on rose roots. *P. zeae* was responsible for 69.6, 36.4, 59.6 and 33.3% reduction in shoot length, shoot weight, root length and root weight, respectively. Affected roots show patchy discolouration in cortical region, which coalesce and lead to root rotting.

(iii) Management

(a) Physical Methods: Hot water treatment of rose seedlings at 50°C for 5 min or 48°C for 10 min eliminates *P. zeae* infestation.

(b) Cultural Methods: Interplanting of roses with African marigold helps in reducing the population of *P. penetrans*.

(c) Chemical Methods: Application of aldicarb at 4 oz/1000 sq. ft. was effective in reducing *P. penetrans* population and in increasing flower production by 16% (Johnson & McClanahan, 1974).

(d) Host Resistance: Ohkawa and Saigusa (1981) found that *Rosa indica* (*cv*. Major) and *R. multiflora* (*cv*. 60–5) were resistant to *P. penetrans* and *P. vulnus*, respectively.

1.1.3.2. Root-knot Nematodes, *Meloidogyne* spp.

M. incognita was recorded on rose (*Rosa indica*) from Aligarh (Alam *et al*., 1973) and *M. javanica* from Delhi (Prasad & Dasgupta, 1964).

(i) Symptoms: Nematodes of the genus *Meloidogyne* attack the roots of a wide range of plants including roses. The damaged root system develops root knots (Fig 1.16), which cause slow growth, wilting, and yellowing of leaves. Nematodes are spread by introducing infested plants or soil to the field, and also can be carried on garden tools. Symptoms of *Meloidogyne* infestation in roses is stunting, slow-growth, pale green leaves and wilting in mild weather.

Figure 1.16. *Meloidogyne* sp. damage on rose roots

(ii) Management

(a) Physical Methods: Hot water treatment of rose roots at 45.5°C for 60 min eliminates root-knot nematode infestation in seedlings.

(b) Chemical Methods: Dipping of bare roots of rose in 0.1% solution of fenamiphos for 30 min gave considerable reduction in root-knot infection (Dale, 1973). Nematodes may be treated using fenamiphos (Nemacur) granules. This treatment will need to be repeated.

(c) Host Resistance: Ohkawa and Saigusa (1981) reported that rose *cv*. Mavetti to be resistant against *M. hapla*. Growing roses grafted onto Fortuniana rootstock will help to reduce problems from nematodes.

(d) Integrated Methods: Pre-treatment of soil with dazomet (at 25 g/m^2) followed by soil amendment with neem cake (at 1 kg/m^2) along with *Pochonia chlamydosporia* (at 2×10^{10} spores/m^2) recorded reduced root galling and increased flower yield (Nagesh & Janakiram, 2004).

1.2. CARNATION

Carnation (*Dianthus caryophyllus*) flower is valued for its excellent keeping quality, wide array of colours and forms, and ability to rehydrate after continuous transportation. Moderate climatic control measures that are economical can deliver quality carnations at the internationally competitive prices year round. In India, Carnation and gerbera are commercially grown mostly under cover around big cites like Solan, Shimla, Mandi, Kullu, Chandigarh, Ludhiana, Guragaon, Kalimpong, Pune, Bangalore and Delhi. Carnation has tremendous potential for export market.

1.2.1. Pests

1.2.1.1. Thrips, *Frankliniella schultzei*

(i) Damage: The young and adult stages mainly attack growing buds, flowers and feed on sepals and petals causing discolouration (silver streaks) and distortion of affected floral parts. Damage to young buds will not allow buds to open leading to considerable flower loss.

Adult female thrips are narrow-bodied, light-brown or grey and 1–2 mm long. Males are smaller and yellow. Both sexes have two pairs of narrow delicate wings fringed with long hairs. The wings lie along the back when not in use. The colourless eggs are inserted into the tissue of petals, sepals, blossom stems, stamens, pistils, calyx cups and young leaves.

The nymphs feed on the stamens, pistils and petals. When full grown they crawl to the ground and pupate about 50 mm under the surface. Generation times vary from 10 days to more than a month, depending on the temperature.

Thrips have mouth parts adapted for rasping and sucking. Unlike bugs they do not have tubular mouthparts and are therefore unable to penetrate deeply into plant tissue. Feeding by thrips is usually confined to soft recent growth or flowers where they rasp the surface cells and suck the sap as it oozes out. Petals and anthers turn brown and shrivel. Normal opening of flowers may be affected if thrips enter before the flowers open. Feeding and egg laying on the pistil may cause it to wither.

Egg laying in the tender tissue of young leaves may cause the tissue around the eggs to die and fall out leaving small, irregularly shaped 'shot-holes'. In addition to damage done by feeding, crops may also be downgraded by fecal deposits.

(ii) Management

Spraying of 0.1% acephate or 0.015% ethofenprox or 0.007% imidachloprid or 0.05% cartap hydrochloride at the initial stage of infestation controls thrips damage. Soil drenching with 0.01% chloropyriphos reduces further incidence by killing thrips present in the soil.

1.2.1.2. Bud Borer, *Helicoverpa armigera*

(i) Damage: Infestation of bud borer is observed from May to September on carnation. Bud worms are the larvae of noctuid moths. Female moth lays cream coloured eggs singly on young buds. Larvae bore into growing points by making holes in buds and damage flowers by feeding on developing petals and reproductive parts resulting in petal drop. They also feed on the young shoots and foliage. They make characteristic round holes in buds and flower heads. The caterpillars are about 40 mm long when fully grown, have various shades of yellow, green, pink or brown, with dark flecking and longitudinal paler and darker stripes.

(ii) Management

Collection and killing of grown up larvae by putting them in 0.1% chloropyriphos solution reduces borer population. Spraying of 0.07% endosulfan or 0.05% methyl parathion or 0.01% ethofenprox when eggs are observed on buds controls damage by the pest effectively.

1.2.1.3. Aphids, *Myzus persicae*

(i) Damage: Aphids are small plump-bodied insects that feed by sucking sap. Several species of aphids attack carnations but the most common is green peach aphid *Myzus persicae*. They secrete copious amounts of honeydew, which favours growth of sooty moulds, and they reduce plant vigour by sucking sap. Their most important effect is the transmission of viruses while feeding.

(ii) Management

Effective management of aphids can be obtained by spraying dichlorvos or methomyl each at 0.1% or cypermethrin at 0.04%. Spraying of 0.07% imidachloprid or 0.05% cartap hydrochloride also gives very good control of the pest.

1.2.1.4. Red Spider Mite, *Tetranychus urticae*

(i) Damage: This is a devastating pest on carnation. Infestation is severe during summer. Mite colonies are found on upper surface of leaves beneath leaf folds and are covered by webs and they spread to entire plant in severe cases. More population is seen on lower part of the plant and economic threshold is 0.5 mite/leaf (Jesitor, 1976). Damaged leaves and buds (Fig 1.17) become discoloured and dry. Stunted growth and drying of plant are the symptoms in case of heavy damage.

The adult female is about 0.5 mm long. The adult male is about 0.3 mm long. In the summer the adult females are the feeding stages which are yellowish-green, with two pronounced dark spots, one on either side of the body. In males these spots are less conspicuous and the body is smaller and tapered towards the rear. Both sexes possess two reddish eye-spots.

Figure 1.17. Mite infestation on carnation bud

Under artificial long day length or in warm situations such as glasshouses, the mites may continue feeding and reproducing through the winter. Outdoors, when temperatures and day length decrease in autumn, the males die and the adult females stop feeding and change colour to orange-red. They usually migrate to sheltered sites where they congregate to overwinter. When the temperatures increase in spring, the females leave their shelters and begin to feed, losing their red colour in the process. They lay up to six eggs a day and usually lay a total of 70 or more. Males develop from unfertilized eggs and females develop from fertilized eggs.

Egg hatch takes from three to 10 days depending on the temperature. The young mites mature in from four to 12 days and may live for about three weeks. In hot weather the complete life cycle may be as short as two weeks and overlapping generations may occur. Damage occurs when nymphs and adults feed by piercing the cells of leaves and sucking the contents, causing collapse and death of the cells. The mite is a profuse web-spinner in its active stages and eggs are laid amongst the webbing.

Heavily infested leaves become pallid and may become bronzed and shriveled. Defoliation may occur.

(ii) Management

Cutting and burning of severely infested and dried plant parts to be undertaken. Washing of plants with water dislodges the webs. A mixture of dicofol and tetradifon at 300 g/ha or proparigate at 150 g/ha were found effective against adults, eggs and nymphs (Pondolfo, 1984). Spraying of pongamia oil in combination with annona oil at 2% effectively checks mite population (Jhansi Rani & Jagan Mohan, 1997). Spraying of triazophos at 0.06% or ethion at 0.05% or dicofol at 0.05% at 15 days interval is also effective.

Cvs. Isaq, Alcardi and Sunrise were reported resistant to mite attack.

1.2.2. Diseases

1.2.2.1. Wilt, *Fusarium oxysporum* f. sp. *dianthi*

(i) Symptoms: Initial symptoms are foliar yellowing and production of crook neck shoots. Leaves and shoots wither and become brownish. Stem when cut open shows brown discolouration at the vascular region.

(ii) Management

Soil fumigation with methyl bromide and treating the beds with benomyl before planting are effective. After transplanting, drenching with 0.1% carbendazim is quite effective.

Bioagents like *Trichoderma harzianum, Bacillus subtilis, Streptomyces* sp. and non-pathogenic isolates of *Fusarium oxysporum* are reported to give good control of Fusarial wilt.

Carnation *cvs*. 'Barbi', 'Candy Maj', 'Capello', 'Carbasio', 'Improved Lilac', 'Juanita', 'Lady Di', 'Light Pink Marble', 'Maiko', 'Maj Pink', 'Meiling', 'Melody', 'Moonlight', 'Picotee Orange', 'Scarlett Elegance', 'Siri #1', 'White Melody' are reported to be highly resistant, while 'Ace', 'Apache', 'Barsemi Yasmino', 'Big Red', 'Comache', 'Corona', 'Exquisite Select', 'Felicia', 'Fiesta', 'Georgia Ann', 'Jolievette', 'Lucy Carrier', 'Maman', 'Mei-Sciang', 'Orchid Beauty', 'Pallas', 'Shiro', 'Silvery Pink', 'Sweetheart', 'White Elegance' are moderately resistant to Fusarium wilt.

1.2.2.2. Foot Rot, *Phytophthora* sp., *Pythium* sp., *Rhizoctonia solani, Sclerotinia scleortiorum*

(i) Symptoms: The infection results in rotting of stem at soil level, resulting in wilt. Leaves gradually get discoloured and start drying and turn bottom upwards.

(ii) Management

Soil treatment with bioagents or benomyl or iprodion against *R. solani* and Al-Fosetyl and metalaxyl against *Phytophthora* sp. and *Pythium* sp. are effective.

1.2.2.3. Basal Rot, *Sclorotium rolfsii*

(i) Symptoms: Rotting of stem at soil level extending to leaves, cottony growth of the fungus mycelium is commonly seen on infected portions. White to dark brown spherical sclerotial bodies are clearly visible at advanced stage of infection.

(ii) Management

Regulation of soil moisture and avoiding excessive nitrogen to the soil helps to reduce the disease incidence. Soil drenching with chlotrothalanil/iprodion/dithionon at 2 g/litre is quite effective in controlling and further spread of the disease.

1.2.2.4. Leaf Spot, *Alternaria dianthi*

(i) Symptoms: Initially, small and purple lesions are seen on leaves which later turn grayish brown. The infection is generally noticed on lower leaves just above the ground level.

(ii) Management

Foliar application of mancozeb or chlorothalanil or propineb or dithionon or iprodion is effective.

1.2.2.5. Carnation Mottle Virus (CarMV)

(i) Symptoms: CarMV is the most common and widespread virus disease of carnations. Most carnation cultivars infected with this virus do not express symptoms. In other cases, when the symptom is expressed, the infected plant may have a yellow mottle on the leaves. Flower quality and yield are reduced. "Virus free" plants have broader leaves and more vigorous growth. CarMV is highly infectious and is rapidly spread from plant to plant by foliage contact or during handling. Research has shown that up to 75% of carnations initially free of CarMV became infected within six months when grown in the same glasshouse as carnations infected with CarMV.

A flexuous rod-shaped virus measuring 718 nm × 12 nm was isolated from carnation. It causes veinal chlorosis, leaf mottling, chlorotic spots, distortion of flowers and stunting of plants. The virus belongs to the corlavirus group.

(ii) Management

The only current method for controlling virus diseases is through the use of virus tested material in conjunction with strict hygienic practices. To prevent infection of this virus-tested material, the carnation grower should enforce hygiene regulations (for example, washing hands with hot soapy water before handling plants, aphid-proof glasshouses, sterilisation of cutting and propagating instruments, etc). It is essential to educate workers about the spread of virus diseases.

Table 1.4. Disease management

Disease (causal agent)	Symptoms	Survival of pathogen and effect of environment	Control
Alternaria leaf spot and branch rot *(Alternaria dianthi, A. dianthicola)*	Gray-brown leaf or petal spots with purple margins; black spore masses form in spots; branch rot starts at nodes and girdles stem.	In infected plants and debris; airborne spores; moist conditions for 8 to 10 hours required for infection.	Provide good air circulation and keep humidity low; irrigate in early morning; do not wet foliage with irrigation water; during periods of high humidity, protect plants with iprodione, mancozeb, or chlorothalonil.
Bacterial wilt *(Pseudomonas caryophylli)*	Sudden wilting of tops or individual branches; basal stem cracks; roots		

Table. 1.4. (Contd.)

Disease (causal agent)	Symptoms	Survival of pathogen and effect of environment	Control
	may be rotted; vascular discoloration in stems is yellowish to brown; the outer layer (epidermis) separates easily from stem, which is sticky to the touch.	Infected plants and debris; bacteria spread in water; favored by high temperatures.	Use disease-free cuttings and treat soil as above. Bacteria can spread quickly; avoid use of cutting dips and avoid splashing water; break cuttings from stock plants; disinfect tools.
Calyx rot *(Pleospora herbarum)*	Rot starts at tip of calyx and progresses towards base.	In old leaves, stems, and debris; favored by high humidity.	Same as Greasy blotch; harvest regularly.
Fairy-ring leaf spot *(Cladosporium echinulatum)*	Conspicuous tan spots with concentric rings; margin of spot may be red; dark spores form in spots.	Infected plants and debris; airborne spores; favored by wet weather.	Provide good air circulation and keep humidity low; irrigate in early morning; do not wet foliage with irrigation water; protect foliage with mancozeb.
Fusarium bud rot *(Fusarium tricinctum)*	Outwardly normal buds are brown and decayed inside; cottony white growth and plump white mites may be visible.	Fungus carried by grass mite, *Pediculopsis graminum.*	Destroy infected buds; control mites; do not bring field-grown carnations into greenhouse; control weeds outside growing area.
Fusarium stem rot *(Fusarium avenacearum, F. culmorum, F. graminearum)*	Stem rotted at soil line and high up on plant; roots and base of stem rotted; tops wilt and die. Pink cushions of spores may form at base of plant on decayed tissues; common as a cutting rot.	In soil and plant debris; spores spread in water; also favored by warm, humid conditions; also favored by high N fertilization and high N : K ratio.	Use clean cuttings and rotting medium; spray cuttings in rooting medium with thiophanate-methyl.
Fusarium wilt *(Fusarium oxysporum* f. sp. *dianthi)*	Yellow, wilted branches frequently occur on one side at first; vascular	Soil and infected plants; spores spread in water; favored by soil temperatures 70°F and	Plant resistant cultivars; use disease-free cuttings planted in treated soil; steam soil

Table. 1.4. (Contd.)

Disease (causal agent)	Symptoms	Survival of pathogen and effect of environment	Control
	discoloration is dark brown; root system usually remains intact; in late stages, stem develops a dry, shredded rot; infected parts die.	above; restricted below 60°F.	in raised beds at 140°F for at least 30 minutes; solarize soil or fumigate with methyl bromide alone or with chloropicrin; residual bromides should be leached by irrigation before planting; adjust soil pH to 6.5 to 7.0.
Gray mold *(Botrytis cinerea)*	Woolly gray fungal spores form on soft brown, decayed blossoms and can move into plant parts wherever blossoms touch them.	In plant debris; airborne spores; favored by high moisture conditions and low temperatures.	Remove old flowers from growing area; maintain horizontal air movement; lower humidity where possible; mist blooms with iprodione or fenhexamid.
Greasy blotch *(Zygophiala jamaicensis)*	Greasy-appearing spots on leaves with radiating weblike margins; pimpling of infected areas.	In infected plants and debris; favored by high humidity; not too common.	Provide good air circulation and keep humidity low; irrigate in early moring; do not wet foliage with irrigation water.
Phialophora wilt *(Phialophora cinerescens)*	Gradual wilting of plants; leaves become straw colored; not one-sided as Fusarium wilt may be; brown discoloration of vascular system; little or no tissue rotting in late stages; uncommon.	Soil and infected plants; spores spread in water; favored by cool soil temperatures.	Use disease-free cuttings and treat soil as above.
Phytophthora stem rot *(Phytophthora parasitica)*	Stem rotted at soil line; may be mistaken for Rhizoctonia stem rot.	In soil and plant debris; favored by warm, moist soil and poor drainage.	Steam or fumigate soil; drench at planting with mefenoxam.

Table. 1.4. (Contd.)

Disease (causal agent)	Symptoms	Survival of pathogen and effect of environment	Control
Pythium root rot *(Pythium* spp.)	Plants are stunted, particularly in lower, poorly drained areas. rootlets rotted.	Soilborne; common in most soils; favored by poor drainage, low spots, excessive irrigation.	Steam or fumigate soil (see Fusarium wilt); drench plants with mefenoxam periodically, depending on severity of disease; make first application at planting.
Rhizoctonia stem rot*(Rhizoctonia solani)*	Stem rotted at soil line; rot progresses from the outside; entire plants wilts and dies; dark fungal strands and sclerotia may be visible with a hand lens.	Soilborne; plant debris; favored by warm, moist conditions.	Steam rooting medium and soil. Use PCNB before transplanting, or spray base of transplant with iprodione.
Rust *(Uromyces diantha)*	Small pustules of powdery brown spores; spores are airborne.	Carried over only on living plants; favored by moist conditions.	Use resistant cultivars or protect plants in problem areas with myclobutanil or mancozeb.

Virus/Bacterial Diseases	Symptoms	Host range and natural spread	Control
Mottle	Faint leaf mottle or no symptoms; common in virtually all carnation cultivars.	Handling and cutting knife; not transmitted by aphids; may be spread in drainage water.	Reduce spread of mottle by disinfecting tools between blocks of plants and several times a day; obtain virus-free plants.
Etched ring	Rings usually oval or elongated; rarely concentric on older leaves and stems; symptoms may be slight in young cuttings; no obvious effect on plant vigor.	Aphids; not spread by handling.	Obtain virus-free plant; control aphids.

Table. (Contd.)

Virus/Bacterial Diseases	Symptoms	Host range and natural spread	Control
Necrotic fleck	Reddish purple necrotic flecks, streaks, or spots appear in leaves; symptoms are masked at low temperatures.	Aphids.	Obtain virus-free plants; control aphids.
Ring spot	Small 0.5- to 1-inch (1- to 2-cm) rings; sometimes concentric; chlorosis, mottling, and distortion of young leaves; plants obviously stunted.	Cutting knife and handling. Insect vectors unknown.	Same as mottle.
Vein mottle	Young leaves exhibit a vein clearing, which develops into chlorotic spots and patterns that mostly follow veins; symptoms tend to disappear on old leaves.	Aphids; not spread by handling or cutting knife.	Obtain virus-free plants; control aphids
Bacterial wilt *(Pseudomonas caryophylli)*	Sudden wilting of tops or individual branches; basal stem cracks; roots may be rotted; vascular discoloration in stems is yellowish to brown; the outer layer (epidermis) separates easily from stem, which is sticky to the touch.	Infected plants and debris; bacterial spread in water; favored by high temperatures.	Use disease-free cutting and treat soil as above; bacteria can spread quickly; avoid use of cutting dips and avoid splashing water; break cuttings from stock plants; disinfect tools.

1.3. GERBERA

Gerbera (*Gerbera jamesonii*), commonly known as Transvaal daisy, Barberton daisy or African daisy, is an important flower grown throughout the world for beds, borders, pots and rock gardens. The flowers are of various colours and suit very well for floral arrangements. The cut blooms also have long vase life. In India, it is distributed in the temperate Himalayas from Kashmir to Nepal at altitudes of 1300 to 3200 meters above sea level. Gerbera belongs to the family Asteraceae and is native to South African and Asiatic regions. To meet the quality standards of export market, it has to be grown under naturally ventilated low cost polyhouses.

1.3.1. Pests

1.3.1.1. Whitefly, *Bemisia tabaci*

(i) Damage: The pest is active during moderately hot and humid conditions (February-July). Small and fragile white adults lay eggs on lower surface of young leaves. Nymphs and adults suck cell sap from lower side of leaves resulting in discolouration, wilting and drying of matured leaves and curling of young leaves. In case of severe infestation, sooty mould develops on leaves on honeydew secreted by nymphs. Affected plants are retarded in growth and produce deformed flowers with short stalks.

(ii) Management

- Remove and burn the heavily infested leaves to check pest build up.
- Spray triazophos 40 EC at 1.5 ml/l or methomyl 40 SP at 1.5 g/l or acephate 75 SP at 1.5 g/l alternating with pongamia oil 1% for management of nymphs.
- Spray 0.05% dichlorovos or deltamethrin against adults.
- Spraying of permethrin, fenvalerate and amitraz each at 0.1% reduced whitefly population by more than 90% (Thachuk *et al.*, 1983).
- Install yellow sticky traps for monitoring and trapping adults.
- Excessive dampness and dark conditions should be avoided.
- Clean cultivation and use of insect proof nets are essential.
- Oil sprays are to be given during evening inside polyhouse for better efficacy. Annona and pongamia oil at 2% were found very effective with more than 90% mortality of nymphs (Jhansi Rani & Jagan Mohan, 1997).
- Release of parasitoid, *Encarsia formosa* at 6 adults/m^2 is recommended for control of whiteflies (Nucifera & Calabretta, 1986).

1.3.1.2. Leaf Miner, *Liriomyza trifolii*

(i) Damage: Attack of leaf miner starts when plants are at 3 to 4 leaf stage. The incidence of leaf miner is severe in March-June. Adults lay eggs by puncturing the tender leaves. Maggots mine the leaves by feeding on mesophyll by leaving characteristic serpent shaped mines (Fig 1.18). Severely damaged leaves show extensive mining, become brittle, turn brown and dry up. The affected plants become stunted in growth and produce weak and small flowers.

(ii) Management

- Remove and destroy severely mined leaves.

- Spray triazophos 40 EC at 1.5 ml/l or abamectin 1.9 EC at 0.4 ml/l against larvae.
- Spray with dichlorvos 76 EC at 1 ml/l or decamethrin 2.8 EC at 1 ml/l for managing adults.

Figure 1.18. Leaf miner infestation on gerbera

- Installing yellow sticky traps coated with chlopyriphos 20 EC at 2.5 ml/l gives considerable protection and encourages parasitization by *Diglyphus isaea* (Calabaretta & Nucifera, 1985).
- Spraying of 1% pongamia oil or 4% neem seed kernel extract is very effective.
- Intercropping with field bean (*Vicia faba*) which acts as trap crop is effective.
- Release of eulophid parasitoid, *Diglyphus intermedius* at 1000 at weekly intervals gives considerable reduction in leaf miner population.

1.3.1.3. Thrips, *Thrips palmi*

(i) Damage: Heavy infestation of thrips is observed on gerbera during September-November and March-May. Adults and nymphs colonize along midrib on upper surface of leaves and petals of growing flowers and suck the sap. Thrips cause damage by sucking sap from leaves, buds and flowers. Damaged leaves become brittle and curl up. Affected buds and flowers get discoloured and distorted in shape (Fig 1.19). Discolouration, development of brown streaks and severe distortion of petals are seen on the damaged flowers.

(ii) Management

- Spray acephate 75 SP at 1.5 g/l or dimethoate 30 EC at 2 ml/l along with 0.5% pongamia oil 2–3 times at 15 days interval with onset of new flush.
- Apply fipronil 5 SC at 1.5 ml/l in case of severe infestations.
- Drench the soil with chlorpyriphos 20 EC at 5 ml/l for killing pupae in the soil.

1.3.1.4. Aphid, *Myzus persicae*

(i) Damage: Pest incidence is found during cool and cloudy months (November-January). Aphids mainly attack growing flowers. Both nymphs and adults colonize on lower side of petals and spread to entire flower. They suck the sap at the feeding site resulting in discolouration, distortion and wilting of flowers (Fig. 1.19). Development of sooty mould on honey dew secreted by aphids makes flowers look sickly with black dots.

Figure 1.19. Thrip infestation on flowers of gerbera

(ii) Management

- Spray dimethoate 30 EC at 2 ml/l or oxydemeton methyl 25 EC at 2 ml/l at 10–15 days interval.
- Effective control of the aphid can be obtained by spraying dichlorvos or methomyl each at 0.1% or cypermethrin at 0.04% (Tusandi, 1984). New insecticides like 0.07% imidachloprid or 0.05% cartap hydrochloride also give very good control of the pest.
- Spray pongamia or neem oil at 1%.
- If the incidence is severe, spray imidacloprid 200 SL at 0.4 ml/l.

1.3.1.5. Bud Caterpillar, *Spodoptera litura*

(i) Damage: Larvae cause severe damage to flower buds before they open as the larvae are protected by bracts on both open and greenhouse grown gerbera plants. Damage results in petal drop, reduces quality of flowers due to excreta and cause severe flower loss.

(ii) Management

A mixture of deltamethrin at 15 g and triazaphos at 30 g/litre spray applied at 150 g formulation/750 m gave 97% mortality and treatment with *Bacillus thuringiensis* offers effective control (Inserra & calabretta, 1985). Methyl parathion, quinalphos and methomyl each at 0.05% sprays protect plants from damage by *S. litura*. Spreading of poison bait (10 kg rice bran + 1 kg jaggery + 0.5 kg chlorpyriphos) where damage is noticed and installing pheromone traps after first showers is quite helpful in bringing down the population of *S. litura*.

1.3.1.6. Red Spider Mite, *Tetranychus ludeni*

(i) Damage: Severe attack of the mite is observed during February-April on gerbera. Colonies of red coloured mites are seen on both sides of the matured leaves and spread to flowers in case of heavy infestation. Nymphs and adults feed on the leaf sap resulting in bronzing and drying of damaged leaves. Severely damaged plants produce few and very small flowers which dry up consequently.

(ii) Management

Thinning of old and heavily infested leaves prevents severe attack. Weekly sprays of 0.15% dichlorvos or 0.1% methomyl are recommended by Kerenyi-Nemestothy and Tusandi (1985). Recent acaricides like 0.0025% vertimec or 0.025% diafenthiuron or 0.05% amitraz or 0.015% flufenuxuron provide highly effective control of the mite (Jhansi Rani & Jagan Mohan, 1997). Annona, pongamia or jatropa oil all at 1% cause considerable mortality of mites.

1.3.2. Diseases

1.3.2.1. Powdery Mildew, *Erysiphe cichoracearum, Oidium erysiphoides* f. sp. *gerbera*

(i) Symptoms: This is one of the most important diseases of gerbera. The infection takes place as mealy white powder coating on the foliage leading to malnutrition of flowers. The leaves become placid, wither and defoliate gradually. The plants start showing decline and suckering is inhibited.

(ii) Management

Foliar application of sulphur compounds and carbendazim, dinocap, triadimefon, fenerimol, penconazole and triforine offer good control.

Foliar sprays of K_2HPO_4 at 4.3 g/l were found to be effective in reducing conidial production and thus controlled the disease (disease incidence of 5% as compared to 62% in control) which was on par with fungicidal spray (Hexaconazole at 0.1%).

1.3.2.2. Root Rot, *Phytophthora cryptogea*

(i) Symptoms: The infection occurs just at the soil level on the collar portion of the stem. The leaves often turn yellow and the entire plant wilts. Rotting extends from the stem towards the leaf, along the petioles, and also through roots.

(ii) Management

Treat the soil in the root zone of plants with either metalaxyl or captan and copper oxychloride.

1.3.2.3. Wilt, *Fusarium oxysporum* f. sp. *dianthi*

(i) Symptoms: The infected plant shows yellowing of lower leaves. The collar portion shows vascular discolouration.

(ii) Management

Rouging and soil application of carbendazim or benomyl is recommended. Since the disease is facilitated by nematodes, these should be controlled first, followed by treatment for *Fusarium*.

1.3.2.4. Foot Rot, *Pythium* spp., *Sclerotium rolfsii, Rhizoctonia solani*

(i) Symptoms: The plants show reduced growth, smaller leaves and flowers with weak stems. Leaves show purple discolouration, with rotting of both feeder and main roots.

(ii) Management

Regular application of copper oxychloride or iprodion or dithionon or captaf to soil may help to control the disease. Soil application of FYM along with *Trichoderma harzianum* is helpful for initial suppression of the disease.

1.3.3. Nematode Pests on Carnation and Gerbera

The root-knot nematode is recognized as the major limiting factor in successful production of carnation and gerbera crops.

1.3.3.1. Root-knot Nematode, *Meloidogyne incognita*

The root-knot nematode, *M incognita* is one of the serious limiting factors in commercial cultivation of carnation and gerbera under polyhouse conditions. Nagesh and Parvatha Reddy (1994) reported that *M incognita* was responsible for 27 and 31% loss in flower yield of carnation and gerbera, respectively.

(i) Economic Importance and Losses: Most of the highly fetching exotic cultivars of Carnation and Gerbera from Europe have shown 40 to 60% mortality in polyhouse beds due to root-knot nematode infection in and around Bangalore (Nagesh & Parvatha Reddy, 1996a).

(ii) Symptoms: The root-knot infected Carnation and Gerbera plants exhibit stunted growth, leaf yellowing and premature dropping and root galling (Fig 1.20).

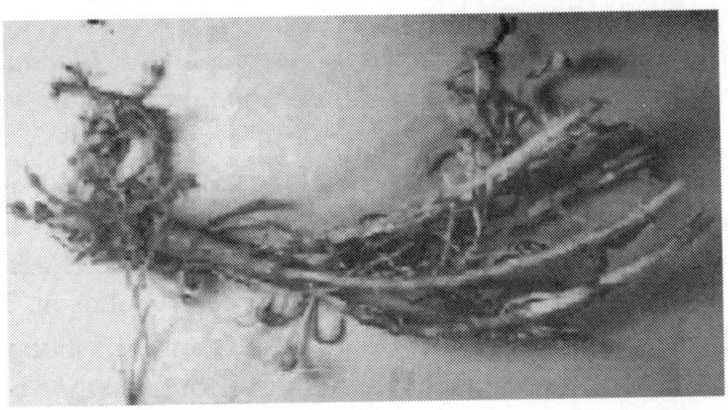

Figure 1.20. Root galling on carnation due to root-knot nematodes

(iii) Interaction with Other Pathogens: *M. incognita* interacts with soil-borne fungi such as *Fusarium oxysporum* f. sp. *dianthi* and *Rhizoctonia solani* in causing wilt/root rot complex in Carnation and Gerbera.

(iv) Management

(a) Cultural Methods: Application of neem cake at 1 kg per m^2 gave effective control of *M. incognita*.

(b) Chemical Methods: Application of carbofuran at 2 kg *a.i.* per ha gave effective control of *M. incognita*.

(c) Biological Methods: An experiment carried out for the biological control of root-knot nematode *M. incognita* in carnation (*Dianthus caryophyllus*) showed that, though the growth parameters were found to be at par in plants treated with *Paecilomyces lilacinus* and *Pochonia chlamydosporia*, significantly higher reduction of root galling was achieved with *P. chlamydosporia*. Thus, it was concluded that under polyhouse conditions, *Paecilomyces lilacinus* and *Pochonia chlamydosporia* were better bio-agents for the control of root-knot nematode compared to *T. harzianum* (Shylaja, 2004).

(d) Host Resistance: Carnation *cvs*. Desco, Castelaro, Kappa, Rara, Logupink, Target and Antalia were highly resistant to *M. incognita*.

Screening of the four commercially available varieties of carnation for their resistance to root-knot nematode, *M. incognita,* revealed that *cv*. Ivonne was the most susceptible while, Randez was moderately tolerant. The other two varieties, Dark Randez and Sunshine were observed to be moderately susceptible (Shylaja, 2004).

(e) Integrated Methods

Bioagents and Botanicals: Integration of neem cake at 0.5 kg per plant with *T. viride* at 100 g per m^2 was effective in the management of root-knot nematodes and increased flower yields.

Integration of *P. lilacinus/T. harzianum* at 0.5 l per m^2 (aqueous spore suspension containing 2×10^4 spores per ml) with neem cake at 0.5 kg per m^2 or fenamiphos at 2 g *a.i.* per m^2 increased plant growth parameters and flower yield of carnation and gerbera. The above treatments also increased root-knot egg parasitization by the parasitic fungi (Nagesh & Parvatha Reddy, 1996a) (Fig 1.21).

Three antagonistic fungi viz., *T. harzianum*, *V. lecanii* and *P. lilacinus* at 2×10^4 spores/ml in combination with neem cake reduced *M. incognita* population in both soil and roots of carnation.

Application of *P. lilacinus* at 0.5 g/kg of soil along with neem cake at 1.0 t/ha efficiently suppressed the nematode population and checked its build up, enhancing the plant growth parameters resulting in better production with increased flower stalk length and flower diameter in carnation. The plants also came to flower early (Johnson, 2000).

Arbuscular Mycorrhizal Fungi and Bioagents: Combined inoculation of AMF and *Pseudomonas fluorescens* had positive effect on root-knot nematode control on carnation (Anusuya & Vadivelu, 1999).

Figure 1.21. Management of root-knot nematodes on carnation by integration of bioagents and botanicals. Left–Untreated; Right–Treated

Bioagents, Botanicals and Chemicals: Pre-plant treatment of beds with dazomet (at 40 g/m^2) followed by the application of neem cake (at 1 kg/m^2 15 days later) along with antagonistic fungi, *P. chlamydosporia/ P. lilacinus* (at 2 × 10^{12} spores/m^2) significantly reduced root-knot nematode population (*M. incognita*), mortality of plants and suppressed the nematode infection for 2 years. The antagonistic fungi established better in the beds treated with dazomet. The above treatment also reduced root galling, nematode multiplication rate and plant mortality, and increased spike/stem length, flower yield and root colonization with the bioagents (Nagesh & Parvatha Reddy, 2005).

1.3.3.2. Root-knot Nematode, *Meloidogyne incognita* and Wilt, *Fusarium oxysporum* f. sp. *dianthi* Disease Complex

(i) Symptoms: In an experiment carried out to study the role of root-knot nematode, *M. incognita* in predisposing carnation to *Fusarium* wilt, it was observed that the appearance of the wilt symptoms were accelerated when *M. incognita* was inoculated two weeks prior to *F. oxysporum* f. sp. *dianthi*. The rate of wilting was observed to be 2.1 and 4.6, respectively during 12th and 25th week of observation, while the root galling index (RGI) values were recorded to be 4.15 and 4.75, respectively. It was observed that maximum mortality was recorded when *M. incognita* was inoculated 2 weeks prior to *F. oxysporum* f. sp. *dianthi* which was 33.4% and 79.2%, respectively, at 12th and 25th week of observation. The plant growth was also reduced significantly due to prior inoculation of *M. incognita* (Shylaja, 2004).

Ferraz and Lear (1976) studied the interaction of four plant parasitic nematodes (*Circonemoides curvatum, Pratylenchus dianthus, Rotylenchus robustus* and *Meliodogyne hapla*) and *F. oxysporum* f. sp. *dianthi* in carnations. They reported that plants wilted earlier and more severe symptoms were present with the nematode-fungus combination than with the fungus alone. All nematodes except *R. robustus* showed synergistic interaction with *Fusarium*.

(ii) Management

(a) Integrated Methods: The studies carried out to evaluate combination of bio-agents for the biological control of wilt (*F. o.* f. sp. *dianthi*) and root-knot nematode (*M. incognita*) disease complex in carnation, revealed that a combination of *P. chlamydosporia* + *P. lilacinus* gave the best control of wilt and root knot disease complex (Shylaja, 2004).

1.4. CHRYSANTHEMUM

Garden chrysanthemum (*Dandrathema grandiflora*), also known as 'Queen of the East', is a highly versatile and accommodating ornamental with a wide range of type, size and colour in flowers. In India, large flowered varieties are grown for cut flower, making garland, wreaths and 'veni', religious offerings and for bedding and potting purposes. Its commercial cultivation is being done in Maharashtra, Rajasthan, Madhya Pradesh and Bihar. Small flowered varieties are commercially grown for loose flowers in Tamil Nadu, Karnataka and Maharashtra. In India, chrysanthemum is grown in about 4,000 ha.

1.4.1. Pests

1.4.1.1. Aphid, *Myzus persicae, Macrosiphoniella sanborni*

(i) Damage: The most common aphids infesting chrysanthemums are the green peach aphid, *Myzus persicae*, and the chrysanthemum aphid, *Macrosiphoniella sanborni*. The green peach aphid is a green to greenish-yellow in colour and the chrysanthemum aphid is a shiny mahogany in colour.

Greenish-black nymphs and chocolate-brown adults suck the cell sap from growing shoots and lower surface of leaves (Fig 1.22). Damage results in loss of vigour, yellowing and premature leaf fall and stunted growth of plant. Honey dew secreted by aphids favours development of black sooty mould on heavily infested leaves. In severely infested plants, flower buds reduce in size and remain unopened.

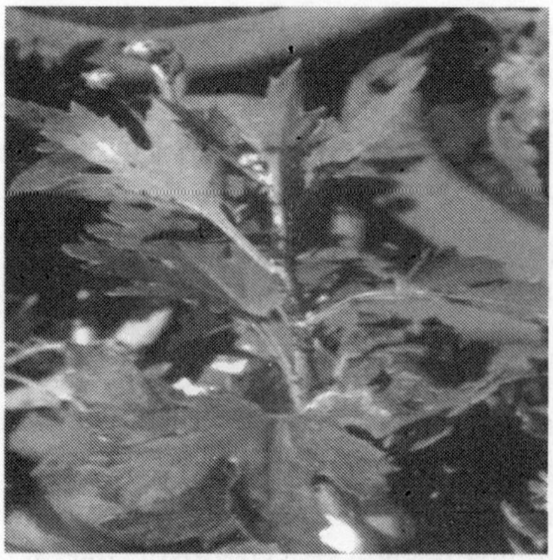

Figure 1.22. Aphid infestation on leaves and flower of chrysanthemum

Aphids feed on chrysanthemum by puncturing and sucking plant fluids from the veins. These insect pests cause damage by secreting large amounts of honeydew which collects on leaves and buds. Honeydew encourages the growth of sooty mould which is unsightly and inhibits photosynthesis by reducing the amount of light reaching the leaves.

Only pregnant females survive after winter. In spring, females reproduce without mating and produce nymphs (immature stages) which form social broods and feed near their mothers. Winged females develop and disperse and aphid populations build up rapidly with spring growth.

(ii) Management

- Spray pongamia or neem oil at 20 ml/l or 4% pongamia kernel extract at 10–15 days interval.
- Spray dimethoate 30 EC at 2 ml/l or oxydemeton methyl 25 EC at 2 ml/l or imidacloprid 200SL at 0.4 ml/l or acephate 75 SP at 1.5 g/l.
- Application of *Verticillium lecanii* (Microgermin) at 3 g/l.
- Release of 1 day old *Chrysoperla carnea* larvae at 50:1 ratio (aphid: chrysopid) eliminated aphid population (Scopes, 1969).

1.4.1.2. Thrips, *Microcephalothrips abdominalis; Frankiniella* spp., *Thrips* spp.

(i) Damage: Thrips have been found in all chrysanthemum growing areas. This major insect pest is responsible for millions of dollars worth of crop damage world-wide due to its resistance to many insecticides and efficiency at spreading tomato spotted wilt virus.

Thrips usually live and feed on flowers or new plant growth, such as buds and young leaves. Thrips have rasping and sucking mouth parts which cannot penetrate deep into plant tissue, so they usually feed on soft-recent growth or flowers by rasping or scraping surface cells and sucking the cell contents. Damage is not always obvious after feeding but becomes more obvious as the affected flowers, leaves or fruit grow and distort.

The thrips life cycle is mostly continuous and all stages of thrips can be found throughout the year. Female thrips live for 30 to 45 days and can produce around 150 to 300 eggs.

More than seven species of thrips are reported on chrysanthemum of which the above mentioned species are common. Slender white nymphs and black adults feed on tender leaves causing silvering, mottling and distortion of leaves. Damaged flowers look discoloured, withered and dried due to scorching (Fig 1.23).

(ii) Management

- Spray dimethoate 30 EC at 2 ml/l at 15 days interval.
- Drench the soil with chlorpyriphos 20 EC at 5 ml/l.
- If the incidence is severe, spray ethofenprox 10 EC at 1 ml/l or fipronil 5 SC at 1.5 ml/l.
- Covering of young buds with polythene bags with holes prior to blooming protects flowers from attack by thrips (Wang & Lin, 1984).

Figure 1.23. Thrip damage on flowers of chtysanthemum

1.4.1.3. Leaf Folder, *Hedylepta indicata*

(i) Damage: This is a serious pest of chrysanthemum under field conditions. Pale white eggs are laid in groups on lower side of leaves which hatch in 10 days. There are 6 larval instars. Green coloured larvae with white stripes and brown head fold the leaves and feed on chlorophyll resulting in skeletonization and drying. Pupae are seen inside the rolled leaves. The larvae also damage flowers leading to heavy loss occasionally.

(ii) Management

Cutting and burning of rolled leaves with larvae and pupae reduce the pest population. Spraying of 0.01% ethofenprox or 0.01% fenvalerate followed by methyl parathion or quinalphos or profenofos each at 0.05% gives effective control of leaf folder.

1.4.1.4. Leaf Miner, *Liriomyza trifolii*

(i) Damage: The incidence of leaf miner is severe in March-June. Eggs are deposited in leaf tissue and larvae mine leaves by feeding on mesophyll leaving characteristic trails (Fig 1.24). Severely damaged leaves show extensive mining which turn brown and dry up.

(ii) Management

- Remove and destroy severely mined leaves to reduce further damage.
- Spray triazophos 40 EC at 1.5 ml/l or methomyl 40 SP at 1.5 g/l or abamectin 1.9 EC at 0.4 ml/l against larvae.
- Spray with dichlorvos 76 EC at 1 ml/l or decamethrin 2.8 EC at 1 ml/l for managing adults.
- Install yellow sticky traps coated with chlopyriphos 20 EC at 2.5 ml/l to trap adult population.
- Dipping of chrysanthemum cuttings in 3% margosan (neem formulation) for 2 hr and foliar spray of the same botanical at 0.4% reduce the number of pupae and adults (Lindquest *et al.*, 1986).

Figure 1.24. Leaf miner infestation on chrysanthemum leaf

- Insect growth regulatory compounds like cyromazine, methoprene, fenoxycarb and avermectin reduce leaf miner population (Parella *et al.*, 1983).
- Removal and destruction of weeds which act as alternate hosts is very important to protect the crop from leaf miner damage.
- The parasitoid, *Diglyphus* sp. at 1000/week effectively checks larval stages (Jones & Parella, 1986).

1.4.1.5. Bud Borer, *Helicoverpa armigera*

(i) Damage: Larvae tunnels its way into the buds and feeds on developing flower parts. The pest can be detected by the presence of holes on buds and fecal matter at entrance of such holes (Fig 1.25).

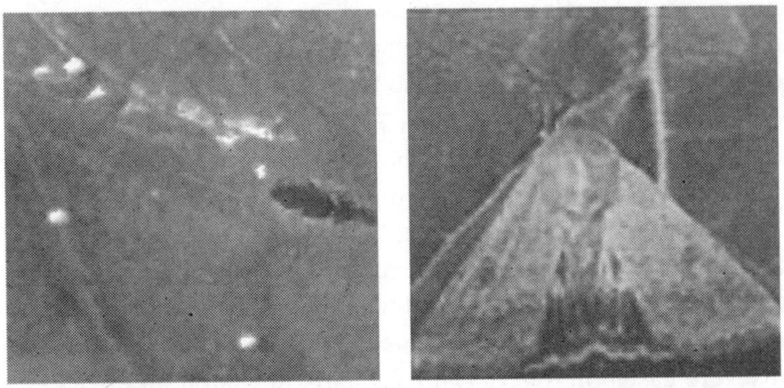

Figure 1.25. Bud borer infestation on chrysanthemum

Bud worms are the larvae of nocturnal (active at night) moths. Bud worm larvae feed mainly on very young terminal shoots, young foliage, flower buds and heads. Evidence of feeding can be seen by characteristic round holes in buds and flower heads.

Fully-grown caterpillars are about 40 mm long and may be various shades of yellow, green, pink or brown, with dark flecking and dark markings or stripes along their bodies. Caterpillars feed for two to three weeks in warm weather. When they have eaten enough, they leave the plant and burrow about 100 mm into the soil where they pupate.

Adult moths have a wingspan of about 40 mm and have grey-brown to reddish-brown fore wings with darker markings. The hind wings are pale at the base and dark at the tips. Moths are inactive during the day and fly at night to lay eggs and feed on nectar. Each female can lay up to 1000 eggs which are deposited singly on new leaves.

(ii) Management

NSKE 4% sprays offers protection to the crop from the borer damage. Other control measures are same as in rose bud borer.

1.4.1.6. Red Spider Mite, *Tetranychus urticae*

(i) Damage: Chrysanthemums are damaged when nymphs and adults pierce the leaves and suck the contents of cells causing them to collapse and die. Heavily-infested leaves become lifeless and may become bronzed and shriveled. Plants may lose their leaves and die.

Females lay up to six eggs a day, totaling 70 or more. Eggs are minute, globular and almost transparent pale-yellow-white. Males develop from unfertilized eggs and females develop from fertilized eggs. Egg hatch takes from three to 10 days depending on the surrounding temperature. Newly-hatched larvae have six legs, are pale yellow-white, minute and oval. When they moult, larvae become eight-legged nymphs which mature after another two moults. Young mites take four to 12 days to mature and live for about three weeks. In hot weather, the complete life cycle may be as short as two weeks and overlapping generations may occur.

Adult females are about 0.5 mm long and adult males are about 0.3 mm long. When feeding in summer, adult females are a yellowish-green with two pronounced dark spots, one on each side of the body. In males, these spots are less obvious and the body is smaller and tapered towards the rear. Both sexes also have two reddish eye-spots on the body.

Red spider mite is a profuse web-spinner in its active stages. Under artificial long-day length or in warm situations, such as glasshouses, spider mite may continue to feed and reproduce through winter. Outdoors, when temperatures and day length decrease in autumn, males die and adult females stop feeding and change colour to orange-red. They usually migrate to sheltered sites where they assemble for overwintering. When temperatures increase in spring, females leave their shelters, begin to feed, and lose their red colour.

Mite colonies are usually seen on the upper surface of leaves and population spreads to entire plant in severe cases of attack. Damage starts in March and reaches a peak during

April-July. Affected leaves become discoloured, wilt and fall off while damaged shoots dry up.

(ii) Management

- Spray of dicofol 18.5 EC at 2.5 ml/l or dimethoate 30 EC at 2 ml/l or triazophos 40 EC at 1.5 ml/l or profenphos 50 EC at 1 ml/l.
- After 15 days repeat the spray with pongamia or neem oil at 1%.
- Removing and burning of heavily infested shoots checks further build up of mites.

1.4.2. Diseases

Although the list of diseases affecting Chrysanthemums is long, they are relatively trouble-free given full sun, fertile well-drained soil, and adequate watering. Maximizing plant vigour by appropriate cultural practices is an important disease management strategy.

1.4.2.1. *Septoria* Leaf Spot, *Septoria chrysanthemi*

(i) Symptoms: The lower leaves are infected first. The spots are at first yellowish then become dark brown and black. Severe infection may result in the infection of entire leaves and they prematurely wither and drop off from the plant. The disease develops more rapidly in cool-wet weather. Spores of the fungus are spread by splashing water.

(ii) Management

Regularly clean up and destroy infected plant debris and hand pick symptomatic leaves from lightly infested plants. Avoid splashing water onto plant foliage and water early in the day to allow foliage to dry quickly. In severe cases, fungicides such as chlorothalonil, mancozeb, myclobutanil, propiconazole, or thiophanate methyl may be applied.

The disease is effectively checked by fortnightly spraying of 0.1% carbendazim. Other fungicides such as captan, mancozeb or zineb also provide satisfactory control of the disease.

1.4.2.2. *Alternaria* Leaf Spot, *Alternaria* sp.

(i) Symptoms: The pathogen cause small dark brown spots with yellow surroundings. The spots merge and leaves begin to rot. Severe infection results in withering of leaves, with dead leaves hanging to the stem for some time.

Symptoms first appear as yellow spots which turn brown to black. Spots often occur on lower leaves first and can coalesce into large necrotic areas and finally death of the entire leaf.

(ii) Management

Spraying mancozeb or copper oxychloride each at 2 g/litre of water gives effective control of the disease.

1.4.2.3. Wilt, *Verticillium albo-atrum*

(i) Symptoms: Wilt is the most serious disease of chrysanthemum. Symptoms of *Verticillium* wilt often appear only after blossom buds have formed; young vigorous plants may be symptomless. Foliage becomes yellow and wilted, sometimes only along leaf margins and on one side of the plant. Leaves begin to die from the base of the plant upward and often remain attached. Stems may exhibit dark streaks in the vascular system. Infected plant remains stunted and fails to produce flowers. This disease is favored when cool weather is followed by hot.

(ii) Management

Start with pasteurized growing media and pathogen-free cuttings. Most cultivars are resistant. Avoid susceptible cultivars including 'Bright Golden Ann', 'Echo', 'Glowing Mandalay', Mountain Peak', 'Paragon', 'Pert', 'Puritan', and 'Wedgewood'.

The pathogen survives in the soil and hence continuous cropping in sick soil should be avoided. Drenching soil around plant base with 0.2% carbendazim, benomyl or pentachloro nitrobenzene (PCNB) helps in reducing the disease incidence.

1.4.2.4. *Fusarium* Wilt, *Fusarium oxysporum* var. *trachaephilum*

(i) Symptoms: The first signs of this disease are yellowing of foliage, stunting, and wilting often along one side of plant. Plants may appear water stressed and foliage may become brown and die. Stems show a reddish brown discolouration of the vascular system. *Fusarium* is spread in contaminated soil and from infected cuttings and is favoured by warm temperatures, high relative humidity, overwatering, and poor drainage.

(ii) Management

Start with pathogen free cuttings or plants and pasteurized growing media. Adjust pH to 6.5 to 7.0 and use nitrate nitrogen fertilization. Avoid highly susceptible cultivars such as 'Bravo', 'Cirbronze', 'Illini Trophy', 'Orange Bowl', 'Royal Trophy', and 'Yellow Delaware'.

1.4.2.5. Gray Mold, *Botrytis cinerea*

(i) Symptoms: Gray mold may occur on petals, leaves, or as a stem canker as brown, water-soaked spots. Infected plant parts may be covered with gray to brown, powdery masses of spores. Senescing tissues are most susceptible. Gray mold is favored by extended periods of cloudy, humid, and wet weather.

(ii) Management

Practice good sanitation including removing senescing flowers and leaves. Avoid wetting flowers when watering and do not over crowd the plants. Provide good air circulation and keep humidity low by a combination of heating and venting. Apply preventive fungicides as soon as disease is detected. Fungicides such as chlorothalonil, mancozeb, copper sulfate pentahydrate, fenhexamid, azoxystrobin, and thiophanate methyl are registered for *Botrytis* control. Be sure to rotate applications among chemical classes as fungicide resistant strains of *Botrytis* have been reported.

1.4.2.6. Flower Blight, *Stemphylium lycopersici*

(i) Symptoms: The infection is noticed on calyx and petals in the form of necrotic brown spots. Early infection on buds may prevent blooming. Infected flowers wither and die. These fungi become serious when temperature and humidity are high.

Stemphylium lycopersici causes small, necrotic, light brown to dark brown lesions which may coalesce and cause blossom death. The pathogen is favoured by wet conditions and temperatures between 60° to 85°F.

(ii) Management

Rogue and dispose of infected plants. Provide good air circulation and do not over crowd plants. Avoid overhead irrigation and keep plants dry.

The disease can be controlled by spraying 0.1–0.2% mancozeb or zineb at 15 days interval. Benomyl at 0.1% sprays have also been reported to be quite effective.

1.4.2.7. Powdery Mildew, *Erysiphe cichoracearm*

(i) Symptoms: As the name implies, it is characterized by a white to ash-gray powdery growth on leaves and occasionally stems. Foliage may become puckered or distorted; severely infected leaves will shrivel and die. The disease is most serious during hot, humid weather. Unlike most fungal diseases, free water is not required for powdery mildew infection; high humidity encourages disease development.

(ii) Management

Powdery mildew can be avoided by proper plant spacing, good air circulation, low relative humidity, and adequate light levels. Apply preventive fungicides at the first sign of disease with myclobutanil, triadimefon, propiconazole, sulphur, potassium bicarbonate, or thiophanate methyl.

1.4.2.8. Rust, *Puccinia chrysanthemi, P. horiana*

(i) Symptoms: Two species of *Puccinia* causes rust on chrysanthemums, *P. chrysanthemi* and *P. horiana*. *P. chrysanthemi* is most common in late summer and is characterized by dirty-brown pustules and yellowish-green spots on upper surfaces of leaves. *P. chrysanthemi* causes minor damage in the field and is uncommon on greenhouse plants. Severe infestation may damage large areas of leaves and lead to defoliation and reduced flower production.

P. horiana causes chrysanthemum white rust. Symptoms include white, pinkish or brownish pustules produced on underside of leaves with white, yellow, to pale-green lesions on upper leaf surfaces. Chrysanthemum white rust results in leaf distortion, discolouration, defoliation, and plant death. White rust is primarily a disease of greenhouse crops; when it occurs outside, direct sunlight and low humidity kill the spores.

(ii) Management

Chrysanthemum varieties resistant to rust include 'Achievement', 'Copper Bowl', 'Escapade', 'Helen Castle', 'Mandalay', 'Matador', 'Miss Atlanta', 'Orange Bowl', and 'Powder Puff'. Infected plants to be destroyed to prevent disease establishment

1.4.2.9. Root Rot, *Pythium* sp., *Phytophthora* sp.

(i) Symptoms: Infected plant parts (root, stem and leaves) show sudden wilting.

(ii) Management

Provision of good drainage conditions to prevent water logging, soil solarization by covering with black polythene mulch during summer and soil drenching with mancozeb at 2 g/litre of water is effective for the management of the disease.

1.4.2.10. Bacterial Blight, *Erwinia chrysanthemi*

(i) Symptoms: Symptoms of bacterial blight extend beyond plant leaves to include water-soaked lesions on stems, darkening and death of buds and stems, blackening of terminals, wilt and collapse of upper portions of the plants. Infected cuttings may have brown to black decay at their base. Bacterial blight pathogen survives in crop debris and is favoured by surface moisture, high temperature, and high humidity. It is easily spread through infested tools, hands, or plants.

(ii) Management

The disease can be managed by starting with pasteurized growing media, use of pathogen-free cuttings, reducing humidity and increasing air circulation, avoiding wetting of the foliage, practicing good sanitation, and regularly inspecting the crop and disposing of infected plants.

Table 1.5. Disease management

Disease (causal agent)	Symptoms	Survival of pathogen and effect of environment	Control
Bacterial blight *(Erwinia chrysanthemi)*	Water-soaked lesions; pith becomes jellylike; tops turn black and exude drops of liquid; stem may break or split.	In plant debris; favoured by high temperatures (80° to 90°F), absence of free water, and high humidity.	Use disease-free cuttings reduce humidity in growing areas; dip cuttings in streptomycin.
Cottony rot *(Sclerotinia sclerotiorum)*	Stems rotted; flower rot is similar to gray mold; cottony fungal mass may occur on rotted tissues; black sclerotia may form inside or outside stems.	Sclerotia in soil; spores produced from sclerotia are airborne but infect only through flowers and dead tissues; favored by high humidity.	Same as for gray mold; also, treat soil with PCNB before planting and spray foliage with iprodione or thiophanate-methyl.
Crown gall *(Agrobacterium tumefaciens)*	Irregular or round galls on stems and sometimes leaves.	Soil and galls; infection favored by moist conditions.	Destroy infected plants.

Table. 1.5. (Contd.)

Disease (causal agent)	Symptoms	Survival of pathogen and effect of environment	Control
Foliar nematode *(Aphelenchoides ritzema-bosi)*	Dark green, angular spots in leaves develop progressively upward from base of plant; leaves turn yellow; these nematodes are rarely important in California.	Adults can survive up to 3 years in dead leaves; spread by splashing water; leaves need to be wet for infection to take place.	Use disease-free plants; discard infected plants; avoid overhead irrigation; control weeds; submerge infected plants in hot water (115°F) for 10 minutes.
Fusarium wilt *(Fusarium oxysporum* f. sp. *chrysanthemi F.o.* f. sp. *tracheiphilum)*	Unilateral chlorosis of one or more leaves near the apex, followed by curvature of the stem towards the affected side; as the disease progresses, there is a general chlorosis and wilt and stunting of leaves; the vascular system becomes reddish brown.	Soilborne and carried in cuttings; favored by high soil temperatures (80°F).	Use disease-free cuttings in clean soil; treat soil (see Verticillium wilt). Adjust pH of soil to 6.5 to 7.0 and use nitrate nitrogen; avoid planting highly susceptible cultivars ('Bravo', 'Cirbronz' 'Illini Trophy' 'Orange Bowl', 'Royal Trophy', 'Yellow Delaware').
Gray mold *(Botrytis cinerea)*	Brown, water-soaked spots on petals; woolly gray fungal spores from on decayed tissues; rotting of lower leaves; fungus may enter and and girdle stem.	In plant debris; favored by high humidity, low temperatures (50° to 60°F), and water on plant.	Keep humidity low; avoid overhead irrigation; protect foliage with a fungicide, especially lower dense foliage; mist blooms with chlorothalonil, iprodione, or fenhexamid.
Hollow stem *(Erwinia carotovora)*	Pith of rooting cuttings deteriorates and collapses; affected tissues are brown; surviving plants do not grow satisfactorily and pith collapse may extend upward involving several internodes; 'Red Torch', 'Tempo', and 'Tempter' are very susceptible.	Bacteria may be present in vascular bundles of symptomless chrysanthemums; bacteria also present in undecomposed debris; favored by high temperatures and high moisture.	Use disease-free cuttings; reduce humidity in growing areas; streptomycin dips may be helpful.

Table. 1.5. (Contd.)

Disease (causal agent)	Symptoms	Survival of pathogen and effect of environment	Control
Powdery mildew *(Erysiphe cichoracearum)*	White powdery growth on leaves and stems; found mainly on older leaves.	Airborne spores produced only on living plants; favored by high humidity, crowding of plants, and cool weather.	Spray with piperalin to eradicate existing infections; protect foliage with myclobutanil, fenarimol, or triadimefon.
Pythium root rot and basal stem rot *(Pythium* spp.) caused by root rot;	Girdling black lesions occur near soil line; plants stunted as a result of reduced root system plants may die.	Soilborne fungus; spores spread in water or in soil; favored by excess soil moisture and poor drainage.	Treat soil as for Verticillium wilt. Drench plants with mefenoxam.
Ray blight *(Phoma chrysanthemi)*	Basal leaf and stem rot; below ground stem infection may cause a one-sided distortion and necrosis of foliage; blackish rot of petals (ray blight) may extend into flower stalk.	In mum refuse; waterborne spores; favored by rainy weather.	Use disease-free cuttings; avoid wetting foliage and keep humidity low; protect foliage with chlorothalonil.
Rhizoctonia stem rot *(Rhizoctonia solani)*	Stem rotted at soil line; plants stunted; dark fungal strands may be visible with hand lens; may kill plants.	Soilborne fungus; favored by warm, moist conditions.	Avoid deep planting; spray base of transplants with thiophanate-methyl, chlorothalonil, or iprodione after planting, or treat soil with PCNB before planting.
Rust *(Puccinia tanaceti)*	Small pustules of powdery, chocolate-brown spores on undersides of leaves and on stems.	Airborne spores produced only on living plants; free moisture necessary for infection. Principally a field disease.	Use resistant cultivars; protect foliage with triadimefon or mancozeb before rust starts to build-up; avoid wetting foliage; keep humidity low.
Septoria leafspot *(Septoria obesa, S. chrysanthemi)*	Irregular or circular, brown or black dead spots develop progressively upward	In plant debris and in soil debris for 2 years; spores spread in splashing water;	Protect foliage with a fungicide at first sign of disease; greenhouse: avoid wetting foliage;

Table. 1.5. (Contd.)

Disease (causal agent)	Symptoms	Survival of pathogen and effect of environment	Control
	from base of plant; tiny black fungal fruiting bodies in centres of spots.	favored by wet weather.	keep humidity low.
Verticillium wilt *(Verticillium dahliae)*	Yellowing and wilting of foliage, may be one-sided; leaves die and dry upward from the base of the plant.	Soilborne for many years; carried in cuttings and root divisions; favored by cool weather followed by hot weather during flowering.	Use resistant cultivars and pathogen-free plants; fumigate with chloropicrin-methyl bromide combination; in sunny climates, soil solarization might be considered.
White rust *(Puccinia horiana)*	Whitish pustules on the lower leaf surface; on the upper surface the infection is evident as pale-green to yellow spots up to an inch in diameter. Raised, waxy, pink-colored pustules are formed on lower leaf surface. As they mature and produce spores, they turn whitish in color.	This microcyclic rust survives on living chrysanthemum foliage; teliospores in the leaf pustules germinate in place producing basidiospores that are airborne and infect by direct penetration; favored by the same conditions as ordinary rust (cool, damp weather), except that direct sunlight destroys airborne basidiospores; Principally a green-house disease.	Not all cultivars are susceptible; white rust is under an eradication program in California; once the disease is confirmed, CDFA requires a control program of 3 to 5 applications of myclobutanil (Eagle or Systhane); the number of applications depends on whether or not the crop is in an infested or non-infested area or if it is propagative material; repeated treatments with myclobutanil can cause stunting of growth, so don't apply it weekly throughout the winter if white rust is not present and treatment not required by law.

Table. 1.5. (Contd.)

Disease (causal agent)	Symptoms	Survival of pathogen and effect of environment	Control
			Growers in nonregulated areas can rotate with mancozeb (Dithane) and chlorothalonil (Daconil) as a protectant treatment program; combine fungicide treatments with destruction of infected plants and crop free periods of 8 weeks; propiconazole, previously recommended as the treatment for eradication programs, may cause phytotoxic symptoms on some cultivars; infected plants or flowers cannot be moved or sold and cleanup of infestations in quarantined areas must be done to CDFA's satisfaction.

Virus or virus-like disease	Symptoms	Host range and spread	Control
Aspermy (Tomato aspermy virus)	Flower distortion and reduction in flower size; color break in florets of red, bronze, and pink cultivars; foliar symptoms not apparent; some cultivars are symptomless.	Transmitted by handling, cutting tools, and vegetative propagation; also transmitted by aphids.	Use disease-free cuttings; control aphids; remove infected plants.
Chlorotic mottle (Chrysanthemum chlorotic mottle viroid)	Mottling followed by complete chlorosis; may be confused with nutritional problems; symptoms somewhat masked under low light and cool temperature conditions below 70°F.	Spread by handling, cutting tools, and vegetative propagation.	Use disease-free cuttings; remove infected plants.

Table. (Contd.)

Virus or virus-like disease	Symptoms	Host range and spread	Control
Mosaic (Chrysanthemum mosaic virus)	Mosaic mottling of foliage; dwarfing and deforming of leaves, stems, and flowers; symptoms variable, depending upon strain of virus and cultivar.	Transmitted by vegetative propagation and by aphids.	Obtain disease-free plants; control aphids.
Spotted wilt (Tomato spotted wilt virus, Impatiens necrotic spot virus)	Frequently one-sided in plant; ring patterns on leaves of some cultivars; leaf distortion and necrosis; dark necrotic streaks on stems; flowers may be distorted and with some necrosis.	Thrips-transmitted; not spread by cutting knife, but can be spread by vegetative propagation; many weeds and perennial ornamental plants act as reservoirs of virus (dahlia, calla, nasturtium, mallow, knotweed, plantain, and others).	Eliminate nearby susceptible ornamental plants and weeds; control thrips inside and outside growing areas; use virus-free cuttings.
Stunt (Chrysanthemum stunt viroid)	General stunting of plants. Foliage may be pale with upright, young leaves; flowers are smaller than normal and some cultivars may flower 7 to 10 days early.	Viroid is easily transmitted by handling, cutting knives, vegetative propagation, etc. not spread by aphids or thrips; viroid has a wide host range; symptomless in some plants.	Obtain disease-free plants from a propagation specialist using an indexing program.
Bacterial blight *(Erwinia chrysanthemi)*	Water-soaked lesions; pith becomes jellylike; tops turn black and exude drops of liquid; stem may break or split.	In plant debris; favored by high temperatures (80° to 90°F), absence of free water, and high humidity.	Use disease-free cutting; reduce humidity in growing areas; dip cuttings in streptomycin.

1.4.3. Nematodes

The lesion and the bud and leaf nematodes are major problems in chrysanthemum cultivation.

1.4.3.1. Lesion Nematode, *Pratylenchus coffeae, P. chrysanthus*

Edward *et al.* (1969) reported *P. chrysanthus* associated with root rot of chrysanthemum.

(i) Symptoms: Rashid and Khan (1975) reported that *P. coffeae* is responsible for heavy root damage which subsequently led to poor growth of chrysanthemum. Stunting of plants with premature yellowing and drying of leaves are common above ground symptoms. Flower size is also reduced.

(ii) Histopathology: *P. coffeae* infects and remains confined to the cortical cells. Parenchymatous cells are completely destroyed due to nematode feeding.

(iii) Interaction with Other Pathogens: The presence of *P. coffeae* has been found to aggravate the disease severity of *Pythium aphanidermatum* and *Rhizoctonia solani* (Hassan, 1988). Edward *et al.* (1969) reported *P. chrysanthus* associated with root-rot of chrysanthemum.

(iv) Management

(a) Integrated Methods: Application of neem cake enriched with *T. harzianum* at 1 t/ha is effective in reducing the nematode population.

1.4.3.2. Bud and Leaf Nematode, *Aphelenchoides ritzemabosi*

A. ritzemabosi is a major pest and causes considerable damage to the foliage of chrysanthemum, zinnia, salvia, aster and dahlia (Gill & Sharma, 1976).

(i) Symptoms: In early infection, *A. ritzemabosi* feeds ectoparasitically on buds and growing points causing retardation of growth and distorted, deformed leaves with roughened feeding areas. Later it invades leaves and feeds on and destroys cells of parenchymatous tissue of the mesophyll resulting in leaf spots or blotches, easily seen first on the underside; gradually these areas turn brown and then black as infestation spreads. The veins hinder nematode movement within leaves and thus the characteristic symptoms take the form of interveinal discolouration (Fig 1.26). These are pale green at first, gradually becoming yellow and then dark brown or black. The infested leaves shrivel, hang down as the leaf dies. The nematodes escape through stomata and migrate in surface water films to infect terminal flower buds which produce deformed and under-sized blossom.

(ii) Life Cycle: In chrysanthemum leaves, a female lays about 25–30 eggs in a compact group. Eggs hatch in 3–4 days and the larvae take 9–10 days to reach maturity. The life cycle takes 10–13 days (Wallace, 1960).

(iii) Management

(a) Physical Methods: Hot water treatment of stools at 46°C for 5 min eliminates nematode infestation.

Sufficient chrysanthemum stools were immersed in hot water at 48°C to bring the temperature down to 46.6°C, no further heating being necessary, and stools were removed after 5.5 min when the temperature was down to 46.1°C.

(b) Cultural Methods: Thorough cleaning up and burning of all infected plant material is essential in all control programmes. Plants propagated from top cuttings selected from plants growing under dry conditions offer an opportunity to secure relatively clean stock for foundation plantings. Avoiding wetting and overlapping of leaves of adjacent plants, banding the base of stems with petroleum jelly or tree-banding grease and destroying all plant remains and fallen leaves are some of the practices recommended for reducing the nematode inoculum. For outdoor plantings, the ground can be kept free of nematodes by a weed free fallow of 2–3 months during winter.

Figure 1.26. Chrysanthemum leaves infected with *Aphelenchoides ritzemabosi* (Courtesy: Sasser, 1971)

(c) Chemical Methods: Dipping of chrysanthemum stools in 0.03% parathion for 20 min was most effective for eliminating *A. ritzemabosi*. Foliar application of methyl parathion, chloropyriphos and quinalphos each at 0.05% were effective in the control of *A. ritzemabosi* infecting zinnia and chrysanthemum (Gill, 1981; Gill & Walia, 1980). Drenching with 0.02% thionazin at 6.8 litres/m^2 is also effective.

(d) Host Resistance: Chrysanthemum *cvs*. Amy Shoesmith, Orange Beauty and Orange Peach Blossom are comparatively resistant.

1.4.3.3. Root-knot Nematodes, *Meloidogyne* spp.

M. arenaria and *M. javanica* have been reported on chrysanthemum (Sen & Dasgupta, 1977; Chandwani & Reddy, 1967).

(i) Management

(a) Biological Methods: Talc and pesta granules of *P. lilacinus* applied at 10 and 15 kg/ha to root-knot infested field reduced root galling (RKI = 1.4 to 2.7), nematode multiplication rate (Rf = 1.4 to 2.12) and enhanced bioagent propagules in rhizosphere (700–1070), bioagent infectivity (32–52%) and chrysanthemum flower yield by 23 to 28% (Nagesh *et al.*, 2003).

1.4.3.4. Sting Nematode, *Belonolaimus longicaudatus*

(i) Interaction with Other Pathogens: Chrysanthemum roots inoculated with *B. longicaudatus* and *Pythium aphanidermatum* develop symptoms of root rot earlier and more extensively than those inoculated with the fungus alone.

1.5. GLADIOLUS

Gladiolus (*Gladiolus* spp.) is very much liked for its majestic spikes containing attractive, elegant and delicate florets. These florets open in sequence over a longer duration and hence have a good keeping quality of cut spikes. It is cultivated for cut flowers, garden and interior decoration, making bouquets, bedding, rockeries and pots. West Bengal, Maharashtra, Uttar Pradesh, Punjab, Haryana and Andhra Pradesh are the major gladiolus-growing states.

1.5.1. Pests

1.5.1.1. Thrips, *Taeniothrips simplex*

(i) Damage: Thrips are minute insects that feed by rasping plant tissues and sucking the sap. Gladiolus thrips cause serious damage to late flowering plants and to stored gladiolus corms. The leaves of infested plants develop a bleached and silvered appearance, the flower spikes are stunted, flowers may not open, or are distorted and the petals are marked with small pale flecks. Damage to dark-coloured varieties is most noticeable and even a few flecks are enough to considerably reduce the value of the flower (Fig 1.27).

Figure 1.27. Thrip damage on flowers and corms of gladiolus

Young nymphs feed inside the leaf tubes and buds. If the plants are left in the field for a long time after flowering, the thrips enter the corms and continue feeding. Stored corms are attacked by thrips, which cause a sticky exudate. The infested corms later become hard and scabby (Fig 1.27). When the damaged corms are planted out, the thrips begin feeding on the young roots, and growth from the corm may be seriously affected.

Thrips can overwinter at any stage of their life cycle on stored corms or on plants growing in greenhouses. Although the gladiolus thrips can maintain a population outside during the summer, they cannot overwinter outside in areas where the temperature consistently falls below 10°C as a result the monitoring of corms in storage is extremely important.

The eggs are opaque, white, kidney-shaped, and about 0.34 mm long and 0.2 mm in diameter. Gladiolus thrips emerge as creamy yellow larvae, but soon turn dark brown, except for the tips of the legs which are lighter. The wings are brown above and lighter below and appear to be darker with a grey band when folded over the back. Gladiolus thrips are about 1.5 mm long and extremely slim. Males are smaller and lighter in colour than the females.

The life cycle of the thrip is very much determined by the weather and in warmer conditions several generations can be produced in a short time hence the apparent sudden epidemic.

(ii) Management

- Spraying of 0.1% acephate or 0.5% dimethoate or oxydemeton methyl when initial symptoms are noticed and at 2 weeks interval checks damage.
- Storing of infested corms at 2°C for 6 weeks and later treating them in hot water at 46°C, completely kills thrips on corms (Conijin & Goren, 1990).
- Acephate at 0.15%, primophos methyl at 0.1% and cypermethrin at 0.05% were found effective as dipping agents for the control of thrips on corms (Labnovisky & Mynett, 1990).

1.5.1.2. Cut Worm, *Agrotis segetum*

(i) Damage: The pest attack is severe in hilly regions. Female moth lays eggs near ground level on plant parts. Larvae feed on emerging shoots at night. Grown up clay coloured larvae cut the plants at ground level. Plants are vulnerable to attack up to 3-leaf stage. Cut worms also damage underground corms and developing spikes in severe cases.

(ii) Management

- Deep ploughing during summer exposes pupae to predators.
- Spread the poison bait consisting of rice bran, jaggery and chlorpyriphos (10: 1: 0.5).
- Spray methyl parathion 50 EC at 1 ml/l or quinalphos 25 EC 2 ml/l or acephate 75 SP at 1 g/l.

1.5.1.3. Leaf Caterpillar, *Spodoptera litura*

(i) Damage: Caterpillars (Fig 1.28) can also be a pest and when present will often attack flower buds chewing a hole through the entire bud and thus the petals are held inside. In addition they may also emerge from the tips of a spike in tight buds generating further holes which will detract from the beauty of the plant.

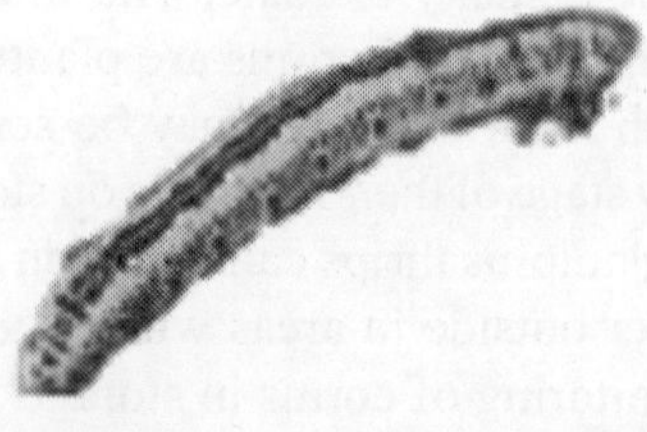

Figure 1.28. Caterpillar pest on gladiolus

Ash coloured eggs are laid in groups of 200–300 on lower side of leaves. Early instar larvae feed on lower surface of leaves by scraping and skeletonises them. Greenish brown mature larvae feed voraciously during nights on leaves, flowers and spikes. Pupation takes place in the soil.

(ii) Management

- Collect and destroy egg masses and leaves with early larval instars.
- Deep ploughing in summer for exposing pupae to predators.
- Spray quinalphos 25 EC at 2 ml/l or chlorpyriphos 20 EC at 2.5 ml/l in combination with diflubenzuron 25 WP at 2 g/l.
- Spraying 4% NSKE protects plants from damage.

1.5.1.4. Mealybug, *Ferrisia virgata*

(i) Damage: The infestation begins in the field on underground corms during dry conditions and is carried on to the storage. Nymphs and adults damage corms by sucking the sap resulting in shriveling and drying of affected corms.

(ii) Management

- Spray methyl parathion 50 EC at 1 ml/l (Krishnaiah *et al.*, 1975) or dimethoate 30 EC at 2 ml/l or acephate 75 SP at 1.5 ml/l at fortnightly intervals.
- Spray pongamia or neem oil 2%.
- Spray *Verticillium lecanii* at 3 g/l.

1.5.1.5. Slugs and Snails

(i) Damage: Both slugs and snails can cause damage but in most cases it is purely cosmetic. However, a severe attack on the leaves (Fig 1.29) can affect the plants ability to produce a new corm for the next season. Slugs have the potential to destroy the growing shoot of a newly planted corm but once established the plant should be safe from their clutches. Snails can affect the florets and make them unsuitable for exhibition.

1.5.1.6. Mite, *Tetranychus equatorius*

(i) Damage: Colonies of mite are seen on undersurface of leaves. Eggs are spherical and translucent in colour. Protonymph and deutonymph are green with dark specks on the dorsum while adults are carmine coloured. Nymphs and adults damage tissue and suck the oozing sap from leaves resulting in white spots. Damaged leaves look discoloured, gradually wilt and finally drop. Mites also infest bulbs during storage.

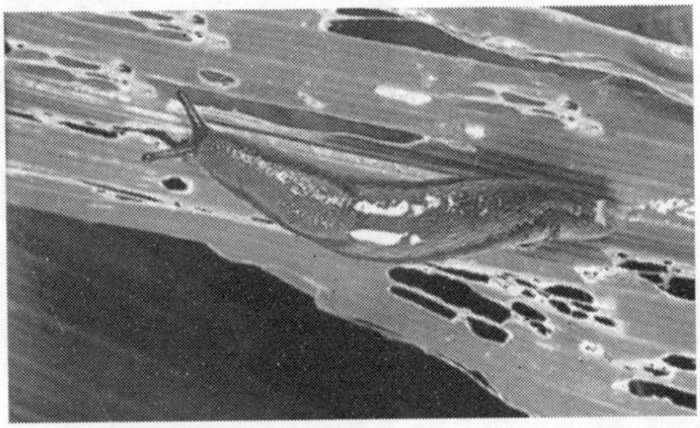

Figure 1.29. Snail damage on gladiolus leaf

(ii) Management

- Spraying of dicofol or frofenphos each at 0.05% or 0.06% triazophos causes significant mortality of the mites.
- Spraying of neem or pongamia oil 1% also gave considerable control of mites.

1.5.1.7. Bulb Mite, *Rhizoglyphus echinopus*

(i) Damage: This slow moving mite is about 0.5 mm long, globular, and yellowish-white with brownish legs. Infested corms produce stunted plants with yellow and distorted leaves. Early infestations are found around the basal plate of the old corm. Roots are destroyed first and stems are attacked later. Corms can be completely destroyed by the combined action of the mites and micro-organisms that invade the damaged tissue.

(ii) Management

Hot water treatment will kill the mites, but good sanitation is very important when digging up, storing, or planting out corms.

1.5.2. Diseases

1.5.2.1. Wilt, *Fusarium oxysporum* f. sp. *gladioli*

(i) Symptoms: This disease is caused by the fungus *Fusarium oxysporum* var. *gladioli* and is often referred to as Fusarium yellows or Fusarium Wilt. The symptoms are yellowing of the leaves starting with the older outside ones first and this is accompanied by the apparent stunting of newer leaves. In addition, the spike itself will often be stunted and faded in colour. The plant may also have blackened areas at the base which spread onto the corm (Fig 1.30). Eventually, if left, the whole plant will wilt and when lifted for destruction, (for this is the only course of treatment), the interior of the corm when opened will appear marbled with a brown colour.

Figure 1.30. *Fusarium* infection on gladiolus

(ii) Management

Where *Fusarium* wilt occurs, fertilize sparingly: nitrogen should be applied in the nitrate form rather than as ammonia or urea. A soil pH 6.6–7.0 is preferred. Harvest corms in dry weather if possible and cure them rapidly using forced air at 80–100°F if available. Inspect stored corms and rogue out suspects.

Planting healthy corms in clean soil, crop rotation, soil solarization and hot water treatment of corms at 50–55°C for 10 min were effective in controlling wilt disease. As a prophylactic measure, corms should be treated in 0.1% benomyl, carbendazim and thiabendazole solution before sowing. Captan at 0.2% can be used in combination with the above fungicides to minimize the chances of development of resistant strains of the pathogen. Solarization of field soil by mulching with transparent polythene sheets (100 gauze) raised the soil temperature which was lethal to wilt pathogen. *T. harzianum*/*T. viride* either alone or in combination with compost (mushroom, vermin, paddy straw) reduced wilt effectively (58 to 92%). Soil application of *T. harzianum, T. viride, T. koningii, T. virens* increased germination (95.6 to 100.0% compared to 79.0% in control), decreased the plant mortality (2.33 to 8.70% compared to 28.67% in control), increased plant height (76 to 85 cm compared to 51 cm in control), increased spike length (65 to 72 cm compared to 46 cm in control) and increased the number of cormels/plant (15 to 22 compared to 13 cm in control). Under field conditions, the disease can be effectively controlled by application of neem cake at 1.25 tonnes/ha along with *T. harzianum* before planting, followed by soil drenching with 0.1% carbendazim one month after planting.

Gladiolus *cvs.* IARI Selection-1, *Psittacinus* hybrid and Suchitra were found resistant to wilt under Pune conditions.

1.5.2.2. Corm Rot, *Pencillium gladioli, Rhizopus arrhizus, Fusarium* spp., *Curvularia* spp., *Stromatinia* spp., *Botrytis* spp.

(i) Symptoms: The fungi cause heavy loss in field as well as in storage by causing dry rot, hard rot, soft rot, etc. Most of these rots develop as a result of mechanical injury to corms during harvesting and storage. Generally, corms harbour more than one pathogen and symptoms are difficult to separate out pathogen-wise.

Corms that have been damaged in some way during the harvesting operation, including damage caused by the removal of the mother corm, or have not been cured properly whilst drying can often attract a *Penicillium* based rot. This rot manifests itself by the appearance of a fluffy blue or green mould which grows around and over the area of the wound.

(ii) Management

Most of these rots can be prevented by avoiding injury to corms during harvesting and storage. They should be stored in cool and dry containers. Badly infected corms should be discarded.

The disease can be controlled effectively by hot water treatment and by corm dipping in carbendazim or captan or chlorthalanil all at 0.2% solution. The corms should be dried after treatment.

1.5.2.3. Stromatinia Rot, *Stromatinia gladioli*

(i) Symptoms: This corm disease is caused by a fungus called *Stromatinia gladioli* and is most prevalent during periods of cool wet weather and as such can be a particular problem in northern European climates. The first signs of disease are usually on the growing plant and are evident by the yellowing of leaves which eventually shrivel and die. Evidence of the disease on the corms is the appearance of reddish brown markings which are sunken and if the corm is dissected, dark markings are seen as a marbling emanating from the centre (Fig 1.31).

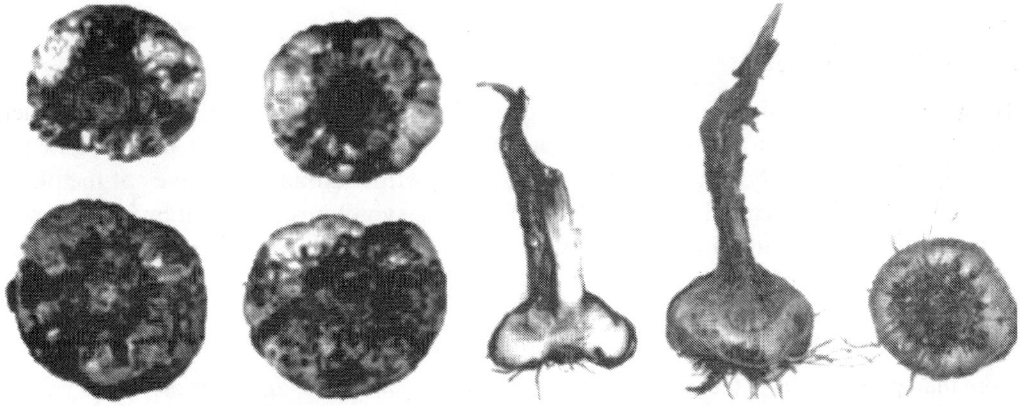

Figure 1.31. *Stromatinia* rot of gladiolus corms

(ii) Management

A particular problem with this disease is that the fungus produces an over wintering particles called sclerotia. As a result diseased corms should never be planted and as a precaution corms should always be harvested in dry conditions. If *Stromatinia* is suspected in growing plants that soil should not be replanted in subsequent years as the disease is capable of passing from soil to plant.

1.5.2.4. Rust, *Uromyces transversalis*

(i) Symptoms: The gladiolus rust is a disease of quarantine importance in Europe and the United States. Because it attacks mainly hybrid cultivars it would have serious consequences if it became established in greenhouses or nurseries. The *sori* or spore producing bodies are unusual as they develop across the width of the leaf rather than longitudinally along the veins as in most monocots. At first the sori are covered by the leaf epidermis which soon splits, releasing the uredospores as shown below (Fig 1.32):

Figure 1.32. Rust on gladiolus leaf

1.5.2.5. *Botrytis* Blight

(i) Symptoms: This disease is usually associated with the later part of the season when days are still warm but nights are cool and there is a large amount of moisture in the atmosphere. It is usually compounded by a lack of air flow around the leaves of the plant, especially evident if planted too closely together. The symptoms are small brownish spots on the leaves which often merge to form larger areas which eventually may develop into a mouldy grey growth.

1.5.2.6. Virus Diseases

More than seven virus diseases have been recorded in gladiolus, and of these, bean yellow mosaic virus (BYMV) and cucumber mosaic virus (CMV) are the most economically important and widespread.

(i) Symptoms: Gladiolus plants infected with BYMV and/or CMV often have pale-yellow mottling of the leaves (Fig 1.33). Symptoms also include "flower break" and "colour break" of flowers, but not in white, cream or yellow cultivars (Fig 1.34). BYMV can reduce the production of gladiolus corms by about 33%. Infected plants may have a shortened life span and corms are also more susceptible to fungal attack.

Figure 1.33. Mosaic on leaves which is typical of BYMV in gladiolus

Figure 1.34. Severe colour break in red variety which is typical of a dual infection with CMV and BYMV.

(ii) Host Range: Both viruses are able to infect a wide range of different plant species, including many common weeds, pasture legumes, other bulb species, and vegetables.

(iii) Spread and Source of Infection: BYMV and CMV can infect a number of plant species apart from gladiolus, including pasture legumes and weeds which provide a field source of these viruses. The most common source of infection in gladiolus is from existing gladiolus crops that are infected with the viruses. Both viruses are spread from plant to plant by aphids which transmit these viruses in only 30 seconds. Aphids will also feed on corms and hence gladiolus corms are susceptible to virus spread during storage.

(iv) Management

The best control strategy is to plant corms from virus tested stock. To minimize the risk of infecting healthy stock:

- Use suitable insecticides for the control of aphids on both the cultivated crop and the stored gladiolus corms.
- Rogue any plants which show symptoms and destroy this material.
- Plant virus-tested gladiolus material at least 1 km (if possible) from virus-infected gladiolus crops.

1.5.3. Nematodes

Root-knot nematodes are recognized as the major limiting factors in successful production of gladiolus crop.

1.5.3.1. Root-knot Nematodes, *Meloidogyne* spp.

The root-knot nematodes are major limiting factors in the production of gladiolus.

(i) Economic Importance and Losses: Parvatha Reddy *et al.* (1979) reported *M. incognita* causing heavy damage to gladiolus.

(ii) Symptoms: Severe galling on roots results in yellowing of leaves which subsequently leads to stunted growth. The nematode invades roots, daughter corms and cormels which develop after flowering.

(iii) Survival: The nematodes would survive in corm tissue in the soil as a source of inoculum for next season (Parvatha Reddy *et al.*, 1979).

(iv) Interaction with Other Pathogens: Incidence of gladiolus scab caused by *Pseudomonas marginata* was found to increase the population of *M. javanica* (Elgoorani *et al.*, 1976).

(v) Management

(a) Physical Methods: Hot water treatment of corms at 58°C for 30 min eliminates root-knot nematode infection.

(b) Cultural Methods: Nematode-free planting material should be used for planting. Crop rotation or intercropping with marigold helps in reducing the nematode population.

(c) Chemical Methods: Dipping of gladiolus corms in thionazin or fensulfothion solution (0.5 g *a.i.* per litre) gave reduced root-knot nematode infestation (Overman, 1970). Application of Vorlex (35 gal/acre) by broadcast method also resulted in better flower yield (Overman, 1967).

(d) Integrated Methods: Soil application of neem cake enriched with *T. harzianum* at 1 t/ha is effective.

Application of 5 tonnes of FYM enriched with *P. fluorescens* (with 1×10^9 cfu/g) per ha significantly reduced *M. incognita* by 61–76%, *R. reniformis* by 65–70% in the roots of gladiolus and increased the flower yield by 19–22%.

1.5.3.2. Root-knot Nematode, *Meloidogyne incognita* and Wilt, *Fusarium oxysporum* f. sp. *gladioli* Disease Complex

(i) Management

(a) Biological Methods: Application of *P. fluorescens* effectively controlled the disease complex leading to significant improvement in plant growth and flowering (Mustafa & Khan, 2004).

(b) Integrated Methods: Gladiolus plants treated with *P. lilacinus* + *T. harzianum* + neem cake and *P. lilacinus* + *T. viride* + neem cake combinations not only controlled *M. incognita* infection but also *Fusarium* wilt (*F. oxysporum* f sp. *gladioli*) till the harvest of flower spikes. The corms and cormels obtained from plants treated with these combinations were free from *Fusarium* infection. Fungal colonization of galled roots was maximum in *P. lilacinus* + *T. viride* combination (94%) followed by *P. lilacinus* + *T. harzianum* combination (69%). Similarly, the parasitization of eggs was maximum in *P. lilacinus* + *T. viride* combination (44%) followed by *P. lilacinus* + *T. harzianum* combination (38%), while egg mass parasitization was same in both the combinations (58%) (Nagesh *et al.*, 1998).

1.6. TUBEROSE

Tuberose (*Polianthes tuberosa*) is one of the most important bulbous ornamentals grown both for cut and loose flowers, and also for the extraction of its highly valued natural flower oil. It is also grown for garden decoration in pots, beds and borders. It is commercially grown in Karnataka, West Bengal, Maharashtra, Punjab, Andhra Pradesh and Tamil Nadu. At present, the total area under tuberose cultivation in the country is estimated at about 3,000 ha.

1.6.1. Pests

1.6.1.1. Aphids, *Aphis* spp.

(i) Damage: Black coloured aphids colonize on growing spikes at base of the buds and suck the sap. Damaged buds and flowers fall off while affected spikes look withered and droop. Black sooty mold develops on the buds in case of heavy infestations.

(ii) Management

- Spray dimethoate 30 EC at 2 ml/l or oxydemeton methyl 25 EC at 2 ml/l at 10–15 days interval.
- Spray imidacloprid 200 SL at 0.4 ml/l or cartap hydrochloride 50 SP 1 g/l, if the incidence is serious.

- Pongamia or neem oil sprays at 1% at 10–15 days interval are also effective against the pest.
- Use of commercial formulations of *Verticillium lecanii* at 15 days interval reduces aphid population.

1.6.1.2. Red Spider Mite, *Tetranychus* sp.

(i) Damage: Mite colonies are found on lower surface of leaves which are covered by dirty webs. Affected leaves get discoloured and drop. Attacked plants lose vigour, get stunted in growth, wilt and dry if damage is severe. Damage will be severe if the plants are affected in the early stage.

(ii) Management

- Cutting and burning of severely infested plant parts.
- Thorough spraying of water on the plant dislodges the webs.
- Spray dicofol 18.5 EC at 2.5 ml/l or ethion 50 EC at 1 ml/l or profenphos 50 EC at 1 ml/l followed by a spray of pongamia oil 1%.

1.6.1.3. Thrips, *Thrips* sp.

(i) Damage: Thrips cause damage by sucking sap from leaves and growing buds resulting in browning and distortion of damaged leaves. Damaged leaves become brittle and develop brown streaks while affected buds get distorted in shape and dry up. Severely damaged plants get retarded and do not produce flower spikes. Sometimes the pest is associated with transmission of bunchy top disease in which inflorescence gets malformed.

(ii) Management

- Spray oxydemeton methyl 25 EC at 2 ml/l or dimethoate 30 EC at 2 ml/l or acephate 75 SP at 1 g/l along with 0.5% pongamia oil at 15 days interval at appearance of early symptoms

1.6.1.4. Weevil, *Myllocerus* spp.

(i) Damage: Ash coloured adult weevils feed on leaves from edges and produce characteristic holes and also damage tender shoots. Grubs feed on roots and tunnel into bulbs causing wilting of attacked plants. Full grown grubs pupate in earthen cells below the ground level.

(ii) Management

- Treatment of soil with 0.1% carbaryl or malathion dust before planting control grubs.
- Sprays of fenthion or carbaryl at 0.1% protect the foliage from adults damage.
- Neem oil at 1% effectively checks feeding by the weevils.

1.6.1.5. Bud Caterpillar, *Helicoverpa armigera*

(i) Damage: Cream coloured eggs are deposited on leaves, buds and flowers. Larvae bore into buds and feed on flowers resulting in considerable flower loss.

(ii) Management

Effective control of the caterpillar can be obtained by a spray of 0.01% fenvalerate or 0.05% ethofenprox followed by 0.07% endosulfan or 0.05% methyl parathion at 10 days interval (Jhansi Rani & Jagan Mohan, 1997).

1.6.1.6. Mealybug, *Ferrisia virgata*

(i) Management

Application of 2% pongamia oil provides good control of mealybugs.

1.6.2. Diseases

1.6.2.1. Leaf Spot/Blight, *Alternaria polyantha*

(i) Management

Spraying of 1% Bordeaux mixture is recommended to control the disease. Tuber treatment with *Trichoderma harzianum* and *Pseudomonas fluorescens* at 4 g/kg tuber and seedling dip in *T. harzianum* suspension before planting resulted in effective control of blight and increased the plant vigour.

1.6.2.2. Hippeastrum Mosaic Virus

(i) Symptoms: A potyvirus measuring 750 nm × 12 nm has been reported on tuberose. Unlike other potyviruses, it can not be transmitted by aphids.

1.6.3. Nematodes

Root-knot and bud and leaf nematodes are recognized as the major limiting factors in successful production of tuberose crop.

1.6.3.1. Root-knot Nematodes, *Meloidogyne* spp.

Jayaraman *et al.* (1975) found that root-knot nematodes as the major factors in tuberose decline.

M. incognita, M. javanica and *M. arenaria* have been reported as the major limiting factors in successful tuberose cultivation in Tamil Nadu (Sundarababu & Vadivelu, 1988b), while *M. incognita* and *M. javanica* were potential pests in Karnataka (Khan & Parvatha Reddy, 1992) (Fig 1.34).

(i) Economic Importance and Losses: *M. incognita* was responsible for 13.25, 9.87, 14.30, 13.78 and 28.58% reduction in plant weight, number of flowers, spike length, spike weight and number of bulblets, respectively (Khan & Parvatha Reddy, 1994).

(ii) Symptoms: Affected plants exhibit stunting, yellowing and drying up of leaves and rotting of bulbs (Jayaraman *et al.*, 1975) (Fig 1.35). Further, the emergence of side shoots from the bulbs was also affected and the numbers were conspicuously less. In severely infected plants, the emergence of spike is suppressed and 65% reduction in top growth occurred (Sundarababu & Vadivelu, 1988b).

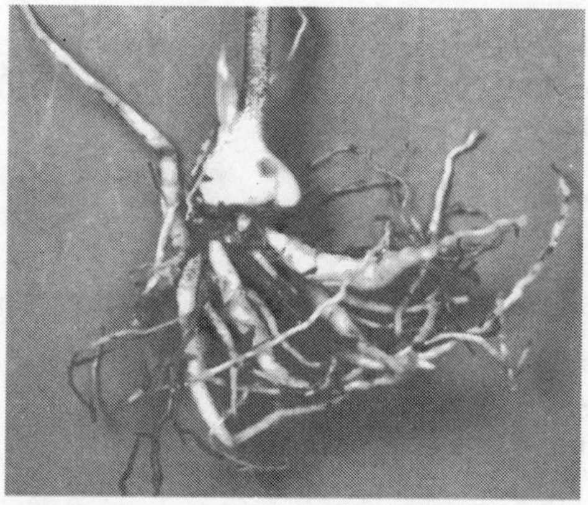

Figure 1.35. Root-knot nematode on tuberose

(iii) Management

(a) Physical Methods: Hot water treatment of bulbs at 49°C for 60 min eliminates root-knot nematode infection.

(b) Cultural Methods: Soil application of neem cake at 1 t/ha is effective.

(c) Chemical Methods: For the control of *M. incognita* on tuberose, aldicarb, fensulfothion and phorate each at 1.5 to 2.5 kg *a.i.* per ha were effective and increased flower yield. Application of carbofuran at 2 g *a.i.*/plant also checked the nematode effectively.

(d) Host Resistance: Tuberose *Cv.* Shringar (Single type) was found resistant, while Suvasini (Double type) was tolerant to *M. incognita* (Nagesh *et al.*, 1995).

Among the six varieties of tuberose (*Polianthes tuberosa*) screened for their resistance to root-knot nematode *M. incognita, cv.* Mexican Single was tolerant under screen house experiments in sterilized soil, while under field conditions it was found to be moderately tolerant. Both, under screen house and field conditions, *cv.* Vybhav and Suvasini were found to be moderately susceptible, while *cv.* Shringar and Pearl Double were found to be moderately tolerant under screen house conditions, but moderately susceptible under field conditions. In case of *cv.* Prajwal, it was observed to be moderately susceptible under screen house conditions, but susceptible under field conditions (Shylaja, 2004).

(e) Biological Methods: The experiment carried out for the biological control of root-knot nematode (*M. incognita*) in tuberose showed that, *P. chlamydosporia* and *T. harzianum* treatments gave significantly higher control of nematodes than *P. lilacinus* (Shylaja, 2004).

(f) Integrated Methods

Arbuscular Mycorrhizal Fungi and Botanicals: Integration of arbuscular mycorrhizal fungi (AMF) such as *Glomus mosseae* or *G. fasciculatum* with neem cake or/and aldicarb gave effective management of root-knot nematodes infecting tuberose (Khan & Parvatha Reddy, 1994).

Bioagents and Botanicals: Integration of *P. lilacinus* with neem cake gave effective control of *M. incognita* on tuberose. Treatment of tuberose bulbs with neem cake extract mixed with *P. lilacinus* spores significantly reduced *M. incognita* infection and multiplication besides stimulating the plant growth. Soil application of neem cake enriched with *T. harzianum* at 1 t/ha is also effective.

Split application (at planting and 45 DAP) of neem cake at 1 kg/1.5 m^2 in combination with *P. lilacinus* (at 25 g/1.5 m^2 containing 18 × 10^8 spores/g) significantly reduced root galling, soil and root population of *M. incognita* and its multiplication rate. The above treatment also significantly increased number of spikes, spike length and weight, number of flowers/spike, parasitization of egg masses and root colonization by *P. lilacinus* (Nagesh *et al.*, 1998).

Bulb dip treatment with 5% neem leaf extract mixed with spores of *P. lilacinus* (2 × 10^4 spores/ml) for 30 min. combined with soil drench with 5% neem leaf extract (at 50 ml/plant) 30 DAP gave significant increase in plant growth, flower yield and parasitization of eggs and egg masses and least root galling and reproduction factor (Nagesh *et al.*, 1997).

Two Bioagents: Combined application of *P. chlamydosporia* + *T. harzianum* gave significantly higher control of *M. incognita* on tuberose (Shylaja *et al.*, 2004). Soil application of *T. harzianum* (10^9 cfu/g) + *P. lilacinus* (10^6 cfu/g) gave effective control of root-knot nematodes.

1.6.3.2. Bud and Leaf Nematode, *Aphelenchoides besseyi*

(i) Management

(a) Physical Methods: Cent percent mortality of *A. besseyi* was obtained by immersing tuberose bulbs in hot water bath maintained at 50°C for 30 min (Khan *et al.*, 2004).

(b) Host Resistance: Tuberose *cvs*. Suhasini, Shringar, Prajwal and Hyderabad Double were tolerant to *A. besseyi* (Khan *et al.*, 2004).

(c) Integrated Methods: Pre-soaking of tuberose bulbs in water overnight followed by hot water treatment at 50°C for 20 min + spraying of the crop twice with monocrotophos 36 EC at 0.15% (at sprouting and 30 days after first spray) in the first year and 3 sprays in second and third year crop at 15–20 days interval was found most effective in reducing % nematode infestation as well as disease indices (Khan *et al.*, 2004).

1.6.3.3. Root-knot Nematode, *Meloidogyne incognita* and Wilt, *Fusarium oxysporum* f. sp. *dianthi* Disease Complex

(i) Symptoms: In an experiment undertaken to assess the impact of *Fusarium* wilt-root knot disease complex on the flower yield in tuberose, it was observed that the disease complex drastically reduced the yield of flowers in *P. tuberosa,* thereby bringing down the production of flowers in this commercial crop. It was found that the damage was maximum in presence of both pathogens *viz;* *M. incognita* and *F. oxysporum* f. sp. *dianthi*, when compared to the presence of either one of these pathogens (Shylaja, 2004).

(ii) Management

(a) Integrated Methods: In course of the experiments carried out to evaluate combination of bio-agents for the control of wilt (*F. o.* f. sp. *dianthi*) and root-knot nematode

(*M. incognita*) disease complex in tuberose, the best result was obtained in plants treated with *P. chlamydosporia* + *T. harzianum* (Shylaja, 2004).

1.7. CROSSANDRA

Crossandra (*Crossandra undulaefolia*) is mainly cultivated commercially for its loose flowers in Tamil Nadu, Karnataka and Andhra Pradesh. It is also grown in Sri Lanka, Tropical Africa and Madagascar. It is a perennial herb and belongs to the family Acanthaceae.

1.7.1. Pests

1.7.1.1. Brown Scale, *Saissetia nigra*; White Scale, *Orthezia insignis*

(i) Damage: Yellowish nymphs and dark brown adult scales are seen in large numbers under leaves (Fig 1.36), on shoots, petioles and stem. Severe infestation results in stunted growth and leaves turn yellow and drop. Honey dew secreted by *Orthezia* sp. favours development of sooty mould which harms the plant growth.

Figure 1.36. Scale infestation on leaves of crossandra

(ii) Management

- Removal and burning of infested portion of plants prevents further spread.
- Application of carbofuran at 1.0 kg *a.i.*/ha and spraying of 1% pongamia oil in combination with 0.05% methomyl or 0.01% acephate or fenthion provide effective control of scales on crossandra.
- Spray ethion 50 EC at 1 ml/l or chlorpyriphos 20 EC at 2.5 ml/l or dimethoate 30 EC 2 ml/l in combination with malathion 50 EC at 1 ml/l.
- The above chemical spray should be followed by 1% pongamia oil.

1.7.1.2. Spike Borer, *Helicoverpa armigera, Archips epicyrta*

(i) Damage: Spike borer attack is seen from October to February. Eggs are laid singly on tender spikes. Hatched larvae feed on spikelets, flowers and developing seeds. Early instar larvae bite holes into spikelets. Matured larvae feed on spikelets, developing buds and tender seed. *Archips epicyrta* often make silken webs on spikelets. Pupae are formed outside the spikes and brown adult emerges in 10 days (Vadivelu *et al.*, 1976).

(ii) Management

- Collection and destruction of grown up larvae and webbed spikes reduces borer population.
- Spray endosulfan 35 EC at 2 ml/l or methyl parathion 50 EC at 1 ml/l in combination with diflubenzuron 25 WP at 2 g/l two to three times at 15 days interval for protecting spikes from borer.

1.7.1.3. Crossandra Bug, *Gynencia affinis*

(i) Damage: Crossandra bug is very specific on this crop. Bugs are small, greenish with brown markings. The entire development from egg to adult takes 23–28 days. Both nymphs and adults suck sap resulting in brown lesions. Severely infested plants fail to produce flowers. Population is high during winter although the pest occurs through-out the year.

(ii) Management

- Spray quinalphos 25 EC at 2 ml/l or acephate 75 SP at 1.5 g/l at or carbaryl 50 WP at 2 g/l.
- Give subsequent spray with neem oil 1% (10 ml/l) to protect foliage from bug damage.

1.7.1.4. Midge, *Contarinia* sp.

(i) Damage: Attack of the midge is more during July-November. Black coloured adult fly lays eggs singly within buds of the spikes. Orange maggots feed on ovaries of developing buds and pupate inside. Due to damage, the buds fail to develop and rot in severe cases.

(ii) Management

- Soil application of carbofuran 3G at 1 kg *a.i.*/ha.
- Spraying of triazophos 40 EC at 1.5 ml/l or ethion 50 EC at 1 ml/l or acephate 75 SP at 1.5 ml/l.

1.7.2. Diseases

1.7.2.1. Foot Rot or Root Rot, *Phytophthora nicotianae*

(i) Symptoms: This disease leads to sudden death of plants. In young seedlings, symptoms appear as lesions on rootlets followed by rotting of the entire rootlet. On the collar region, peculiar berry root can be seen. Leaves show pink discolouration and drooping. In advanced stage of infection, wilting of whole plants can be noticed.

(ii) Management

Growing seedlings in raised beds drenched with captan, application of neem cake along with *Trichoderma harzianum* helps to control the disease. Soil drenching with Al-Fosetyl for 2–3 times at monthly intervals during monsoon is effective.

1.7.2.2. Wilt, *Fusarium* sp., *Rhizoctonia solani*

(i) Symptoms: The characteristic symptoms of the disease are drooping of leaves, wilting of plants and rotting of roots.

(ii) Management

Drenching the plants with captaf at 2 g/litre of water is effective in controlling the disease.

Table 1.6. Pest/disease management in crossandra

Crossandra Pests/Diseases	Chemicals and concentration
Leaf eating caterpillars	Endosulfan at 2 ml or carbaryl at 3 g/l of water
Thrips	Dimethoate at 1 ml/l of water
Mites	Wettable sulphur at 3 g/l of water
Nematodes	Soil application of carbofuran at 10–15 kg/ha and application of neem cake at 500 kg/ha
Fusarium corm rot	Soil drenching with carbendazim at 2 g/l of water and foliar spray at the same concentration
Leaf spot	Mancozeb at 2 g/l of water

1.7.3. Nematodes

Pratylenchus delattrei, Meloidogyne incognita and *Longidorus africanus* are the major nematode pests of crossandra.

1.7.3.1. Root-knot Nematode, *Meloidogyne incognita*

In India, *M. incognita* was first reported on crossandra by Rajendran *et al.* (1976).

(i) Economic Importance and Losses: The nematode is responsible for 25.62% and 21.64% loss in number of flowers and weight of flowers, respectively (Khan & Parvatha Reddy, 1994).

(ii) Occurrence and Distribution: Root-knot nematodes are found in all crossandra growing areas like Tamil Nadu, Karnataka and Andhra Pradesh.

(iii) Symptoms: Infected plants are stunted with dried peripheral branches bearing smaller chlorotic leaves almost turning to white in later stages (Rajendran *et al.*, 1976). Roots exhibit severe galling. Inflorescences are small and sometimes fail to produce flowers.

(iv) Interaction with Other Pathogens: *M. incognita* interact synergistically with wilt fungi *Fusarium oxysporum* and *F. solani* and results in premature death of crossandra plants

(Khan & Parvatha Reddy, 1992). *F. solani* and *F. oxysporum* are invariably associated with the lesions caused by *P. delattrei* in crossandra (Khan & Parvatha Reddy, 1992; Srinivasan & Muthurksihanan, 1975). Srinivasan (1974) found that wilt symptoms appeared earlier when nematodes were added two weeks prior to *F. solani*.

(v) Management

(a) Cultural Methods: Deep ploughing and fallowing in summer for about a month helps in reducing the nematode population. Planting crossandra after graminaceous crops should be avoided. Incorporation of FYM in soil and intercropping of crossandra with marigold or pangola grass would reduce the root-knot nematode infection.

Intercropping of crossandra with enemy plant like marigold significantly reduced the root-knot and lesion nematode population both in soil and roots (Khan & Parvatha Reddy, 1994).

(b) Chemical Methods: Application of carbofuran at 2 to 3 kg *a.i.* per ha would be beneficial.

(c) Biological Methods: The parasitic fungus, *Paecilomyces marguandi* grown on paddy seeds at 2 g per kg soil gave effective control of root-knot nematode infecting crossandra (Khan & Parvatha Reddy, 1994).

Among four arbuscular mycorrhizal fungi evaluated, *Glomus fasciculatum* and *G. mosseae* colonized crossandra roots better with comparatively lower nematode infection and multiplication, thus promoting better root health and plant growth compared to *G. intradices* and *Aucalospora laevis*.

(d) Integrated Methods

Bioagents and Botanicals: Incorporation *V. lecanii* with neem cake facilitated the effective management of *M. incognita* on crossandra. Application of neem cake enriched with *T. harzianum* at 2 kg/m^2 (2 t/ha) in nursery beds gave effective control.

Combinations of *V. lecanii* and *P. lilacinus* (2×10^4 spores/ml each) with 5% neem leaf extract resulted in significantly higher plant growth parameters and crossandra flower yield (Nagesh & Parvatha Reddy, 1995). Root galling, nematode multiplication factor were least and the parasitization of eggs and egg masses was highest in *V. lecanii* +5% neem leaf extract.

Arbuscular Mycorrhizal Fungi and Botanicals: *G. fasciculatum* and *G. mosseae* in combination with neem cake gave better control of nematodes over carbofuran treatment. The root colonization of AM fungi increased significantly in the presence of neem cake which in turn improved their efficacy in reducing *M. incognita* population in crossandra roots (Nagesh *et al.*, 1998). Integration of neem, karanj and castor cakes with AMF, *G. mosseae* significantly enhanced plant growth parameters and flower yield of crossandra, root colonization and sporulation of AMF. The above treatments also reduced root-knot nematode multiplication and root-galling (Nagesh & Parvatha Reddy, 1997).

Bioagents, Arbuscular Mycorrhizal Fungi and Botanicals: Integration of a bioagent (*V. lecanii*), endomycorrhiza (*G. mosseae*) with botanicals improved the growth of crossandra and reduced the population of *M. incognita. G. mosseae* reduced the requirement of phosphatic fertilizer and favoured the antagonistic potential of *V. lecanii* against *M. incognita*.

Two Bioagents: Combined application of *P. lilacinus* and *Pasteuria penetrans* enhanced plant growth and flower yield of crossandra besides reducing root galling due to *M. incognita* (Nagesh *et al.*, 1995).

Bioagents and Botanicals/Chemicals: Integrated management of root-knot nematodes infecting crossandra was achieved by rational combination of a biocontrol agent (*T. harzianum* or *P. chlamydosporia*) with a nematicide (aldicarb) or oil cake (neem cake) (Khan & Parvatha Reddy, 1994).

1.7.3.2. Lesion Nematode, *Pratylenchus delattrei*

The lesion nematode was first recorded by Srinivasan (1974) on crossandra from Tamil Nadu.

(i) Economic Importance and Losses: Heavily infested plants do not produce tertiary spikes and thereby the flower yield is reduced by nine-fold (Srinivasan & Muthukrishnan, 1975).

(ii) Occurrence and Distribution: *P. delattrei* has been reported on crossandra from Tamil Nadu and Rajasthan.

(iii) Symptoms: *P. delattrei* was frequently associated with crossandra crop exhibiting stunting, chlorosis of leaves and wilting. The leaves showed mottled appearance which turned to brown and becoming pinkish eventually. Later, the leaf tips dried and entire leaves turned yellow and shed. The plants were devoid of side shoots and remained completely defoliated. The roots exhibited brown to black lesions. Heavily infested plants did not produce tertiary spikes and thereby flower yield is reduced (Srinivasan & Muthukrishnan, 1975).

(iv) Host Range: Besides crossandra, *P. delattrei* has been recorded on sorghum, maize, red gram, okra, sunnhemp and wheat.

(v) Life Cycle: The total life cycle of this nematode from egg to adult stage ranges between 46 to 55 days. Eggs are laid inside the roots and all stages from larvae to adult are infective.

(vi) Histopathology: Nematode moves intracellularly and remains in cortical region. Its infection results in dissolution of cell walls which leads to cavity formation in cortical region. Cell wall gets thickened and cell contents become more dense and granular.

(vii) Interaction with Other Pathogens: *Fusarium solani* and *F. oxysporum* are invariably associated with the lesions caused by *P. delattrei* (Khan & Parvatha Reddy, 1992). Srinivasan and Muthukrishnan (1975) also opined that crossandra decline in Tamil Nadu occurs due to *P. delattrei* and *F. solani* complex. Srinivasan (1974) found that wilt symptoms appeared earlier when nematodes were added two weeks prior to *F. solani*.

(viii) Survival: The nematode can survive up to eight months in host-free moist soil, whereas in desiccated soil it could survive up to three months.

(ix) Management

(a) Cultural Methods: Intercropping of crossandra with enemy plant like marigold significantly reduced the lesion nematode population both in soil and roots (Khan & Parvatha Reddy, 1994). Application of farm yard manure has been found to reduce the lesion nematode population in crossandra.

(b) Chemical Control: Nursery bed treatment with phorate, aldicarb, fensulfothion and carbofuran each at 5 g per m^2 gave good and healthy crossandra seedlings free from *P. delattrei* (Vadivelu & Muthukrishnan, 1979).

Root dip treatment for 8 hr in 0.05 to 0.1% solution of aldicarb, carbofuran, fensulfothion, fenamiphos or phorate protected crossandra seedling from *P. delattrei* for 30 days after transplanting (Vadivelu & Muthukrishnan, 1979).

Srinivasan and Muthukrishnan (1976) found that carbofuran at 2.5 kg *a.i.* per ha was effective under field conditions. Application of phorate at 1 g per plant a week after planting checked the lesion nematodes effectively and increased the flower yield (Vadivelu & Muthukrishnan, 1979).

(c) Biological Methods: Application of *T. viride* at 10 g/kg soil was found effective to maximize the flower yield and suppressing the lesion nematode population both in soil and roots (Jothi *et al.*, 2004).

1.7.3.3. Needle Nematode, *Longidorus africanus*

In India, this nematode was reported for the first time on crossandra by Vadivelu *et al.* (1976) from Tamil Nadu.

(i) Symptoms: In crossandra, *L. africanus* causes stunting of plants, branching, swelling and darkening of root tips (Muthukrishnan *et al.*, 1977).

(ii) Host Range: Sorghum, snap bean, lima bean, sugar beets are reported as best hosts, while wheat, cotton, okra, cucumber, egg plant and tomato as good hosts (Kolodge *et al.*, 1987).

(iii) Histopathology: *L. africanus* initiates hyperplasia in parenchymatous cells of peripheral region which subsequently progress towards the centre, forking of roots and clubbing of root terminal. Due to this, meristem is pushed laterally which ultimately leads to formation of darkly stained cells (Muthukrishnan *et al.*, 1977).

(iv) Management

(a) Cultural Methods: Pea, carrot, squash, spearmint and onion are reported to be poor hosts; while cauliflower, cabbage and radish as non-hosts. Either of these crops can be fitted as intercrops or in crop rotation scheme. Rotation of crossandra with spearmint which is a poor host of *L. africanus* would reduce the needle nematode population (Kolodge *et al.*, 1987).

1.8. CHINA ASTER

China aster (*Callistephus chinensis*) is one of the important annual flower crops. It ranks next to chrysanthemum and marigold among traditional flowers. In recent years, it has gained more popularity due to its multifarious uses including as cut flower. The flowers fetches very good price and returns to the farmer when their production coincides with their demand during particular season/festival. It is more popular among small and marginal farmers due to its easy cultivation. It is grown for cut flowers, herbaceous borders, bedding, bouquet and pots and cultivated in most parts of the country. It is commercially grown in Karnataka, Tamil Nadu, Andhra Pradesh, Maharashtra (Pune) and West Bengal. In Bangalore and Pune alone it is being grown in an area of 500 and 400 ha, respectively.

1.8.1. Pests

1.8.1.1. Flower Caterpillars, *Helicoverpa armigera, Phycita* sp.

(i) Damage: Caterpillars feed on the floral parts damaging the flowers by making groves in the receptacles. They feed on ovaries and stamens of flower buds.

(ii) Management

- Spray endosulfan 35 EC at 2 ml/l or methyl parathion 50 EC at 1 ml/l or phenthoate 50 EC at 1 ml/l or fenvalerate 20 EC at 0.5 ml/l or Neem seed kernel extract 4%.

1.8.1.2. Stem Borer, *Platyptilia molopias*

(i) Damage: The adult female plume moths lay white elongate eggs on tender leaves and tips of shoots. Hatched larvae bore into stem and side branches and their feeding results in hollowness and shoot wilt. Caterpillars are white with black head, which feed inside shoots and make them hollow. Damaged shoots bear black excreta on the top. Affected shoots wilt and dry. Larvae pupate inside the stem. Pupae are green in colour (Udayagiri, 1985).

(ii) Management

- Collection and destruction of infested shoots.
- Spray profenphos 50 EC at 1 ml/l or endosulfan 35 EC at 2 ml/l or ethofenprox 10 EC at 1 ml/l and application of carbofuran granules at 1 kg.*a.i.*/ha (Furadan 3 G at 33 kg/ha).

1.8.1.3. Aphids, *Aphis gossypii*

(i) Damage: Aphids mainly infest lower side of flowers and base of petals (Fig 1.37). Nymphs and adults suck the sap from the flowers causing discolouration, deformation and withering.

Figure 1.37. Aphid infestation on flower of China aster

(ii) Management

- Spraying of oxydemeton methyl 25 EC at 2 ml/l or dimethoate 30 EC at 2 ml/l.
- Spray neem or pongamia oil 1% if necessary.

1.8.1.4. Red Pumpkin Beetle, *Aulacophora foveicallis*

(i) Damage: Beetles attack China aster plants soon after transplantation. Adults feed extensively on leaves by making holes and also cut tender shoots thereby affecting the growth of the plants. Grubs feed on under-ground roots and stem leading to wilting and drying of attacked plants.

(ii) Management

- Neem oil 1% or NSKE 4% reduces damage to foliage by beetles.
- Two or three sprays of quinalphos or methyl parathion each at 0.05% or 0.1% carbaryl give good control of the beetle.
- Soil application of phorate granules at 0.5 kg *a.i.*/ha control grubs.

1.8.1.5. Semilooper, *Ctenoplusia albistriata*

(i) Damage: The attack of semilooper is observed during June-July on young plants. Eggs are deposited on leaves. Green coloured semilooper with lines on dorsal and lateral sides feed voraciously on leaves and pupates on the plant causing severe defoliation.

(ii) Management

Collecting and killing of grown up larvae reduces further damage. Spraying of quinalphos or methyl parathion each at 0.05% or endosulfan at 0.07% controls semilooper effectively.

1.8.2. Diseases

1.8.2.1. Wilt, *Fusarium oxysporum* var. *callistephi, Sclerotium rolfsii*

(i) Symptoms: Both the pathogens are widely distributed and gross symptoms are same. The leaves wilt and roots are rotted. Generally the lower leaves show wilting symptoms and gradually wilting progresses upwards. A close examination of the stem base in Fusarial wilt shows a thin coating of pale pink or rose coloured fungal growth which is more prominent especially under humid conditions. In Sclerotial wilt, there is buff to dark brown round bodies with infected network attached to the stem base near the soil level.

(ii) Management

Use of clean soil, avoiding bringing in soil on tools from beds previously infected with wilt fungi and discouraging continuous cropping in infested soil may help in reducing the disease incidence. Soil solarization should be done during summer months. Planting may be done on ridges.

Soil drenching with 0.2% benomyl or carbendazim gave good control of fusarial wilt, while soil drenching with 0.2% PCNB (Brassicol) is effective for sclerotial wilt.

Cvs. Heart of France, Crego, Ostrich, Feather, Stardust and Roment are reported to be resistant to Fusarial wilt.

1.8.2.2. Root Rot, *Pythium* sp.

(i) Symptoms: The main symptoms are yellowing of leaves, stems and roots of affected plants. The infected plants appear to be water-soaked and black.

(ii) Management

Irrigation should be regulated and proper drainage should be provided. The seedling should be transplanted on ridges.

1.8.2.3. Viral Disease

(i) Symptoms: The characteristic symptoms of viral infection include appearance of yellowish tinge on leaves, stunted growth, chlorosis and flower discolouration.

(ii) Management

The infected plants should be collected and destroyed to prevent further spread of the disease.

1.8.3. Nematodes

The root-knot nematodes are recognized as the major limiting factor in the successful production of china aster crop.

1.8.3.1. Root-knot Nematodes, *Meloidogyne* spp.

The root-knot nematodes are a major limiting factor both in nursery and main field (Krishnappa *et al.*, 1980; Khan & Parvatha Reddy, 1992).

(i) Management

(a) Host Resistance: China aster *cv*. Shashank was found resistant, while Poornima was moderately resistant to root-knot nematodes. There were significantly higher total phenols and total proteins and higher activity of polyphenol oxidase in infected roots of resistant variety, compared to that in susceptible variety (Kamini) during early stages of infection. As the infection advanced the activity of phenyl alanine ammonia lyase was higher in the roots of resistant variety (Nagesh *et al.*, 1995).

1.9. MARIGOLD, *Tagetes* spp.

1.9.1. Pests

1.9.1.1. Red Spider Mite, *Tetranychus urticae*

(i) Damage: Mite attack on marigold is observed during August-September. Colonies of mites are found on lower surface of leaves which are covered by dirty webs (Fig 1.38). Both nymphal and adult stages feed on the sap. Affected leaves get discolored and drop. Attacked plants get stunted in growth, wilt and dry if damage is severe.

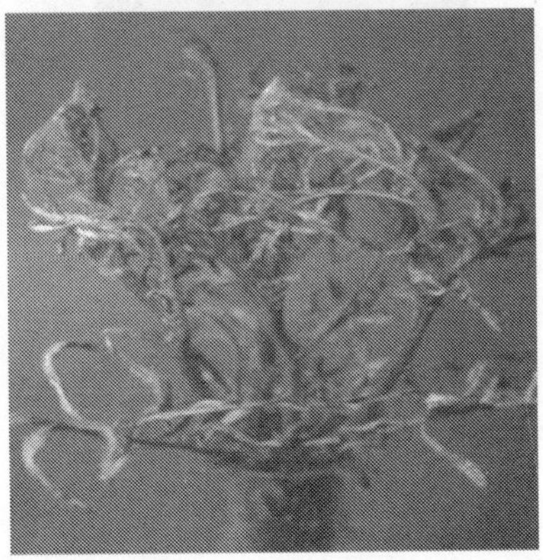

Figure 1.38. Red spider mite damage on marigold

(ii) Management

- Cut and burn the severely infested plant parts.
- Thoroughly spray water on the plant for dislodging the webs.
- Spraying dicofol 18.5 EC at 2.5 ml/l or wettable sulphur 80 WP at 3 g/l or ethion 50 EC at 1 ml/l or profenophos 50 EC at 1 ml/l.
- Follow the above spray with spray of pongamia oil 1%.

1.9.1.2. Aphid, *Aphis gossypii*

(i) Damage: Aphids mainly infest growing tips and floral parts. Nymphs and adults suck the sap from growing parts and the plants give a sickly appearance.

(ii) Management

- Spray oxydemeton methyl 25 EC at 2 ml/l or dimethoate 30 EC at 2 ml/l.
- Spraying of neem or pongamia oil 1% cause considerable reduction in aphid population.

1.9.1.3. Bud Caterpillars, *Helicoverpa armigera, Phycita* sp.

(i) Damage: Marigold is one of the preferred hosts of *H. armigera* and an average of 4–5 larvae are seen per plant (Joginder Singh & Sohi, 1990). Moth lays eggs singly on young buds. Hatched larvae bore into buds and growing flowers and feed on floral parts damaging florets and petals.

Larvae of *Phycita* sp. damage flowers by making groove on the central receptacle and feed on ovaries and stamen filaments (Fig 1.39). Excreta of the caterpillars are seen on damaged flowers.

Figure 1.39. Bud caterpillar on marigold flower

(ii) Management

- Collection and destruction of infested buds, flowers and mature larvae to reduce further damage.
- Spray endosulfan 30 EC at 2 ml/l or methyl parathion 50 EC at 1 ml/l or fenvalerate 20 EC at 0.5 ml/l in combination with diflubenzuron 25 WP at 2 g/l at appearance of eggs on tender foliage checks borer damage.
- Spray neem seed kernel extract 4%.
- Spraying of *Ha* NPV at 250 LE/ha.

1.9.2. Diseases

1.9.2.1. Damping-off, *Pythium* sp.

(i) Symptoms: Brown necrotic spots appear on the germinating seedling first on the root followed by stem. Brown ring forms on the stem at ground level. The roots begin to rot.

(ii) Management

Drenching the nursery beds with mancozeb at 2 g/litre of water gives effective control of the disease.

1.9.2.2. Powdery Mildew

(i) Symptoms: The characteristic symptom of the disease is covering of leaves with white powdery mass.

(ii) Management

Spraying with wettable sulphur at 3 g/litre of water is effective in managing the disease.

1.9.2.3. Leaf, Spot, *Alternaria tagetica, Septoria* sp., *Cercospora tagetis, C. tageticola*

(i) Symptoms: The main symptom of the disease is appearance of brown, brownish-grey and blackish spots on leaves.

(ii) Management

Spraying with mancozeb at 2 g/litre of water is effective in controlling the disease.

1.10. NARCISSUS

1.10.1. Pests

1.10.1.1. Bulb Mite, *Rhizoglyphus echinopus*

(i) Damage: This slow-moving mite is about 0.5 mm long, globular, and yellow-white with brownish legs. Infested bulbs produce stunted plants with yellow and distorted leaves. Early infestations are found around the basal plate of the bulb. Roots are destroyed first and bulb scales and stems are attacked later. Bulbs can be completely destroyed by the combined action of the mites and micro-organisms that invade the damaged bulbs.

(ii) Management

Hot water treatment will kill the mites but good sanitation is very important when digging up or planting out bulbs.

1.10.1.2. Bulb Flies, *Merodon equestris, Eumerus tuberculatus*

Adult bulb flies are stocky, hairy flies, blackish to dark green with pale yellow, orange, or gray markings. Adult bulb flies look like small bumble bees and may also resemble other beneficial species. On warm sunny days, adults often hover around blooming plants, where they feed on pollen and nectar. Bulb fly larvae are plump, wrinkled, dirty yellow, gray, white, or brownish maggots with a short brown or blackish breathing tube at their rear. Maggots of several species of flies infest bulbs of all types, including daffodil, iris, lily, and tulip.

(i) Damage: The adult narcissus bulb fly is about 0.5 inch long and closely resembles a small bumblebee. The larva is yellowish-white and 0.75 inch long. Larvae feed in bulbs, causing yellow foliage and stunted plants. If infested, large bulbs produce fewer leaves and distorted growth. Instead of a large shoot, small leaves may emerge in a ring around the central growing point that has been killed by a larva. Blooms and small bulbs often die. *Merodon equestris* attacks healthy plants; *Eumerus* spp. prefer bulbs and rootstocks that are already diseased or decayed. The narcissus bulb fly attacks amaryllis, daffodil, *Galtonia,* hyacinth, iris, lily, *Leucojum, Narcissus, Scilla* , tulip, and *Vallota.*

(ii) Biology: The female lays eggs near the bulbs during May or early June. The eggs hatch into grubs that move down into the soil and burrow into the base of the bulb to feed there, reducing the bulb's center to a rotten mass. In the process, the embryonic flowers that would have unfolded the following spring are destroyed. After feeding, larvae leave the bulb to pupate in the soil. In the spring, the grubs emerge as adult flies to mate and lay eggs.

(iii) Management

Prevention is the most effective bulb fly control strategy. Plant only pest-free bulbs. Handle bulbs carefully to avoid injury. Remove and destroy infested bulbs and plants as soon as you find them.

Narcissus fly can be evaded by deep planting (10 inches). Plant in open areas exposed to wind. Bulb flies are usually less of a problem in such sites. Covering foliage of plants with a row cover after bloom prevents egg-laying. Plant only firm, healthy bulbs. Soft bulbs may be infested from the previous season. Destroy any soft bulbs to prevent emergence of adult bulb flies.

1.10.2. Diseases

Table 1.7. Disease management

Disease/Causal	Symptoms	Survival of pathogen	Control
Basal rot, *Fusarium oxysporum* f. sp. *narcissii*	Infected plants are stunted with distorted and yellow leaves which usually die; the basal plate is decayed and reddish brown in colour; the white to pinkish fungus is sometimes seen at the base between the scales; eventually the bulb rots.	Fungus survives as chlamydospores in soil for long periods; short rotations are not effective; disease is favoured by warm soils and is limited below 55°F; disease is favoured by high levels of nitrogen fertilization; potassium fertilizer may decrease disease sensitivity.	Avoid excessive fertilization. Dig up bulbs in diseased areas as soon as possible; store bulbs below 64°F; dip bulbs in thiabendazole as soon as possible after digging; hot water treatment (see stem and bulb nematode) is also effective in reducing the disease; rotate out of narcissus for at least 3 years; plant when soil is cool; *Cvs.* Malvern City, Rijinveld's Early

Table. 1.7. (Contd.)

Disease/Causal	Symptoms	Survival of pathogen	Control
			Sensation, White Lion, 'Soleil d'Or' 'Dulcimer' are mod. resistant; cv. St. Keverne is resistant.
Crown rot, *Sclerotium rolfsii*	A bulb rot that is at first wet and later becomes dry and woody; a white fungus mat and small (0.03 to 0.125 inch), round, tan to brown sclerotia occur on decaying bulbs and in surrounding soil.	Disease is favoured by warm weather; fungus survives in soil for long periods (10 years); fungus has a wide host range; can be spread by infected bulbs.	Avoid infested fields for 4 years or longer; deep ploughing is sometimes practiced to bury sclerotia; treat bulbs in hot water (see stem and bulb nematode); PCNB applied in the furrow at planting can be effective.
Fire *Botryotinia polyblastis*	Water-soaked areas on petals that become brown and wither; flowers are attacked first, and later the foliage; foliage spots are small, elliptical, tan and usually near the tips; yellow streaking follows the leaf spots.	Fungus survives as sclerotia in leaf debris; Sclerotia germinate in spring to produce fruiting bodies (apothecia), which produce airborne spores (ascospores); conidia, which are also airborne, cause secondary spread; the disease is favoured by mild, humid conditions.	Remove flowers before fungus produces apothecia in March/April; protect flowers and foliage with a fungicide.
Scorch, *Stagonospora curtsii*	Yellow to brown lesions at leaf tips; elongate reddish brown leaf spots; small black pycnidia form in necrotic areas.	The fungus survives in the neck and between scales; disease is favoured by mild, moist conditions; spores are spread in splashing water; the fungus infects members of the Amaryllidaceae.	Treat bulbs with hot water (See stem and bulb nematode); protect plants in the field with fungicides; make first application as leaves emerge.
Smoulder, *Botrytis narcissicola*	A dark brown lesion first appears on leaf tips; infected leaves may curl when infection occurs on inner edge; masses of	Fungus survives in infected bulbs and as sclerotia in soil; conidia are airborne; favoured by cool, wet weather.	Rotate with other crops; dig bulbs early; Dip bulbs in suitable fungicide such as thiabendazole; if disease is severe in

Table. 1.7. (Contd.)

Disease/Causal	Symptoms	Survival of pathogen	Control
	woolly gray spores (conidia) and small black sclerotia form on diseased tissues, especially near soil.		field, spray with iprodione, fenhexamid, or thiophanate-methl.
Stem and bulb nematode, *Ditylenchus dipsaci*	Leaves are stunted and distorted, and often swollen and thickened near the base; yellow or brown swellings occur in leaf centres and margins; severely infected bulbs are unproductive and may rot.	Nematodes survive in infected bulbs and in bulb debris in fields for about 3 years; opt. temp. for infection and reproduction is 50° to 60°F (Little activity occurs when soil temperatures are lower than 50°F and greater than 68°F); nematodes are spread by irrigation water and equipment; they also survive in weed hosts and can survive desiccation.	Remove infected plants and surrounding plants from the field; clean equipment after use in diseased fields; treat bulbs in hot water; store bulbs at 60° to 64°F before treatment to reduce heat injury; pre-soak bulbs 2 to 3 hours or overnight in water plus a wetting agent at 75°F; increase temperature in morning to 109°F; maintain a constant temp. of 109° to 111°F for 3 to 4 hours; cool and dry blubs immediately.
White mold, *Ramularia vallisumbrosae*	Small, sunken, grey or yellow spots on leaves and green parts; spots enlarge and darken to a yellow-brown with yellow margins; masses of white; powdery spores (conidia) occur on leaves under moist conditions; small, dark sclerotia are produced in older infected tissues.	Fungus survives as sclerotia in dead leaves and on bulbs; sclerotia germinate to produce conidia as leaves emerge; disease is favoured by warm, moist conditions.	Do not replant for 1 year in fields where the disease has occurred; protect foliage with mancozeb; copper fungicides are also helpful.
Narcissus white streak virus	Narrow, dark green to purple streaks, which later become white to yellow-white, appear in leaves and flower	Narcissus is the only host of the virus, which is transmitted by aphids; symptoms do not occur until air	Eliminate infected plants; control aphids.

Table. 1.7. (Contd.)

Disease/Causal	Symptoms	Survival of pathogen	Control
	stalk after flowering; bulb size and yields are reduced by premature senescence.	temperatures exceed 64°F.	
Narcissus yellow stripe virus	Conspicuous light green to yellow streaks and mottling of leaves and flower stalk, which occurs shortly after emergence; sometimes leaves are distorted and a colour-break of flowers occurs; bulb yields are reduced.	Narcissus is the only host of the virus, which is spread by aphids; symptoms appear early in growing season.	Eliminate infected plants; control aphids.

1.11. ORCHID

1.11.1. Pests

1.11.1.1. Armored Scale, *Diaspis boisduvalli*

(i) Damage: This is a serious and persistent pest on orchids. Scales colonize on the lower side of leaves, stems and shoots. They suck the plant juice and create toxic effects on plants resulting in burning and drying of damaged parts.

(ii) Management

Removal of heavily infested leaves avoids spread of scales. Spraying of dimethoate or oxydemeton methyl or ethion each at 0.05% followed by spray of 1% pongamia oil controls the scale. Spray of fish oil rosin soap 0.05 kg in 30 litres of water is recommended for controlling scale population (Abraham & Vatsala, 1981).

1.11.1.2. Mealybugs, *Pseudococcus maritimus, Ferrisia virgata*

(i) Damage: These two species of mealybugs are reported to occur commonly on orchids in India. They have soft and filamentous yellow or pink coloured body covered with a white powdery wax. They suck plant sap and cause deformation of leaves and retardation in growth of the plant. They excrete honey dew which attracts ants and sooty mold develops in case of severe infestation.

(ii) Management

Prompt collection and destruction of heavily infested plant parts reduce further spread of the pest. Spraying of methyl parathion or methomyl each at 0.05% or 0.1% acephate, when crawling of ants on plant is noticed, provide effective control of mealybug. Pongamia oil 2% sprays check mealybug infestation. Release of coccinellid predator, *Cryptolaemus montrouzieri* and a parasitoid, *Leptomastrida abnormis* reduces the mealybug population.

1.11.1.3. Thrips, *Anaphorathrips orchidii*

(i) Damage: Both nymphs and adult stages suck the sap from tender leaves, growing buds and flowers. Affected leaves develop brown streaks while buds and flowers are distorted and discoloured causing flower loss.

(ii) Management

Spraying of oxydemeton methyl or dimethoate or cartap hydrochloride or profenphos each at 0.05% or imidachloprid at 0.007% at 10 days interval, when early symptoms are noticed, effectively checks thrips damage. Misting using deltamethrin at 0.0015% or acephate at 0.05% controls thrips in glasshouse (Timmerman *et al.*, 1982).

1.11.1.4. Snail, *Achatina fullica*

(i) Damage: Attack of the snails is observed during monsoon season on orchids. Adult and young stages devour the plants by feeding on leaves at night. They scrape leaves leaving only veins and cut tender shoots (Fig 1.40).

Figure 1.40. Snail damage on leaves of orchid

(ii) Management

Hand picking of grown up stages and killing them by dropping in 5% salt solution reduce the snail population. Spreading of 3% metaldehyde pellets (Snail Kill) in the field or application of 5% metaldehyde dust on plants gives effective control of the pest. Neem oil 1% or soapnut extract 6% sprays protect foliage from damage.

1.12. ANTHURIUM

1.12.1. Insect Pests

1.12.1.1. Whitefly, *Aleurotulus anthuricola*

(i) Damage: Nymphs and adults colonize under flowers, petiole sheath and leaves; and secrete white, powdery, waxy material. As a result of sucking, the infested parts turn yellow, deform and dry in severe cases. The life cycle of whitefly is completed in 35 days on anthurium sheaths.

(ii) Management

Removal and burning of old and heavily infested leaves check pest build up. Spray of methomyl or acephate each at 0.05% significantly reduced the number of nymphs and pupae (Hat & Hara, 1992). A holistic spray schedule of methomyl or acephate at 0.05% for adults, 0.06% triazophos or methomyl or profenphos each at 0.05% followed by a spray of 1% pongamia oil for nymphs and pupae effectively controls whitefly menace on anthurium.

1.12.1.2. Thrips, *Chactanaphothrips orchidii*

(i) Damage: This is a serious pest of anthurium which causes white streaks and distortion of spathes sucking the sap. Severe injury to flowers is caused during September. Life cycle of thrips is completed in about 27 days. Matured nymphs migrate and pupate in the growing medium.

(ii) Management

Hara *et al.* (1987) recommended 0.01% oxydemeton methyl, dimethoate and chlorpyriphos each at 0.06% for the control of thrips. Spraying of new chemicals like profenphos and cartap hydrochloride each at 0.05% or acephate at 0.01% also provides effective control.

1.12.1.3. Mite, *Brevipalpus* sp.

(i) Damage: Colonies of the mite are seen on the lower side of leaves and inside spathes. Entire plant is covered with mites in case of severe attacks. Feeding by mites causes mottling and wilting of affected parts.

(ii) Management

Successful control of mites can be obtained when 6% flower injury (ETL) is noticed. Foliar application of 0.0025% vertimec or 0.025% diafenthiron followed by profenphos or amitraz each at 0.05% reduces mite population effectively.

Table. 1.8. Management of insect/mite pests in ornamental crops

Crop	Pest	Symptoms	Pesticide/ Dose	Remarks
Rose	Scale insects	Reddish-brown encrustations, loss of vigour and drying of shoots	Carbofuran 3G at 33 kg/ha (Soil application) followed by 10% Pongamia oil	Immediately after pruning.
	Aphids	Cluster of aphids found on tender parts	Dimethoate at 1.5 ml/l; phosphamidon at 0.5 ml/l	As and when found
	Thrips	Silvering, browning and distortion of tender leaves, buds and flowers	Methyl demeton/ dimethoate at 2.0 ml/l	As and when found
	Leaf eating beetles	Damaged flower buds, flowers and foliage	Endosulfan 35 EC at 2 ml/l; monocrotophos at 1.5 ml/l	As and when found
	Leaf & flower eating caterpillars	Damaged flower buds, flowers and foliage	Quinalphos at 2.0 ml/l; monocrotophos at 1.5 ml/l; endosulfan at 2.0 ml/l	As and when found
China aster	Aphids	As in rose	Phosphamidon at 0.5 ml/l; 1% neem oil	As and when found
	Leaf & flower eating caterpillars	As in rose	Quinalphos at 2.0 ml/l; monocrotophos at 1.5 ml/l; endosulfan at 2.0 ml/l	As and when found
Chrysanthemum	Aphids	Yellowing, leaf fall and stunted growth	Dimethoate at 1.0 ml/l; Neem oil/pongamia oil at 10 ml/l	As and when found
	Thrips	Adults and nymphs suck the sap from leaves and flowers causing streaks, browning and distortion.	Methyl demeton at 2 ml/l; dimethoate at 2 ml/l	As and when found
	Leaf caterpillar	Larvae fold the leaves and feed inside causing skeletonization	Fenvalerate at 0.5 ml/l; methyl parathion at 1 ml/l	As and when observed
Crossandra	Scales	Scales suck the sap from leaves, petioles and shoots resulting in yellowing, leaf fall and stunted growth	Furadan 3G at 33 kg/ha; chlorpyriphos at 5 ml/l; neem oil at 10 ml/l	As and when observed
	Spike borer	Larvae bore into spikes and feed on buds & tender seeds.	Quinlaphos/endosulfan at 2 ml/l	As and when observed

Table. 1.8. (Contd.)

Crop	Pest	Symptoms	Pesticide/ Dose	Remarks
Gladiolus	Thrips	Flowers and leaves resulting in silvering, browning & deformation	Phorate 10G at 10 kg/ha; methyl demeton/dimethoate at 2 ml/l	As and when found
	Leaf caterpillar	Early instars skeletonize the leaves while grown up larvae damage spikes	Carbaryl at 2 g/l; quinalphos at 2 ml/l	Collect and destroy egg masses before spray
Tuberose	Thrips	Adults and young thrips suck the sap from leaves and spikes and drying of affected parts.	Monocrotophos at 1.35 ml/l; methyl demeton at 2 ml/l	As and when observed
	Aphids	Aphids suck the sap from growing spikes causing withering and bud drop.	Phosphamidon at 0.5 ml/l; dimethoate at 2 ml/l	As and when observed
Carnation	Red spider mite	Dirty white colonies of mite are found on leaves, buds and growing shoots; affected parts lose vigour, get discoloured and deformed.	Monocrotophos at 1.5 ml/l; metasystox at 2 ml/l; neem seed kernel extract at 4%	As and when observed
	Bud borer	Larvae bore into buds and feed on growing petals by making holes.	Endosulfan at 2 ml/l	As and when observed
Gerbera	Whitefly	Adults and nymphs colonize lower surface of leaves and suck the sap; damaged leaves turn yellow, curl and dry.	Carbofuran 3G at 33 kg/ha; triazophos at 1.5 ml/l	As and when observed
	Leaf miner	Larvae mine into leaves and feed on mesophyll; affected leaves turn brown and dry up.	Triazophos at 1.5 ml/l; neem seed kernel extract at 4%	As and when observed
	Aphids	Aphids attack flowers and suck the sap from petals; damaged flowers become discoloured deformed and wither.	Phosphamidon at 0.5 ml/l; methyl parathion at 1 ml/l	As and when observed

Table. 1.8. (Contd.)

Crop	Pest	Symptoms	Pesticide/ Dose	Remarks
Croton	Mealy bugs		Monocrotophos at 1.5 ml/l	
	Spider mite and gall mite	Dirty white colonies of spider mite cause discoloration, browning and drying of leaves; gall mite damage results in gall and malformation of affected parts.	Monocrotophos at 1.5 ml/l; triazophos at 2 ml/l; wettable sulphur at 3 g/l	Spray of monocrotophos/ Triazophos to be followed by sulphur
	Blossom midge	Maggots feed inside buds and the affected buds show purple coloration	Monocrotophos at 1.5 ml/l; neem oil at 1%	First spray Monocrotophos followed by neem oil
Marigold	Bud borer	Larvae bore into buds and damage petals and head portion.	Endosulfan/quinalphos at 2 ml/l	As and when observed
	Leaf hopper	Curling and purplish red coloration of leaves	Phosphamidon at 0.5 ml/l; monocrotophos at 1.25 ml/l	As and when observed
	Aphids	Terminal buds & flowers get malformed and wither	Phosphamidon at 0.5 ml/l; monocrotophos at 1.25 ml/l	As and when observed

Table 1.9. Biological control of plant parasitic nematodes in ornamental crops

Crop	Nematode	Effective bioagent/s	Reference
Tuberose	*M. incognita*	*P. chlamydosporia, P. lilacinus, T. harzianum*	Shylaja, 2004
Crossandra	*M. incognita*	*P. lilacinus, V. lecanii*	Nagesh & Parvatha Reddy, 1995
	M. incognita	*P. margaundi*	Khan & Parvatha Reddy, 1994
	M. incognita	*G. mosseae*	Nagesh *et al.*, 1996b
Carnation	*M. incognita*	*P. lilacinus*	Shylaja, 2004

2 ORNAMENTAL CROPS UNDER PROTECTED CULTIVATION

Pest management strategies for protected cultivation of ornamental crops need to include use of pesticide residue-free, highly effective, economical and safe chemicals to crops and natural enemies along with plant products and biological control agents. At present, the control methods adopted in ornamental crops include use of very costly and imported chemicals which are often toxic to predators, parasitoids and plants. Development of pest prediction models is also another significant aspect since these models can be simulated in protected cultivation and will help in proper implementation of management programmes. Integrated management methods developed by considering these aspects will provide economically and ecologically sound pest management in ornamental crops.

2.1. ROSE

Major protected rose growing areas include Ambala (Punjab), Bangalore (Karnataka), Hyderabad (Andhra Pradesh), Kodaikanal (Tamil Nadu), Nasik (Maharashtra), New Delhi and Pune (Maharashtra).

2.1.1. Pests

2.1.1.1. Thrips, *Scirtothrips dorsalis*

(i) Damage: Thrip is a serious problem on roses grown in polyhouses almost round the year except during November-December and July-August. Both nymphs and adults suck cell sap from tender leaves, growing buds and flowers (Fig 2.1). Curled leaves with brown streaks appear on the young leaves. When they feed on the young buds, scorchy appearance of sepals and petals is seen. Later the deformed flowers with burnt margins are very common.

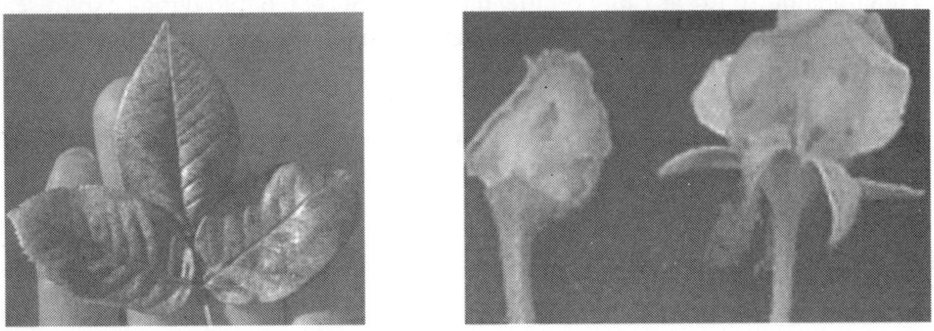

Figure 2.1. Thrips damage on rose leaves and flowers

(ii) Management

- Spray insecticides like imidacloprid 200 SL at 0.4 ml/l or fipronil 5 SC at 1.5 ml/l or acephate 75 SP at 1.5 g/l at 10–15 days interval.
- Drench the beds after raking up with chlorpyriphos 20 EC at 4 ml/l for killing resting pupae.
- Spray jatropha or pongamia oil at10 ml/l.

2.1.1.2. Red Spider Mite, *Tetranychus urticae*

(i) Damage: The mite is found initially in colonies on lower surface of mature leaves (Fig 2.2). In case of severe infestation, population spreads to entire foliage including flowers and buds covering them with webs. Hot and dry conditions (mainly summer) favour mite multiplication. Cream coloured nymphs and red coloured adults with two lateral black spots feed on leaves resulting in discolouration, bronzing and drying. Damage to buds and flowers leads to discoloration, retardation and drying.

Figure 2.2. Red spider mite infestation on leaves of rose

(ii) Management

- Cutting and burning of infested plant parts.
- Proper ventilation and watering to reduce temperatures.
- Spray abamectin 1.9 EC at 0.5 ml/l followed by flufenoxuron 10 DC at 1 ml/l or fenazaquine 10 EC at 1 ml/l or difenthiuron 50 SC at 0.6 ml/l during export season.
- Spray dicofol 18.5 EC at 2.5 ml/l or wettable sulphur 80 WP at 3 g/l or profenofos 50 EC at 1 ml/l or ethion 50 EC at 1 ml/l followed by pongamia or neem or jatropha oil at 5 ml/l during off-season.
- Release predatory mite *Amblyseius tetranychivorus* at 20 mites/plant when *T. urticae* populations are at moderate levels.
- Spray *Verticillium lecanii* formulation at 5 g/l during evening.

2.1.1.3. Whitefly, *Bemisia tabaci*

(i) Damage: The pest is active during January-June. Small and fragile white adults lay eggs on lower surface of young leaves. Nymphs and adults suck cell sap from lower side of leaves resulting in yellowing and dropping of leaves. In case of severe infestation sooty mould develops on honeydew secreted by nymphs and significantly affects the growth of the plants resulting in production of small flowers.

(ii) Management

- Removal and burning of heavily infested shoots.
- Clean cultivation and use of insect proof nets.
- Spray triazophos 40 EC at 1.5 ml/l or methomyl 40 SP at 2 ml/l or acephate 75 SP at 1.5 g/l at 15 days interval alternating with pongamia oil 10 ml/l for nymphs.
- A management package of spraying diclorvos 76 EC at 1 ml/l followed by lambda cyhalothrin 5 EC at 1 ml/l or deltamethrin 2.8 EC at 1 ml/l at 5–7 days interval and installing yellow sticky traps effectively controls whitefly adults.

2.1.1.4. Tobacco Caterpillar, *Spodoptera litura*

(i) Damage: Adult moths lay eggs in groups on lower surface of leaves and damage is seen through out monsoon period in polyhouses. Early instars are gregarious in nature and feed on leaves by scraping chlorophyll. Black coloured mature larvae damage growing buds and flowers during nights resulting in qualitative loss to flowers.

(ii) Management

- Collection and destruction of egg masses and gregarious early instar larvae.
- Spray quinalphos 25 EC at 2 ml/l or endosulphan 35 EC at 2 ml/l at 15 days interval.
- Spray indoxacarb 14.5 SC at 1 ml/l or thiodicarb 75 WP at 1 g/l, if the incidence is high.
- Sprays of *Sl* NPV at 250 LE/ha followed by neem formulations at 1–2 ml/l are also effective.
- Spread poison bait made of wheat/rice bran, jaggery and chlorpyriphos (10: 1: 0.5) for killing grown up larvae.

2.1.2. Diseases

2.1.2.1. Powdery Mildew, *Podosphaera pannosa*

(i) Symptoms: It is a debilitating disease. Though it does not kill the plant, it renders the foliage and flowers unsightly. This is the most important disease on roses grown in both polyhouses and in the open. Grayish white powdery patches are seen on the tender leaves (Fig 2.3), pedicels and flower buds. Younger leaves of 7–10 days old are more susceptible than older leaves. Infected leaves get malformed, flower buds fail to open and plants exhibit wilting symptoms. The infection occurs on the abaxial side of leaves. The luxuriant growth of the fungal mat on leaves results in greater water loss and reduced photosynthetic efficiency.

The fungus has a very short life cycle and spreads at an alarming rate. Unlike in garden roses where the infection occurs mostly in winter months, the disease occurs throughout the year on greenhouse roses. Favoured by warm and dry weather and widely varying day and night temperatures, the fungus grows well at 18–25°C and optimum germination of conidia occurs at 21°C. A night temperature of 15.5°C and RH of 90–99% promote optimum production of conidia and their germination. The fungal spores are spread by wind.

Figure 2.3. Powdery mildew on leaves and stem of rose

The fungus survives in the form of mycelial mat called "Pannose mycelium" on buds, stems and thorns. Water stress predisposes the plants to infection by the fungus.

(ii) Management

Since the disease spreads very fast, the spray schedule should begin before the onset of favourable conditions for fungal growth. As the fungus grows on external surfaces of plants even contact fungicides such as sulphur are quite effective in controlling the disease. However, when the days are hot, application of wettable sulphur should be avoided because of phytotoxicity. Early detection followed by sulphur fumigation is most effective and economical under greenhouse conditions. The most commonly used contact and systemic fungicides are 0.1% carbendazim (Bavistin), 0.1% tridemorph (Calixin), 0.1% dinocap (Karathane), 0.1% triadimefon (Bayleton), 0.1% bitertanol (Baycor), 0.05% hexaconazole (Contaf) and 0.05% myclobutanil (Systhane). Among the more recent compounds, strobilurins such as kresoxym methyl and azoxystrobin which are surface systemic protectants are effective in controlling powdery mildew disease.

Foliar sprays of some inorganic salts namely, sodium bicarbonate, dipotassium hydrogen orthophosphate and potassium dihydrogen orthophosphate at 4.5 g/litre gave better control of powdery mildew compared to fungicides. Other desirable aspects of these inorganics are: (a) their rapid absorption and extreme mobility in the plant tissue, (b) nutrient value, (c) low mammalian toxicity and environmental safety, and (d) low cost.

Some antitranspirants such as polyterpenes and hydrocarbon wax emulsions are also known to be effective in controlling the disease.

The cultural practices that keep the disease under check include minimal use of nitrogenous fertilizers and irrigation through sprinklers or mistifiers.

Among the bioagents, two antagonistic fungi namely, *Ampelomyces quisqualis* and *Pseudozyma flocculosa* can control powdery mildew when their spore suspension are sprayed onto the foliage.

2.1.2.2. Die-back, *Diplodia rosarum, Colletotrichum* sp.

(i) Symptoms: Die-back generally occurs after pruning. The infection occurs at the pruned ends of stems or through wounds caused due to harvesting of flowers or other operations. Damage caused by digger wasps also facilitates the infection. Old plants are more prone to die-back. The infected stems dry from tip downwards, often extending to the main stem and thus killing the entire plant. Browning and rotting of infected stem extends inside the stem even beyond the infected lesions on the stem surface. The fruiting bodies of the fungus are seen as small eruptions around the bases of thorns on the infected stems (Fig 2.4). Plants defoliated due to black spot and downey mildew are more susceptible because of depletion of carbohydrates in the stems. The fungus *Alternaria alternata* is also reported to cause die-back in *Rosa damascene.*

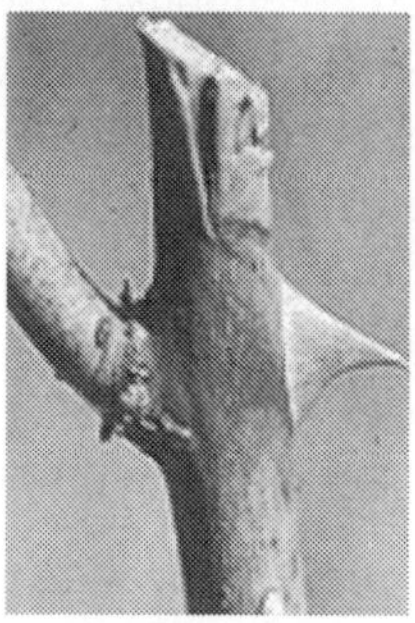

Figure 2.4. Die-back on stem of rose

(ii) Management

Pruning the infected stems 5–10 cm below the margin of infection and pasting the pruned ends with Bordeaux paste or rose paint [copper carbonate, red lead (Pb_3O_4) and linseed oil at 4: 4: 5 ratio] or 0.1% carbendazim or 0.2% copper oxychloride gives effective control of the disease. To avoid digger wasp damage, any of the common insecticides may be mixed with these fungicides. While pruning, a slant cut is advised in order to avoid formation of water droplets on the cut surface since the fungus requires water film for spore germination and infection. Use of sprinklers and mistifiers enhances the spread of the disease.

2.1.2.3. Downy Mildew, *Peronospora sparsa*

(i) Symptoms: Initially the symptoms appear as pinkish brown irregular spots on tender leaves. The fungus causes chlorotic blotches on the upper surface of the matured leaves, which ultimately become necrotic (Fig 2.5). The corresponding areas on the lower surface of the leaves show sparse growth of the fungal mycelium and sporangia that are not easily noticed. The infection on flower buds occurs very early and leads to total drying of flowers. Tender canes on infection show die-back type of symptoms. The infection results in total defoliation and severe flower blight. The infected flowers get malformed and fail to open.

Figure 2.5. Downy mildew on leaves of rose

Downy mildew is very common on greenhouse roses because of the prevalence of high humidity and low temperatures. The disease spreads very fast resulting in total defoliation of plants. The fungus survives through special perennating structures called oospores. Humid, wet and cool conditions favour the spread of the disease. RH of about 85% and temperature around 18°C are optimal for multiplication of this fungus. Under these conditions the fungus is capable of explosive spread through the production of innumerable zoospores which are the dispersal units of the fungus. Since *Peronospora* is an obligate pathogen, movement of the planting material from infected nurseries and polyhouses is the main source of spread of the fungus to new areas.

(ii) Management

The disease can be effectively brought under control by spraying 0.3% fosetyl-AL (Aliette), 0.2% metalaxyl-MZ (Ridomil-MZ) or 0.2% cymoxanil (Curzate-M). Strict phytosanitation is a must in order to keep the disease under check as the fungus can survive in the infected plant parts. All the infected flowers, stems and fallen leaves should be collected and destroyed. Anti-resistant strategies like mixing or alternating fungicides with different modes of action should be followed in order to overcome the fungus developing resistance to fungicides. Reducing the humidity in polyhouses effectively checks the spread of the disease. In nurseries, overcrowding of seedlings should be avoided by providing enough space between them. Irrigation should be so timed so as to facilitate quick drying of the foliage.

2.1.2.4. Grey Mold or Botrytis Blight, *Botrytis cinerea*

(i) Symptoms: *Botrytis* causes water-soaked lesions on leaves (Fig 2.6), which spreads very fast. On flowers, the infection occurs as small water-soaked spots resulting in immature fading and drying of petals, which drop ultimately. The infection is generally seen on inner whorls of petals. During the favourable conditions, the infection spreads very fast resulting in complete blighting of the flowers. In roses, grey mold coincides with the progress of senescence. The fungus affects all growth stages of the crop starting from planting to post-harvest. Humid and wet conditions of the greenhouses favour the growth and multiplication of the fungus. A temperature range of 10–25°C and a RH of above 85% are highly favourable for disease development. During cold storage and transhipment, the condensation on the flowers caused due to temperature fluctuations leads to flower blight. High RH besides promoting prolific sporulation of the fungus is also known to predispose the rose plants for grey mold infection by impeding the uptake of calcium by the plants. Application of calcium to rose buds reduces the severity of post-harvest *Botrytis* blight. The onset of post-harvest senescence of the flowers has a direct correlation with the severity of the disease in storage and transit.

Figure 2.6. Grey mold on leaves of rose

(ii) Management

Many fungicides belonging to benzimidazoles (carbendazim, benomyl at 0.1%), dicarboximides (iprodion) and sterol biosynthesis inhibitors are commonly used for controlling the disease. *Botrytis* has developed resistance to all the important groups of fungicides that were effectively used earlier. Chemicals with anti-senescent activity render the plants more tolerant.

Application of growth regulators like gibberellic acid, paclobutrazol and methyl jasmonate is reported to suppress grey mold in roses. Inhibitors of ethylene biosynthesis such as amino-oxyacetic acid and carbon dioxide also reduced the disease incidence.

Since, *Botrytis* sporulates readily on wounded and senescent tissues; these should be thoroughly removed from the greenhouses and destroyed. Roses under photo-selective plastic have shown a lower disease incidence compared to commercially used plastic. Occurrence of free moisture should be reduced by proper ventilation. Polyfilms that block UV rays and allow blue light are known to reduce the sporulation of the grey mold fungus to a large extent.

Biological agents like *Gliocladium roseum, Myrothecium verrucaria, Trichoderma harzianum* and *Pencillium* spp. are known to be effective in controlling grey mold.

2.2. CHRYSANTHEMUM

2.2.1. Pests

2.2.1.1. Aphid, *Macrosiphoniella sanborni*

(i) Damage: Greenish-black nymphs and chocolate-brown adults suck the cell sap from tender growing shoots and lower surface of leaves (Fig 2.7) during cool and cloudy months. Damage results in loss of vigour, yellowing and premature leaf fall and stunted growth. Heavy infestations affect bud and flower development considerably.

Figure 2.7. Aphid infestation on leaves of chrysanthemum

(ii) Management

- Spray pongamia or neem oil at 10 ml/l at early infestation.
- Spray dimethoate 30 EC at 2 ml/l or oxydemeton methyl 25 EC at 2 ml/l or imidacloprid 200 SL at 0.4 ml/l or acephate 75 SP at 1.5 g/l at 15 days interval, if the aphid incidence is severe.
- Spray *Verticillium lecanii* at 3g/l during evening.

2.2.1.2. Thrips, *Microcephalothrips abdominalis*

(i) Damage: Slender white nymphs and black adults feed on tender leaves causing bronzing and mottling of leaves. Damaged flowers get discoloured, distorted and dry due to feeding by thrips resulting in reduction in quality and yield of flowers (Fig 2.8).

(ii) Management

- Spray dimethoate 30 EC at 2 ml/l or oxydemeton methyl 25 EC at 2 ml/l at 15 days interval.
- Drench the soil with chlorpyriphos 20 EC at 4 ml/l to reduce resting populations.

Figure 2.8. Thrips damage on flowers of chtysanthemum

- Spray imidacloprid 200 SL at 0.4 ml/l or fipronil 5 SC at 1.5 ml/l or acephate 75 SP at 1.5 g/l, if the incidence is severe.

2.2.1.3. Red Spider Mite, *Tetranychus* sp.

(i) Damage: Mites mainly feed on leaves and spread to entire plant in case of heavy infestations. Damage starts in March and reaches a peak during April–June. Affected leaves become discoloured, dry and fall off, while attacked flowers get discoloured and dry.

(ii) Management

- Spray of dicofol 18.5 EC at 2.5 ml/l or triazophos 40 EC at 1.5 ml/l or profenofos 50 EC at 1 ml/l or ethion 50 EC at 1 ml/l at 15 days interval followed by pongamia or neem oil at 10 ml/l.
- Spray *Verticillium lecanii* formulation at 3 g/l.

2.3. CARNATION

2.3.1. Pests

2.3.1.1. Thrips, *Frankliniella schultzei*

(i) Damage: Both nymphs and adults attack buds and flowers. Damaged buds show brown streaks and get distorted in severe cases with burnt margins leading to quality reduction.

(ii) Management

- Spray dimethoate 30 EC at 2 ml/l or oxydemeton methyl 25 EC at 2 ml/l at 15 days interval.
- Drench the soil with chlorpyriphos 20 EC at 4 ml/l to reduce pupae.
- Spray imidacloprid 200 SL at 0.4 ml/l or fipronil 5 SC at 1 ml/l or acephate 75 SP at 1.5 g/l.

2.3.1.2. Red Spider Mite, *Tetranychus urticae*

(i) Damage: It is a devastating pest on carnation grown under protected conditions especially during September-November and February-May. Mite colonies are found on lower surface of leaves, covered by dirty webs. Affected leaves get discoloured and dry up. Infestation spreads to buds (Fig 2.9), flowers and entire plant in severe cases. Attacked plants lose vigour and get stunted in growth and dry in severe cases.

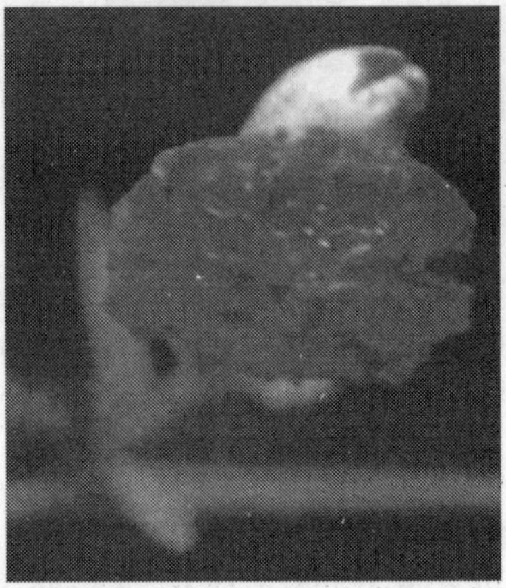

Figure 2.9. Mite infestation on carnation bud

(ii) Management

- Cutting and burning of infested plant parts.
- Thorough spraying of water on the plant dislodges the webs.
- Spray wettable sulphur 80 WP at 3 g/l or profenofos 50 EC at 1 ml/l or ethion 50 EC at 1 ml/l at 15 days interval followed by 10 ml/l pongamia or neem oil.
- Spray abamectin 1.9 EC at 0.5 ml/l followed by flufenoxuron 10 DC at 1 ml/l or fenazaquin 10 EC at 1 ml/l during summer when infestation is high.
- Release predatory mite, *Amblyseius tetranychivorus* at 20 mites/plant.

2.3.1.3. Bud Borer, *Helicoverpa armigera*

(i) Damage: Eggs are laid singly on growing buds during September–October and damage continues up to February–March. Larvae bore into young buds and feed on developing petals making them hallow. Larvae also attack open flowers resulting in dropping of petals and significant yield loss.

(ii) Management

- Collection and destruction of larvae.
- Spraying of quinalphos 25 EC at 2 ml/l or methyl parathion 50 EC at 1 ml/l or endosulphan 35 EC at 2 ml/l at 15 days interval.

- Spray indoxacarb 14.5 SC at 1 ml/l or thiodicarb 75 WP at 1 g/l if the incidence is severe.
- Application of *Ha* NPV at 250 LE/ha followed by neem formulations at 1–2 ml/l or neem seed kernel extract at 40g/l.

2.3.2. Diseases

2.3.2.1. Wilt, *Fusarium oxysporum* f. sp. *dianthi*

(i) Symptoms: Fusarial wilt is widely prevalent in carnation all over the world. The initial symptoms are foliar yellowing and production of 'crookneck' or bent shoots. The leaves and shoots wither and become brownish. Initially the symptoms are confined to a few branches or a part of the plant but ultimately the entire plant shows the symptoms. The symptoms are first noticed during mid-day when the temperature is quite high inside the polyhouses, but they are not visible when the temperature goes down. Stems when cut open show brown discolouration at the vascular region. Sometimes pith and cortex also get discoloured. Finally the stems show shredding of the internal tissue. The disease spreads through contaminated planting material as young rooted cuttings carry the pathogen without expressing the symptoms. Spread of the disease within the greenhouse occurs through root contacts, contaminated soil carried by workers and implements.

(ii) Management

The best way to control wilt of carnation is by integrated approach which includes use of clean stock, soil treatment and phytosanitation. Soil treatment with benzimidazoles (benomyl, carbendazim, thiophanate methyl at 0.1%), pthalamides (captaf, captafol at 0.2%) and sterol biosynthesis inhibitors (prochloraz) before planting is reported to give effective control of wilt.

Biocontrol agents like *Trichoderma harzianum, Bacillus subtilis, Streptomyces* spp. and non-pathogenic isolates of *Fusarium oxysporum* are also reported to give good control of wilt disease.

The substrate pH should be maintained above 6.5 as acidic soils favour the growth of *Fusarium*. During all the stages of crop development, calcium deficiency in the plants is known to increase their proneness to disease.

2.3.2.2. Foot Rot, *Phytophthora nicotianae* var *nicotianae, P. nicotianae* var *parasitica*

(i) Symptoms: *Phytophthora* spp. attacks the roots and crown portion of carnation especially under high moisture conditions. The infection results in rotting of the stem at soil level resulting in wilt. In case of root infections, the leaves gradually get discoloured and start drying from bottom upwards. High soil moisture is the predisposing factor. The pathogen spreads through water and contaminated soil.

(ii) Management

The reduction in soil moisture is an important step in the control strategy. The fungicides useful are benomyl (Benlate at 0.1%) or iprodion (Rovral at 0.1%) or combination product

(Quintal which contains both Bavistin and Rovral in equal proportions) or fosetyl-AL (Aliette at 0.3%) or metalaxyl (Ridomil MZ at 0.2%).

2.3.2.3. Leaf and Blossom Blight, *Alternaria dianthi, A. dianthicola*

(i) Symptoms: Dark brown circular spots occur on the margins of leaves and also on flower buds. The infected parts get blighted. The spores of the fungus are spread through water emitted by misting and fogging devices.

(ii) Management

The disease can be managed through application of pthalamides (Captafol at 0.2%) or chlorothalanil (Kavach at 0.2%) or iprodion (Rovral at 0.1%). Carnation *cv.* "Forca" is free from blight infection.

2.4. GERBERA

2.4.1. Pests

2.4.1.1. Whitefly, *Bemisia tabaci*

(i) Damage: The pest is active during moderately hot and humid conditions (February–July). Small and fragile white adults lay eggs on lower surface of young leaves. Nymphs and adults suck cell sap resulting in discolouration, deformation and withering of leaves. In case of severe infestation, sooty mould develops on honeydew secreted by nymphs affecting the plant growth. Severely damaged plants produce small and deformed flowers with crooked stalks.

(ii) Management

- Excessive dampness and dark conditions should be avoided.
- Removal and burning of infested leaves.
- Spray triazophos 40 EC at 1.5 ml/l or methomyl 40 SP at 2 ml/l or acephate 75 SP at 1.5 g/l followed by pongamia or neem oil at 10 ml/l at 15 days interval against nymphs.
- Spray diclorvos 76 EC at 1 ml/l followed by lambda cyhalothrin 5 EC at 1 ml/l or deltamethrin 2.8 EC at 1 ml/l at 10 days interval against adults.
- Install yellow sticky traps for attracting adults.

2.4.1.2. Aphid, *Myzus persicae*

(i) Damage: Mainly growing buds and flowers are attacked during cool months (October–January). Both nymphs and adults colonize on lower side of petals and suck the sap leading to discolouration, withering and deformation of flowers. Development of sooty mould on honeydew secreted by aphids makes flowers look sticky with black dots.

(ii) Management

- Spray pongamia or neem oil at 10 ml/l at early infestation levels.

- Spray dimethoate 30 EC at 2 ml/l or oxydemeton methyl 25 EC at 2 ml/l or imidacloprid 200 SL at 0.4 ml/l or cartap hydrochloride 50 SP at 1.5 g/l or acephate 75 SP at 1.0 g/l.
- Application of *Verticillium lecanii* at 3 g/l also reduces aphid population.

2.4.1.3. Leaf Miner, *Liriomyza trifolii*

(i) Damage: Incidence starts on newly planted crop at 3–4 leaf stage. Small black flies lay eggs by punching the young leaves. Maggots feed inside leaves on mesophyll by leaving characteristic serpentine mines (Fig 2.10). Damaged leaves bear extensive mining marks; become brittle and dry in case of severe attack affecting the growth of the plants.

Figure 2.10. Leaf miner infestation on gerbera

(ii) Management

- Remove and burn mined leaves.
- Spray of triazophos 40 EC at 1.5 ml/l or methomyl 40 SP at 2 ml/l or profenofos 50 EC at 1 ml/l followed by spray of pongamia or neem oil 1% at 15 days interval.
- Drench the beds after raking up with chlorpyriphos 20 EC at 4 ml/l for killing pupae.
- Spray of deltamethrin 2.8 EC at 1 ml/l for managing adults.
- Install yellow sticky traps.

2.4.1.4. Thrips, *Thrips palmi*

(i) Damage: Thrips cause damage by sucking sap from leaves, buds and flowers from July to February. Damaged leaves develop brown streaks, become brittle and curl up. Affected buds and flowers get discoloured and distorted in shape affecting the quality and yield of flowers (Fig 2.11).

Figure 2.11. Thrip infestation on flowers of gerbera

(ii) Management

- Spray dimethoate 30 EC at 2 ml/l or oxydemeton methyl 25 EC at 2 ml/l at 15 days interval if the incidence is low.
- Drench the soil with chlorpyriphos 20 EC at 4 ml/l to reduce resting stages.
- Spray imidacloprid 200 SL at 0.4 ml/l or cartap hydrochloride 50 SP at 1.5 g/l or fipronil 5 SC at 1 ml/l or acephate 75 SP at 1.5 g/l followed by pongamia or neem oil at 10 ml/l in case of severe incidence.

2.4.2. Diseases

2.4.2.1. Foot Rot, *Phytophthora cryptogea*

(i) Symptoms: Flaccidity of lower leaves is the initial symptom noticed generally. As the disease progresses, the infected plants wilt suddenly. In some cases the leaves turn purple. The infected plants show severe rotting of roots, which retain only the stellar portion because the cortical portion gets sloughed-off. The tissues of the crown and root become soft and brownish. Rotting can be seen extending upward through the petioles and occasionally the flower stalks. Gerbera does not tolerate excess moisture in the soil. The plants under such conditions succumb to infections caused by root rot fungi such as *Phytophthora* and *Pythium*. The pathogen proliferates well under wet weather conditions and propagules spread through contaminated soil and water.

(ii) Management

Adopting strict phytosanitation and soil treatment with metalaxyl-MZ (Ridomil-MZ at 0.2%) or fosetyl-AL (Aliette at 0.2%) help in managing the disease. Regulation of soil moisture becomes decisive in containing the disease. Precautionary steps like preparation of raised beds (25 cm above soil level) and a friable substrate help in reducing the disease incidence. Pre-planting fumigation of beds and preventive application of broad spectrum fungicides help in preventing the infection.

2.4.2.2. Wilt, *Fusarium oxysporum*

(i) Symptoms: Discolouration and loss of turgidity of leaves are the initial symptoms noticed on infected plants. Cutting open of the crown portion shows clear discolouration of the vascular region. Severe wilting occurs under anaerobic conditions of the substrate caused due to excessive soil moisture. The fungus is soil-borne and is also occasionally transmitted through the plantlets which act as symptomless carriers.

(ii) Management

Planting disease-free rooted cuttings is the most important step in managing the disease. The substrate should be sterilized with chemicals or solarization before planting. Rouging and soil application of benzimidazoles (carbendazim, benomyl at 0.1%) combined with pthalamides (captaf at 0.2%) help in managing the disease.

2.4.2.3. Powdery Mildew, *Erysiphe cichoracearum, Oidium erysiphoides* f. sp. *gerbera*

(i) Symptoms: Powdery mildew is a major production problem in gerbera and accounts for considerable economic losses. The infection occurs both on leaves and flowers. The symptoms are seen on leaves and flower stalks in the form of grayish to white patches. The white powdery patches on the leaves reduce the effective leaf area available for photosynthesis and ultimately leading to reduced productivity and poor quality of flowers. The infection also occurs on petals and flower stalks resulting in malformed flowers and brittle stalks. The lesions on leaves and flowers produce fungal spores called conidia in abundance. These are produced under dry and humid conditions that are commonly prevalent in polyhouses. The dry spores are readily spread by the movement of workers and by wind.

(ii) Management

The control measures are same as those given under rose powdery mildew.

2.5. ORCHIDS

2.5.1. Pests

2.5.1.1. Scale, *Diaspis boisduvalli*

(i) Damage: One of the two most common scale found on orchids is soft brown scale. Initially, when in the crawler stage, they look like tiny flat crabs, but quickly mature to resemble little shiny brown, tan or greenish bumps which are attached to leaves and pseudo bulbs. This pest is readily visible in the adult stage, when they form a hard domelike dark brown shell. They also excrete a sticky substance which attracts ants and subsequently develops an ugly grey mold-like fungus.

The other common scale is boisduval scale, also known as white or woolly scale. They also have several stages of development which are difficult to see at first. Boisduval scale first appear as small round raised spots on the leaf or pseudo bulb. These are the females, which later develop a thin protective cover. The larger white males are more conspicuous and you will usually find them in clusters, especially on the underside of leaves. Most likely there are many more hiding in large fuzzy colonies around and under the rhizome as well as under bracts and dry sheathing. This is a very persistent critter and seems to be able to survive repeated treatments by appearing again perhaps even a year or so later.

The young and mature scales suck the juice from the leaves, petioles and shoots resulting in stunted growth and drying of shoots in case of severe attack.

(ii) Management

Gently scrub to dislodge as many scales as you can see. Rinse the entire plant thoroughly with clean water and let it dry. After that, a systemic insecticide specifically geared towards scale should be used according to directions. Applications may have to be repeated several times as recommended.

- Selection of scale free planting material.
- Cutting and burning of infested parts.
- Application of carbofuran 5–7 g/pot.
- Spray malathion 50 EC at 1 ml + dimethoate 30 EC at 1 ml/l or chlorpyriphos 20 EC at 2 ml/l or ethion 50 EC at 1 ml/l followed by pongamia or neem oil at 10 ml/l.

2.5.1.2. Mealy Bug, *Pseudococcus maritimus*

(i) Damage: Both nymphs and adult mealy bugs suck the sap from the plant parts and secrete honeydew leading to sooty mould development. In case of severe infestation, plants look withered with chlorotic and deformed leaves.

(ii) Management

- Prompt collection and destruction of infested parts.
- Crawling of ants on the plants is a sign of mealy bug infestation when spraying should be taken up.
- Spray dichlorvos 76 EC at 1 ml/l followed by methyl parathion 50 EC at 1 ml/l or methomyl 40 SP at 2 ml/l or acephate 75 SP at 1.5 g/l at 15 days interval.
- Pongamia oil or neem oil sprays at 10 ml/l are also effective.

2.5.1.3. Thrips, *Anaphorathrips orchidii*

(i) Damage: Both adults and nymphs suck the cell sap from growing leaves and buds. Affected leaves curl with brown streaks. Damage to developing buds results in discolouration, distortion and drying.

(ii) Management

- Spray oxydemeton methyl 25 EC or dimethoate 30 EC at 2 ml/l at 10 days interval when the infestation is noticed.

- If the incidence is severe, spray cartap hydrochloride 50 SP at 1.5 g/l or fipronil 5 SC at 1.5 m/l or imidacloprid 200 SL at 0.4 ml/l followed by pongamia or neem oil at 10 ml/l.

2.5.1.4. Slugs and Snails, *Achatina fullica*

(i) Damage: They absolutely love orchids and will travel up the spikes to reach and chomp on tender new buds (Fig 2.12). They are difficult to get rid of and even harder to locate during the daytime since they are nocturnal and spend the day hiding out in the sheaths and axils of the plant.

Figure 2.12. Snail and slug damage on leaves of orchid

Adult and young ones devour plants during nights. Snail's attack is seen mainly during monsoon period when high moist conditions prevail. They feed on leaves by scraping, making big irregular holes and also cut tender growing shoots (Fig 2.12).

(ii) Management

Slug pellets may be distributed on the floor of the greenhouse to control slugs. Slugs love beer almost more than new orchid buds! Place a saucer filled with beer near the affected plants, you will find the drunken offenders in the morning. Another method is to discourage slugs and snails from dining on your newly developing blossoms. Wrap a small piece of fluffy cotton around the stem of the inflorescence, below the first buds. Snails won't cross over the cotton and don't appear to have much interest in the stem itself.

- Handpicking and killing of snails by dropping them in 5% salt solution.
- Sprinkling of salt crystals in the paths or around pots.
- Spread Snail kill (3% metalaldehyde pellets) at 50 kg/ha.
- Spray neem oil at 10 ml/l or soap nut extract at 60 g/l on foliage.

2.5.1.5. Spider Mite, *Brevipalpus* sp.

(i) Damage: Spider mite is a serious and devastating pest in the greenhouse. They attack mostly the softer leaved varieties of orchids such as phalaenopsis, dendrobium and cymbidium. They are so miniscule in size that they are difficult to detect with the naked eye. This is why infestation most often is only noticed when leaf damage is obvious.

The first indication is when the leaf colour develops a silvery white sheen which soon turns yellow and the leaf will drop off (Fig 2.13). In severe cases, dirty grey webs may be seen. Red spider mite leaves a reddish/orange residue when rub your thumb over the leaf of a suspect plant. You can do another test by tapping a leaf over a piece of white paper. A good magnifying glass will bring them into disgusting focus.

Figure 2.13. Mite infestation on leaves of orchid

(ii) Management

Spraying of horticultural oil, neem oil, mineral oil or insecticidal soaps is effective against mites. Use of predatory mite, *Phytoseiulus persimilis* is particularly successful in greenhouses. Spider mite is also controlled with the miticides like dicofol (Kelthane) and abamectin (Avid).

2.5.2. Diseases

2.5.2.1. Basal Rot

(i) Symptoms: Basal rot is also commonly known as 'Southern Blight'. It is directly related to watering methods and practices. Either over watering, or water left standing in the crown of plants combined with cool humid situations can bring on this condition.

On soft and broad leaved orchids there is little warning. Basal rot, as the name suggests, starts at the bottom or the crown of the plant and proceeds upward. Therefore, by the time you notice a problem, it is already too late.

(ii) Management

If the condition is discovered on a plant with multiple growths, and only a small portion is afflicted, you can attempt to save the plant by cutting away the diseased portion with a sharp knife and treating the wound with sulphur. Keep on the dry side after the "operation" and hope the rot does not spread.

2.5.2.2. *Botrytis* Blight

(i) Symptoms: This is a fungal disease which affects primarily the flowers of phalaenopsis, cattleya, cymbidium and dendrobium. This condition occurs most frequently in cool, moist conditions where there is little air movement. Symptoms are tiny black spots on the petals which soon grow larger and in extreme circumstances are covered with a grey fuzzy mold. If caught early, take measures to increase night time temperatures to no less than 65°F and provide maximum air circulation.

(ii) Management

Remove and destroy badly infected material such as dead flowers which have fallen to the floor and are covered with mold. Spray flooring with a strong solution of fungicide. Spray plants with a mild solution of physan. This will not reverse the damage already manifest on your bloom, but should dry up the affected areas and halt the further spread of spores.

2.5.2.3. Fusarium Wilt, *Fusarium proliferatum*

The most serious disease caused by *F. proliferatum* on orchids is damping-off and rots of young plants. Hundreds of young seedlings are lost in community pots due to this devastating disease. Rots of these plants kill many and stunt the growth of survivors. As surviving plants mature they are plagued by the recurrence of fusarial diseases during wet periods.

(i) Symptoms: Various parts of orchid plants are attacked by this fungus. On dendrobium plants at least 9 months old, new shoots are blackened by *Fusarium* infections and slow rot develops especially at the leaf tip. Leaves are lost and the entire apical tip of new shoots are killed. *Fusarium proliferatum* causes small (2–5 mm) circular to oval spots with irregular edges. Spots expand very slowly. As the leaf matures, some of the spots develop a black edge, while the centre of the spots becomes sunken. The leaf tissue surrounding the spot is slightly chlorotic or pale green.

With mature dendrobium plants, only the young leaves or wounded leaves are susceptible, while mature unwounded leaves are resistant. The occurrence of leaf spots on mature leaves is the result of infection when leaves were young. The distribution pattern of leaf spots reflect the leaf position of the unopened, young leaves of a new shoot. A characteristic pattern of spots for this disease is a row of spots across the width of the leaf (Fig 2.14). Sunken brown spots also occur.

The fungus infects the leaf sheath causing black to dark brown rots with irregular edges. Rot expands very slowly and after a few months, the centre of sheath blight or rot can be white to tan. Circular to elongate spots also develop on flowers. Spots are tan to deep brown with a dark border (Fig 2.14).

Figure 2.14. Fusarium infection on leaf and flower of orchid

(ii) Biology: *Fusarium proliferatum* produces microscopic, long, canoe-shaped spores called conidia. These asexual conidia have 3 to 7 cells and are produced on specialized hyphae called conidiophores. A second, smaller type of conidia with 1 or 2 cells, called microconidia are also produced. Both spore forms are produced in very large numbers (millions) and germinate on suitable substrates, such as the host and grow into a colony which produces more spores. Growth of *Fusarium* is favoured by moisture and temperatures in the range of 25–31°C.

(iii) Epidemiology: *Fusarium proliferatum* produces many spores on the surface of infected host tissue. Some microconidia are also produced within rotting tissue. White cottony masses on the diseased tissue are spores and mycelia of the fungus. These spores are blown or splashed to nearby plants or other healthy tissue. For many orchid plants such as dendrobium, new shoots are formed below the canopy of the older leaves. Thus fungal spores formed on older shoots easily splash or drip on young shoots emerging from the base of the plant.

Young shoots are very susceptible to infection and rapid disease development. Likewise, fragile young plants in community pots are extremely susceptible to this fungus and many are killed. Since growers commonly salvage all plants that survive, the *Fusarium* will persist on mature plants and cause leaf, sheath, shoot, and flower rots. These plants serve as sources of inoculum, and disease spread within a greenhouse is insured by overhead watering.

Fusarium spores are distributed by wind, splashing water or the movement of spores by contact (hands, clothing, gloves, etc). Insects, snails, and slugs also move spores from diseased to healthy plants. *Fusarium* will survive in diseased plant tissue for many months and spores will also survive in potting media and in the immediate environment of the diseased plants. Greenhouse walls, benches, floors, etc. will be contaminated as long as diseased plants are present.

(iv) Management

Great emphasis must be placed on keeping young, disease-free seedlings healthy. The production of new seedlings should be in a greenhouse separated from the production of mature plants. This seedling-house should not have any mature plants of any kind, especially samples of mature blooming plants.

All diseased leaves, dead shoots and infected flowers need to be gathered and discarded from the nursery. Do not drop infected material to the ground as this provides the fungus a good opportunity to infest the ground for long periods.

Spores of *Fusarium* need water to germinate and penetrate the host. Thus controlling moisture will reduce disease levels. Avoid watering in the early evening as this will expose leaves for a long period of wetness during the night. Water during the day and keep leaves as dry as possible. Increase air movement to reduce humidity levels. Remove weeds and other shrubs that prevent adequate air movement.

Fungicides such as mancozeb will also reduce infection levels.

2.5.2.4. Brown Wet Spot, *Pseudomonas cattleyae*

Pseudomonas cattleyae, also known as 'brown wet spot' occurs mostly in phalaenopsis. However, a similar wet bacteria infection can occur on other orchids as well.

(i) Symptoms: Plants develop brownish black, soft areas usually starting on the underside of the leaves and quickly penetrating to the leaf surface (Fig 2.15). This affliction is a destructive rot which can spread with amazing speed, taking in the entire plant within only a few days. The cause is a combination of cool temperatures, high humidity and inadequate air movement. Prevention is easier than the cure. Keep minimum night temperatures around phalaenopsis at least 60 to 65°F, along with brisk air circulation. Brown wet spot can not develop under such conditions.

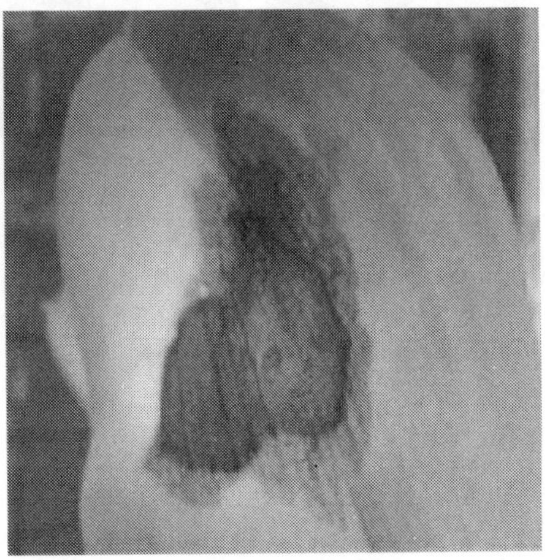

Figure 2.15. Wet blight on orchid leaf

(ii) Management

If the disease has reached the crown portion of the plant, there is no cure even if some of the leaves look healthy. On phalaenopsis, if caught early, the affected portion of the leaf should be cut off (well below the afflicted area) with a new razorblade or some other sterilized cutting tool. Wet rot on other orchids must be treated by lancing and then cutting and/or scraping out the damaged tissue.

Dust the cut surface with sulphur. Try to rapidly dry off the injury by placing the plant before a fan. Isolate from your regular growing area to prevent any spread of the disease.

2.5.2.5. Leaf Spot, *Phyllosticta capitalensis*

Phyllosticta leaf spots and blights are one of the most serious problems in the commercial orchid industry. The quality and quantity of exports shall be affected due to leaf loss in dendrobium fields grown for cut sprays, and leaf rots of potted plants that are shipped.

(i) Symptoms: The fungus causes leaf spots on orchids. On dendrobium, young leaf spots are typically yellow or black, circular, and small (Fig 2.16). These spots remain unchanged for many months. However, the fungal growth within the leaf tissue eventually increases and tan to brown rot develop. Within the next week or two, the entire leaf is killed and the characteristic, black web-like pattern formed by this fungus on diseased leaves.

On dendrobium, flower spots are small, faint, and commonly blue or lavender on purple cultivars (Fig 2.16). When the flowers are harvested or drop from age, a black rot rapidly covers the flowers in 1 or 2 days.

Figure 2.16. *Phyllosticta* leaf and flower spot on orchid

(ii) Biology: *Phyllosticta* produce small, black fruiting bodies (pycnidia) that contain conidia on the surface of infected tissue. These conidia are released when moisture is present and also require moisture to germinate. In addition, black fruiting bodies called perithecia, are also formed, shortly after pycnidia are produced. Perithecia produce sexual spores or ascospores. Ascospores are produced in sacs and there are 8, single-celled spores per sac.

(iii) Epidemiology: In the field or greenhouse, disease spread is due to movement of both conidia and ascospores. Conidia are splashed from older diseased leaves to other leaves nearby. Ascospores are discharged into the air and are carried to long distances. When either spore type lands on young dendrobium tissue, it germinates and forms a small structure called an appressorium. Moisture is needed for this process. The appressorium adheres to the surface of the host tissue. The fungus then grows from the appressorium and penetrates the host. Without moisture the spores fail to germinate and are unable to infect the host.

After *Phyllosticta* penetrates the leaf, for some time there are no symptoms or signs of the pathogen on the infected leaves. It takes 3 to 6 weeks before a small faint spot can be observed on the infected leaf. Many more weeks are required before a clearly defined yellow spot appears. Because of this, many growers are surprised to see spots developing many months following plant purchase.

In the field or in commercial greenhouses, there is sufficient sunlight for the plant to inhibit the expansion of *Phyllosticta*, confining the fungus to the yellow spot stage. However, when light levels are reduced the plant loses its ability to inhibit *Phyllosticta*. The fungus starts to grow rapidly and large rots develop. Leaves are lost and only the cane or pseudostem remains. This is a major problem for growers who ship infected potted plants. Within shipping boxes, darkness and high humidity favor disease development and plants arrive with severely diseased leaves. Similarly, light levels in garden shops are also too low to prevent disease development.

(iv) Management

Since many orchids are propagated by seeds or cloning, it is possible to start new plants without the fungus. These plants should be protected by: (a) isolating them and keeping them away from mature plants that could have pathogens; growing new plants in a small, separate greenhouse or glasshouse would be ideal; (b) preventing the movement of ascospores to new plants by keeping them in a separate greenhouse or in a section separated by solid plastic; (c) reducing the number of ascospores by removing all rotted leaves and especially those in pots, and on the bench or ground.

For larger plants and for those already infected: (a) remove all rotted or diseased leaves; clean benches, pots, and walkways by removing all dried sheaths, leaves, old stems etc., since the fungus survives in diseased plant material; (b) trim old dead stems and flower spikes; (c) apply fungicides such as mancozeb to kill fungal spores.

Since moisture is needed for disease spread by asexual spores (conidia), for germination of the spores, and for host penetration, moisture control is very important. Potted plants grown under solid cover will be nearly free of this disease. Water plants early in the day to allow leaves to dry as opposed to just before dark, when foliage would remain wet for many hours. Remove tall bushes around fields to encourage air movement in the field and reduce long periods of high humidity.

Application of fungicides such as mancozeb will also reduce infection levels. However, these fungicides are not effective, once the fungus has penetrated the host. Thus, disease prevention through cultural practices and fungicide application is crucial.

2.5.2.6. Black Rot, *Phytophthora* sp.

(i) Symptoms: Black Rot is a particularly aggressive infection of Cattleyas caused by the fungus, *Phytophthora*. A new shoot suddenly turns black: the rot moves rapidly, killing the rest of the plant. A whitish 'bloom' of fungus spores may be seen on the diseased tissue. Heat-stressed orchids are more susceptible. This disease is more prevalent during wet weather: it is spread primarily by splashing water.

(ii) Management

Disease spread primarily by splashing water. Shelter plants from rain. Isolate infected plants. Burn diseased material. Spray or dust with appropriate fungicide.

2.5.2.7. Black Leg/Dry Rot, *Fusarium* sp., *Rhizoctonia* sp.

(i) Symptoms: Dendrobiums and Vandas growing in waterlogged potting or bedding materials are susceptible to a slow but inexorable rot of the roots and stem from below. The causative fungi are *Fusarium* and *Rhizoctonia*. Pseudo bulbs become spongy and discoloured. The leaves, especially in Vanda, will turn yellow and drop off, one by one, until none are left and the plant dies. Plants can be salvaged. Uninfected pseudo bulbs of Dendrobiums can be removed to allow keikis to form. If a Vanda has many aerial roots, sever the stem above the line of infection. The disease is caused primarily by poor culture.

(ii) Management

This disease is caused primarily by poor culture. Fungicides are not recommended. Use potting/bedding materials appropriate to the plant type. Replace potting materials before they become stale and waterlogged. Burn infected material.

2.5.2.8. Bacterial Soft Rot

(i) Symptoms: Soft rots are difficult to diagnose but whatever the causative agent, they can be devastating to an orchid collection. Orchids will be more vulnerable to infection if they are over fertilized, given insufficient light and ventilation, and if they are permitted to remain wet especially in the crown. Rots are a problem during the wet season, also after storms when plants are bruised and torn by high winds.

(ii) Management

Shelter susceptible plants from rain. Be vigilant for rot during the wet season.

2.5.2.9. Bacterial Spot/Rot, *Pseudomonas* sp, *Erwinia* sp.

(i) Symptoms: Soft, brown, smelly, fluid-filled blisters are produced on leaves and in the crown of *Phalaenopsis*, *Paphiopedilum*, and *Catasetum*. It is highly contagious. Crown rot will quickly kill a plant. Leaf spots can be excised and the plant saved.

(ii) Management

Withhold water. Improve ventilation. Remove and burn the infected tissue. Be careful not to break the blisters. Consider the fertilizer being used: reduce the nitrogen and increase the potassium component. Pot *Phalaenopsis* so that the crown drains freely. Use Physan-20 at 7.5 ml in 5 litres as a preventive spray.

2.5.2.10. Virus Diseases

The three most important viruses for the orchid grower are undoubtedly Cymbidium Mosaic Virus (CyMV), Tobacco Mosaic Virus (TMV) and Tomato Spotted Wilt Virus (TSWV).

Figure 2.17. Tomato Spotted Wilt Virus (*Cattleya* hybrid)

(i) Foliar symptoms: Chlorotic streaks in *Cymbidiums*, to black necrotic spots and necrotic line patterns with sunken patches in *Phalaenopsis*, *Cattleya*, *Dendrobium* and many other genera (CyMV, TSWV) (Fig 2.17). In lavender-flowered Cattleyas, the foliage patterns are associated with pronounced red or purple pigmentation. The leaf patterns appear as spots, chevron patterns or irregular line patterns of chlorotic or red-pigmented tissues (TMV) (Figs 2.17 and 2.18).

Figure 2.18. Tobacco Mosaic Virus (*Cymbidium* hybrid)

(ii) Flower symptoms: The flowers may show necrotic streaks within a few days of the flower being open. However, flower symptoms usually show up only after about two weeks and are especially noticeable in white Cattleya flowers (CyMV).

Flower colour breaking is consistent with a TMV infection. Streaks of dark pigmentation are often observed in predominantly pink to lavender-flowered orchids. Colour breaking should, however, not be confused with white streaks on lavender-flowered cattleyas that are caused by a genetic abnormality.

(iii) Management

- Virus diseases are managed by prevention. Once a plant is infected with a virus it is virtually impossible to get rid of it and the plant will have to be destroyed with the pot and potting medium.
- Do not propagate from infected plants.
- Pest control is essential to reduce the risk of virus infections.
- Weeds can be an important source of infection both as reservoirs of viruses and as hosts for vectors.
- Disinfect all cutting tools and pots between each plant that is handled as viruses can be transmitted in sap from infected plants. Flaming them or dipping them into a 10% hypo chloride solution can disinfect the cutting tools. Hands should be washed with soap.
- Avoid dusting the greenhouse floor, or the garden, with tobacco dust to get rid of snails, slugs or ants. The tobacco dust may contain virus particles, which can be a source of infection. Smokers should not smoke and handle orchids at the same time.

2.6. ANTHURIUM

2.6.1. Pests

2.6.1.1. Whitefly, *Aleurotulus anthuricola*

(i) Damage: White and waxy-coated adults and nymphs colonize under surface of the leaves and petiole sheath. The pest sucks the sap from infested parts causing yellowing, wilting and drying of affected plant parts.

(ii) Management

- Remove and burn infested leaves.
- Spray methomyl 40 SP at 2 ml/l or acephate 75 SP at 1.5 g/l alternating with triazophos 40 EC at 1.5 ml/l at 15 days interval for managing nymphs.
- Dichlorvos 76 EC at 1 ml/l or malathion 50 EC at 1 ml/l spray checks adult flies.
- Install yellow sticky traps for attracting and killing adults.

2.6.1.2. Thrip, *Chaetanaphorothrips orchidii*

(i) Damage: Damage starts in March and reaches to a peak in August-September. Thrips produce white or brown coloured streaks and distortion of spathe by sucking cell sap.

(ii) Management

- Spray oxydemeton methyl 25 EC at 2 ml/l or dimethoate 30 EC at 2 ml/l or profenofos 50 EC at 1 ml/l or acephate 75 SP at 1.5 g/l.
- After 15 days give a follow up spray of pongamia or jatropha oil at 10 ml/l.
- Spray imidacloprid 200 SL at 0.4 ml/l or fipronil 5 SC at 1.5 ml/l, if the incidence is severe.

2.6.1.3. Mite, *Brevipalpus* spp.

(i) Damage: Mite colonies are seen on lower surface of mature leaves. Feeding by mites result in specking, mottling and drying of affected leaves in severe cases.

(ii) Management

- Spray of dicofol 18.5 EC at 2.5 ml/l or triazophos 40 EC at 1.5 ml/l or ethion 50 EC at 1 ml/l at 15 days interval followed by pongamia or neem oil at 10 ml/l. If the incidence is severe, spray abamectin 1.9 EC at 0.5 ml/l or flufenoxuron 10 DC at 1 ml/l.

2.6.1.4. Slugs and Snails

(i) Damage: Snails cause damage to young roots. Slugs and snails are of quarantine significance on exported plant material. Slugs escape hot, dry conditions by hiding between the soil surface and any material on the soil during the daytime.

(ii) Management

Control must be 100% for quarantine security. Sanitation is vital to slug control. Removing boards, pots, and any debris from growing areas will reduce slug cover for slug and snail breeding areas. Pruning anthurium plants will allow sunlight to reach the soil. Weather conditions affect slug and snail populations. Hot dry weather or excessive flooding can reduce the populations; while cool, damp weather encourages population growth. Baits are the preferred method of control. Baits consist of a base that is attractive to slugs plus a poison. Use of metaldehyde can prevent the attack of snails.

2.6.2. Diseases

2.6.2.1. Anthracnose, *Glomerella cingulata (Colletotrichum gloeosporioides)*

(i) Symptoms: The infection occurs on young plants soon after planting. Symptoms start as tiny brownish spots on the flower spadix. During high humidity these spots enlarge, appear water-soaked and turn necrotic. Sometimes the entire spadix will turn black as lesions coalesce. The shape of most lesions is, however, angular due to the shape of the spadix tissue. As the disease becomes more severe, masses of orange coloured spores form on necrotic areas. Leaves and spathes are rarely if ever infected. Warm wet weather favours the outbreak of the disease which spreads fast in densely planted beds.

(ii) Management

Sanitation and use of common fungicides (0.1% carbendazim + 0.2% copper oxychloride or 0.1% thiophanate methyl or 0.2% mancozeb or 0.15% kitazin) at fortnightly intervals helps in management of the disease.

2.6.2.2. Black Rot and Leaf Blight, *Phytophthora nicotianae* var. *parasitica, P. citrophthora*

(i) Symptoms: Symptoms appear as black spots with water soaked margins on spathe, spadix, leaves, stalks and petioles. The rotting spreads faster on leaves and spathe through water droplets. The infection spreads rapidly during wet weather with moisture on leaf surface.

(ii) Management

Aliette at 0.3%, dimethomorph at 0.2% and other anti-oomycete fungicides are effective for the management of the disease. Root rot caused by *Pythium* and *Phytophthora* can be controlled by the application of potassium phosphonate 0.3%.

2.6.2.3. Bacterial Wilt, *Ralstonia solanacearum*

(i) Symptoms: The bacterium causes both systemic and localized infections. The symptoms constitute yellowing starting from the margins of outer leaves. The yellowing gradually extends to entire leaf area and finally to inner leaves. The infected plants show systemic yellowing and wilting leading to death of plants. Root and collar rot are also quite common. Often the infection results in necrotic lesions from petiole spreading along the veins. Local infections on leaves cause black necrotic lesions with greasy margins resulting in leaf distortion. On the spathe these spots are angular and irregular. The bacterium is carried through plants without expressing external symptoms. Under humid conditions, the bacterial ooze is produced on infected plants and contributes to spread of the disease through water droplets, garden tools and workers activity. The disease is more severe during summer and under excessive soil moisture conditions.

(ii) Management

The disease can be managed by using disease-free planting stock besides adopting strict phytosanitation.

2.6.2.4. Bacterial Blight, *Xanthomonas compestris* pv. *diffenbachiae*

(i) Symptoms: In the humid tropics, high rainfall coupled with year-round high temperatures increases the severity and spread of bacterial blight disease making it a very important impediment to the cultivation of anthurium.

The infection causes black necrotic spots on the leaves and spathe (Fig 2.19). The spots on leaves coalesce and result in blighting. The bacterium survives in the latently infected plants without producing symptoms. The infection mainly spreads through contaminated planting material. High temperature and humidity enhance the infection. Misting though lowers the temperature, spreads the inoculum effectively. The disease also spreads through contaminated tools. Blackening of the stem and decay of leaf axils are the symptoms of bacterial blight.

Figure 2.19. Bacterial blight on anthurium

(ii) Management

Use of clean tissue cultured plants, following appropriate cultural practices and strict phytosanitation helps in reducing the disease incidence.

3 MEDICINAL PLANTS

In medicinal plant sector, the WHO has estimated that about 80% of the population in developing countries rely on traditional medicines mostly plant drugs for primary health-care needs. Even modern medicines contain about 25% drugs derived from plants. A large number of people are earning their livelihood from different activities of medicinal plants such as collection, cultivation, marketing, processing, etc. The area under cultivation of the above crops is low, but the cash returns are high because of intensive cropping.

In recent times, due to increasing realization of health hazards and toxicity caused by synthetic drugs (side effects), there has been a renewal of interest in the use of plant based drugs throughout the world. A campaign is going on worldwide to utilize more and more plant derived chemicals in the human health care/personal care system. As a result, the demand for plant-based raw materials has increased enormously in both national and international markets. Herbal plants are finding diverse uses as raw materials, not only for medicines but also as biopesticides.

Medicinal crops form one of the important groups due to their demand in various pharmaceutical industries and also to earn foreign exchange by way of export. The potential for earning foreign exchange by India from the exports of medicinal and aromatic plants is estimated over US$ 3,000 million per annum. The export of these plants and their products has a tremendous potential in advanced countries like Europe, USA and Japan. The international market for medicinal plants related trade is estimated at US$ 60 billion/year having a growth rate of 7% per annum. Destination of foreign tourists is now aimed at India for health treatment. Kerala and Gujarat have become their favourite destinations since Ayurvedic treatment is very popular in these two states.

Unless we start cultivating the important medicinal plants, it will not be possible to meet the increased demands. However, cultivation of medicinal plants has to be taken up in polyculture models rather than in monoculture models. There is a huge task to be carried out at every level to make the Indian system of medicine a viable, sustainable and modern. For this task, urgency is felt which can only be established through regular and systematic scientific research on collection, conservation, evaluation and development of good agricultural practices to compete in the world market.

3.1. SOLANUM, *SOLANUM* SPP.

Solanum (*Solanum viarum*), also called as Khasi Kateri, is a steroid-bearing perennial, tall bush distributed all over Assam, Sikkim and Manipur between 1,600–2,000 m elevations. Its berry pulp is rich in solasodine alkaloid, which is a starting chemical for production of steroids. About 22 species including *Solanum viarum* contain an alkaloid 'Solasodine' in their berries. Solasodine serves as an excellent raw material for the manufacture of Cortisone and Cortisonal hormones. It is used in the production of contraceptive pills, corticosteroids and sex hormones. It is cultivated in 3,000–5,000 ha, mainly in Maharashtra and Andhra Pradesh. It is also cultivated in West Bengal, Assam, Madhya Pradesh and Gujarat on a small scale.

3.1.1. Pests

3.1.1.1. Fruit Borers, *Leucinodes orbonalis, Helicoverpa armigera*

(i) Damage: Female moth lay eggs on tender leaves, buds and fruits. *Helicoverpa armigera* larvae feed on foliage initially and later bore into developing fruits by making holes and feed on internal contents. Early instars of *Leucinodes orbonalis* initially bore into tender shoots resulting in wilting and drying of damaged shoots. Grown up stages bore into fruits and feed inside.

(ii) Management

Collection and destruction of bored fruits reduce larval population. Spraying of 0.05% cypermethrin or 0.0015% deltamethrin followed by a spray of 0.07% endosulfan or 0.05% quinalphos or monocrotophos at 0.5 ml/l beginning with flowering stage provides protection to fruits from borer damage.

3.1.1.2. Fruitfly, *Dacus latiformis*

(i) Damage: Adult flies lay eggs inside young fruits by puncturing. Hatched grubs enter into fruit and feed on the pulp. Damaged fruits rot, turn black and drop causing severe fruit loss. Loss is high in rainy season.

(ii) Management

Collection and destruction of fallen fruits reduce fruifly population. Spraying of 0.005% cypermethrin or 0.0015% deltamethrin or 1% neem oil kills the eggs and effectively controls fruitfly damage.

3.1.1.3. Foliage Caterpillars, *Spodoptera litura, Spilosoma obliqua*

(i) Damage: Eggs are laid in clusters on lower surface of leaves. Hatched larvae are gregarious in nature and they damage leaves by scraping resulting in skeletonization and drying. Later instars feed voraciously on leaves causing severe defoliation.

(ii) Management

Collection of egg masses and early instar larvae and killing them effectively reduce the pest population. Use of poison baits consisting of rice bran, jaggery and chlorpyriphos (in 10: 1: 0.5 ratio) kills grown up stages. Spraying of methyl parathion or phenthoate each at 0.05% or acephate at 0.01% or quinalphos at 2.0 ml/l gives good control of the caterpillars.

3.1.2. Diseases

3.1.2.1. Wilt, *Fusarium oxysporum* f. sp. *lacimiti*

(i) Symptoms: The disease generally appears after first showers and is characterized by wilting and sudden death of plants. The disease spreads very fast and with in a fortnight majority of the plants in a field show wilting symptoms and many of them ultimately die irrespective of the age of the plant.

(ii) Management

- Direct sowing of seeds instead of transplanting of seedlings can reduce the disease incidence.
- Early planting during October and November helps to escape the disease.
- Application of PCNB to the field at 300 g/100 m^2.
- Drenching of the basal portion of infected plants and surrounding soil with 0.1% benomyl or captan.
- Mix 2.5 kg *Trichoderma viride* in 50 kg of FYM and apply along plant lines.

3.1.2.2. Powdery Mildew

(i) Symptoms: Powdery white mass appear on leaves.

(ii) Management

Spray wettable sulfur at 3 g/litre of water.

3.1.2.3. Bacterial Blight

(i) Symptoms: Streaks of yellow alternated with green stripes appear on the leaves.

(ii) Management

Dissolve 150 g of streptocycline and 750 g copper sulphate separately in 2.5 litres of warm water. Add both solutions to 750 litres of water to spray one ha.

3.1.3. Nematodes

The root-knot nematode is recognized as the major limiting factor in successful production of khasi kateri crop.

3.1.3.1. Root-knot Nematode, *Meloidogyne incognita*

(i) Symptoms: *M. incognita* infected *S. viarum* plants exhibited severe leaf yellowing and wilting symptoms (Shetty & Reddy, 1984). It was responsible for 33.6 and 11.5 per cent reduction in plant height and root weight, respectively.

(ii) Management

(a) Chemical Control: Bare root-dip treatment of *S. viarum* seedlings with phenamiphos and aldicarb sulfone both at 1000 ppm for 2 hr was effective against *M. incognita* infection (Shetty & Reddy, 1985).

(b) Host Resistance: Among four species of *Solanum* tested for their reaction to *M. incognita, S. torvum* and *S. seaforthianum* gave resistant reaction which was reflected in the reduction in number of galls, egg masses and fecundity of females (Shetty & Reddy, 1985b).

3.2. PERIWINKLE, *CATHARANTHUS ROSEUS*

Catharanthus roseus, commonly known as periwinkle or vinca, is a common bedding plant. It is popular in container gardens and new flower colours produced on compact cultivars, have increased its use in many public and home gardens.

It contains several alkaloids found to be effective in the treatment of leukemia and other diseases such as diabetes, hypertension, etc. The more important alkaloids obtained from Vinca are Leurosine, Vinblastin, Vincristin, etc.

3.2.1. Pests

3.2.1.1. Sphingid Larvae/Oleander Hawk Moth, *Dilephila nerii*

(i) Damage: Smooth, green, stout larvae with horn on dorsal side cause damage to plants by feeding extensively on leaves, buds and flowers resulting in severe defoliation and yield loss. Grown up larvae pupate in soil in earthen cocoons.

(ii) Management

Deep ploughing before planting expose pupae to predators. Collection and destruction of grown up larvae check further attack. Spraying of carbaryl at 0.2% or methyl parathion or chlorpyriphos each at 0.05% in combination with 0.5% neem oil controls damage by larvae.

3.2.1.2. Leaf Eating Caterpillar, *Anomis flava*

(i) Damage: Eggs are deposited on tender foliage. Larvae feed voraciously on leaves and cause severe damage to foliage.

(ii) Management

Spraying of phenthoate or quinalphos each at 0.05% or acephate at 0.01% protects foliage from damage by caterpillars. Reddy *et al.* (1981) found that sprays of 0.07% endosulfan and 0.2% carbaryl were effective against the caterpillars.

3.2.1.3. Blister Beetle, *Mylabris pustulata*

(i) Damage: Adult and young stages of blister beetle attack plants in large numbers during the day and feed on the flowers and buds causing severe flower loss.

(ii) Management

Collection and destruction of beetles by tapping them into insecticidal solution reduce the pest population. Spraying of 0.2% carbaryl or 0.1% malathion provides protection to plants from beetle attack.

3.2.2. Diseases

3.2.2.1. Wilt/Root Rot, *Phytophthora nicotianae*

(i) Symptoms: Wilting of plants is very common (Fig 3.1). The wilt begins from infected leaves or as a result of root infections. Leaf spots are initially small, irregular water-soaked areas easily visible on the under surface of the leaf. These spots expand to rot the entire leaf.

Figure 3.1. Wilt/root rot of periwinkle

Diseased leaves are gray-olive green to brownish. The rot progress from leaves to the stem, and sections or branches of the plant are killed. Yellowing of leaves and black leaf rot also occur near the stem. Movement of the fungus into the primary stem or into the larger stems causes wilting of the entire plant or large sections of the plant.

Root rot is not initially noticeable. Small feeder roots are infected and larger roots become diseased with time as the fungus feeds and destroys the root system. Infected plants may be stunted, lack vigour, or appear to be suffering from nutritional deficiency. On smaller plants, movement of the fungus from the roots to the collar causes rot at the soil line. These plants wilt rapidly since upward movement of water is inhibited. On larger plants, collar rots may progress into a lower branch and gradually the plants are killed. Infected branches commonly have no leaves and only dead twigs remain.

(ii) Biology: *Phytophthora nicotianae* produces microscopic, hyaline (clear) spores called sporangia. These asexual spores are papillate (have a pointed tip), nearly spherical (round), and release zoospores when water is available. These zoospores are able to swim for many hours. They are attracted to host tissue or organic matter. Zoospores eventually stop swimming, encyst (round-up and form a wall) and germinate. If the environment is not wet enough, sporangia germinate by producing fungal threads or germ tubes.

The fungus also produces a spherical spore with thickened walls. These are chlamydospores and do not release zoospores. They germinate by producing germ tubes. These spores are able to survive in soil for many months in the absence of the host.

Phytophthora nicotianae is heterothallic and has two mating types, A1 and A2. When both mating types are grown together, numerous sexual spores, called oospores are produced. Oospores are also thick-walled. The fungus is favoured by warmer temperatures (25 to 28°C) and moisture.

(iii) Epidemiology: This pathogen produces sporangia on the surface of the diseased tissue. The sporangia release zoospores when water is present and zoospores or sporangia are splashed to healthy leaves or other plants. On the host, the germ tube produced by the zoospores will penetrate the host. Penetration of the host and growth of the fungus in the host begins the infection process. Continuous feeding and growth of the fungus results in rot and production of disease symptoms. Sporangia also produce germ tubes that penetrate the host epidermis and initiate the disease process.

Chlamydospores are produced inside the diseased leaves and stems, which fall to the ground. The diseased plant tissue decomposes and the chlamydospores remain in soil for months, accounting for the soil-borne nature of this fungus. Movement of contaminated soil or infected plants spreads the pathogen to long distances.

(iv) Management

Use clean seed and monitor seedlings for any sign of disease. Remove any diseased seedling immediately. Control the moisture levels in the nursery, landscaped areas, or home garden by using covered greenhouses or benches, increasing air movement to reduce leaf wetness, preparing the soil by adding compost and insuring good drainage. Removal of diseased plants and plant parts reduces the amount of spores that spread the disease.

Fungicides are available that reduce disease levels. Mancozeb is a good contact fungicide and can reduce the number of leaf spots. Metalaxyl is an effective systemic fungicide that will protect leaves and new shoots.

3.2.2.2. Die-back, *Colletotrichum dematium*

(i) Symptoms: The dieback has been found to affect the crop during the monsoon. The characteristic symptoms of the disease are withering of the terminal bud of the young plants followed by wilting and drooping of leaves. In later stages, the entire branches dry up and wither tip downwards. The fruiting bodies develop on the buds of the withered branches which become grayish in colour.

(ii) Management

Mulching between the rows at seedling stage with dry spent straw of citronella or palmarosa reduces the incidence of dieback to a considerable degree. Cultivation of a new strain WH-40 developed at CIMAP Centre, Bangalore, which has high level of field resistance, minimizes the loss due to dieback.

Spraying of carbendazim or benomyl each at 0.1% at fortnightly intervals is effective for the disease management. Spraying 0.2% mancozeb or 0.3% captafol at fortnightly intervals is also effective.

3.2.2.3. Little Leaf (Mycoplasma)

(i) Symptoms: Occasionally some plants have been found to suffer from little leaf disease due to mycoplsma infection resulting in stunted growth and rosetting of leaves on the plant. The disease spreads through insect vectors.

(ii) Management

The disease can be checked from spreading by uprooting and destroying the affected plants. The disease can be minimized by controlling the vector by spraying Metacid or Nuvacron at 15 days interval when the infection is prevalent.

3.3. ISABGOL, *PLANTAGO OVATA*

It is an important cash crop of Gujarat. In Gujarat, it is cultivated in 33,600 ha area producing 30.742 tonnes of seed. Isabgol is also grown in Bihar (Sasaram), Gujarat (Ahmedabad, Banaskantha, Bhavnagar, Jamnagar, Junagadh, Khaira, Kutch, Mehsana, Rajkot), Haryana (Rewari), Madhya Pradesh, Rajasthan (Sirohi) and Uttar Pradesh.

The husk contains xylose, arabinose, galacturonic acid, rhamnose and galactose. Husk is a safe laxative, beneficial in habitual constipation, chronic dysentery and diarrhea. The seed and husk are used to cure inflammation of the nervous membrane of gastrointestinal and genito-urinary tracts, duodenal ulcer, gonorrhea and piles. It can also be used as cervical dilator for termination of pregnancy. Mucilaginous husk absorbs toxins from the gut and helps in excreting them from the body. It acts as a demulcent and anti-irritant in soothing the intestinal ulcers and reduces the serum cholesterol. Seed contains protein (17–19%) and yellow semi-drying oil (5%) containing linoleinic acid (0.2%), linoleic acid (47.9%), oleic acid (36.7%), palmitic acid (3–7%), stearic acid (6.9%), liguoceric acid (0.8%).

3.3.1. Pests

3.3.1.1. Aphid, *Aphis gossypii*

(i) Damage: Yellowish green and brown coloured aphids colonize the plants on tender shoots and lower surface of leaves. Both adult and young stages suck the sap causing curling, withering and retardation of growth of the infested plant parts (Fig 3.2).

(ii) Management

Spraying 1.0% neem or pongamia oil effectively reduces aphid population. Dimethoate or oxydemeton methyl or fenthion each at 0.05% were found effective against the aphids (Prem Sagar, 1992).

Figure 3.2. Aphid infestation on isabgol

Several chemical products such as dimethoate 0.2%, oxydemeton-methyl 0.025% (Metasystox 25 EC), methomyl 0.05% (Dunet 12.5 WP), acephate 0.05% (Acephate 75% SP) and neem based products such as Kranti neem spray (0.3 and 0.5%), Neem gold (0.3 and 0.5%) and crude extract of neem seed kernel (1 and 2%) were compared for their efficacy on *Aphis gossypii* on *Plantago ovata cv*. Jawahar Isabgol by Upadhyay and Mishra (1999). They reported that the best treatment in control of pest was oxydemeton-methyl with a higher benefit cost ratio of (16.9: 1). Among neem products 2% crude extract treatment resulted in the next highest benefit cost ratio (13.4 : 1).

3.3.1.2. Seed Beetle, *Lasioderma serricorne*

(i) Damage: This is a serious pest of seeds in storage. Both grubs and adults cause damage by feeding on the internal contents of the seed.

(ii) Management

Seed treatment with 2% neem/pongamia oil protects them from beetle damage. Seed treatment with 0.05% chlorpyriphos or 0.07% endosulfan also protects seed from beetle damage.

3.3.1.3. White Grubs and Termites

(i) Damage: Sometimes, white grubs and termites damage the isabgol crop by cutting the roots.

(ii) Management

These pests can be controlled by broadcasting 65% lindane at 1.25 kg/ha before the last ploughing.

3.3.2. Diseases

3.3.2.1. Damping-off, *Pythium ultimum, Rhizoctonia solani*

The crop suffers from both the phases of damping-off *i.e.* pre-emergence and post-emergence. Older plants are less susceptible than the young seedlings.

(i) Symptoms: In pre-emergence phase, the young seedlings are killed before they reach the surface of the soil. Germinating seeds undergo complete rotting. In post-emergence phase, new emerging plants are affected first. The initial infection appears as a slightly darkened spot on collar region of the seedlings. The infection extends rapidly and the invaded cells collapse and the seedling is over-run by the fungus and dies shortly. The disease normally appears if the prevailing temperature is high with high relative humidity at the time of sowing. Unseasonal rain after seeding the crop also helps in development of this disease.

(ii) Management

Treatment of seeds with metalaxyl or fenaminosulf or captafol each at 5 g/kg of seeds protect the seedlings. Seed treatment with *Trichoderma harzianum* at 5 g/kg of seeds protected the seedlings from damping-off disease.

3.3.2.2. Downy Mildew, *Peronospora plantaginis, P. alta*

The downy mildew disease causes enormous quantitative and qualitative losses. Many times, it becomes a limiting factor in successful cultivation of isabgol. The disease is responsible for 30% loss in seed yield.

(i) Symptoms: *Peronospora plantaginis* develops chlorotic areas on the upper surface of the leaves and ashy-white frost-like mycelial growth on the lower surface of the leaves. Later, the infected leaves become chlorotic and brownish in colour followed by curling, crinkling and drying of the plants (Fig 3.3). On the other hand, the fungus *P. alta* develops chlorotic streaks, extending along with midrib of the leaves. Profound downy growth is observed on the lower leaf surface. Ultimately the whole leaf turns necrotic and in severe cases, the entire plant gives blighted appearance. The abundant oospores formation causes the thickening of the affected leaves.

Figure 3.3. Downy mildew on isabgol

The disease is both seed and soil-borne. The sporangia upon germination, produces germ tubes and penetrates through stomata. Sporangia and sporangiophores emerged through stomatal opening before death of the tissues. Studies on sporangial germination under field conditions revealed that the temperature of around 20°C and relative humidity of around 89% favour sporangial germination between 2.00 a.m. and 8.00 a.m. and also at 10.00 p.m. Maximum disease developed during the 75 to 95 days age of the crop.

(ii) Management

(a) Cultural Methods: Seed rate higher than 2.5 kg/ha increased the disease severity in November sown crop and reduced the yield significantly. However, December sown crop generally escapes from the disease.

(b) Chemical Methods: Overnight seed soaking before sowing in Aureofungin at 0.25 g/10 litres of water followed by 3 sprays of Aureofungin at 15 g/ha controlled the disease effectively. The first spray is to be given immediately after appearance of the disease and should be repeated twice at an interval of 8–10 days.

Seed treatment with Apron SD at 5 g/kg of seed coupled with 3 sprays of 0.2% captafol or 0.05% metalaxyl or 0.3% mancozeb effectively controlled the disease. The sprays are to be initiated just after the appearance of the disease and should be repeated at 10 days interval. Seed treatment with metalaxyl at 4 g/kg seed and 3 sprays of metalaxyl at 2 g/l of water at 15 days interval starting from first appearance is best for disease control. Similarly seed treatment with 0.3% mancozeb + three sprays of 0.2% ridomil resulted in maximum seed yield.

Bordeaux mixture (6-3-50) at 600 to 800 l/ha should be sprayed 30 DAS followed by 2 or more sprays at an interval of 15–20 days.

3.3.2.3. Powdery Mildew, *Erysiphe cichoracearum*

(i) Symptoms: Powdery mildew normally appears at the time of flowering. The disease is characterized by small whitish or grayish spots on the leaves which gradually enlarge and cover the whole plant surface with talcum-like powdery mass.

(ii) Management

The disease can be controlled by spraying with 0.2% karathane or wettable sulphur as soon as the disease appears.

3.3.2.4. Wilt, *Fusarium* sp.

(i) Symptoms: The disease is manifested either as pre-emergence damping-off or as a late season wilting of plants as they start to produce seed scales. Wilting, which first occurs in the outer leaves, is preceded by a change in leaf colour from green to silver. Wilting of the whole plant is rapid; leaves may become prostrate within 36 to 48 hours after incipient wilting. Discolouration of the tap root (commencing at the tip) takes place accompanied by the presence of large quantities of hyphae in xylem elements, vascular discolouraion and cortical root rot.

(ii) Management

As a prophylactic measure, spraying of 6-3-100 Bordeaux mixture or copper oxychloride or mancozeb or zineb each at 0.2% commencing from 30 DAS and repeated twice at 15–20 days interval is effective for the management of wilt.

3.4. ASWAGANDHA, *WITHANIA SOMNIFERA*

Ashwagandha (*Withania somnifera*) is a plant of immense medicinal importance growing all over North-Western and Central India. Roots of this plant are the major source of alkaloids and steroidal lactones (withanoids), which are intensively used in various pharmaceutical industries. It has adaptogenic, immuno-modulator, aphrodisiac, anti-stress and mildly sedative properties. Its roots yield valuable drugs which are employed in rheumatic pain, inflammation of joints, nervous disorders, cough, cold, epilepsy, and cardio and nerve tonic. At present, 4,000–5,000 ha area is under its cultivation mainly in Madhya Pradesh (Manasa Neemuch and Jawad tehsil of Mandsaur district) and neighbouring districts of Rajasthan (Kota & Charu).

3.4.1. Pests

3.4.1.1. Epilachna Beetle, *Henosepilachna vigntioctopunctata*

(i) Damage: Clusters of eggs are laid on lower surface of leaves. Grubs and adults fed on leaves by scraping resulting in skeletonization, drying of leaves and defoliation of attacked plants.

(ii) Management

Collection and destruction of eggs, grubs and adults bring down the population. Spraying of 0.2% carbaryl or 0.1% malathion or 0.01% fenthion protects foliage from the beetle damage. Spraying 2% neem oil effectively prevents feeding of the beetle.

3.4.1.2. Mealybug, *Coccidohysterix insolitus*

(i) Damage: White and wax coated nymphs and adults infest lower side of leaves in colonies and suck the sap. Affected leaves get deformed; turn yellow and drop-off in severe cases. Honey dew excreted by bugs favours development of sooty mold which affects the plant growth.

(ii) Management

Removal and destruction of severely infested plant parts avoid further spread. Spraying of 0.1% acephate or methomyl or fenthion each at 0.05% followed by a spray of 2% pongamia oil checks mealybug infestation.

3.4.1.3. Ash Weevil, *Myllocerus viridanus*

(i) Damage: Adult weevils cause damage to leaves by eating from edges in a characteristic manner resulting in severe foliage loss if populations are very high. Grubs feed on roots leading to wilting and drying of the plants.

(ii) Management

Tapping of adults into insecticide solution reduces their population. Spraying of 0.2% carbaryl 0.05% fenthion protects foliage from weevil attack. Soil application of malathion dust or drenching of soil with 0.1% chlorpyriphos kills the grubs.

3.4.2. Diseases

3.4.2.1. Seedling Blight

(i) Symptoms: Seedling blight is known to be the major disease of ashwagandha. The disease reduced the plant population drastically by seedling mortality, which ultimately reduced the yield.

(ii) Management

Seed treatment with 0.2% PCNB (Brassicol) (27.5% mortality), 0.3% mancozeb (31.5% mortality) and 0.2% delan (32.75% mortality) gave significant reduction in seedling mortality and increased the dry root yield.

Seed treatment with *Trichoderma viride* at 4 g/kg of seed along with metalaxyl at 6 g/kg of seed showed minimum seedling mortality (18.7%).

3.4.2.2. Leaf Blight

(i) Symptoms: Leaf blight is a serious disease of ashwagandha at Malwa region of Madhya Pradesh. The symptoms appeared as small, light brown spots on leaves and flowers which enlarged, coalesced with each other and gave a blighted appearance.

(ii) Management

Spraying of Cuman-4 and Rovral both at 0.3% were found effective against this disease.

3.4.2.3. Die-back

(i) Symptoms: The pathogen caused necrosis of tender twigs from the tip backwards. The entire branch or the entire top of the plant withered away. The dead twigs were water-soaked, turn brown, become grayish or straw coloured in advanced stage of the disease.

(ii) Management

Spraying of mancozeb at 3 g/litre of water gives effective control of the disease.

3.4.3. Nematodes

3.4.3.1. Root-knot Nematode, *Meloidogyne incognita*

The root-knot nematode, *M. incognita* is an important limiting factor in the successful cultivation of ashwagandha.

(i) Symptoms: The infected plants were chlorotic, stunted, less branched with fewer and smaller leaves. Roots of such plants were severely galled.

(ii) Management

(a) Integrated Methods: Integration of neem cake with *T. harzianum,* vermicompost with *T. harzianum*, and cow urine with *T. harzianum* considerably reduced the root-knot nematode development and enhanced plant growth and yield. Maximum root-knot suppression was noticed in vermicompost with *T. harzianum* followed by mentha distillate with *Glomus aggregatum*. Highest increase in plant yield was recorded when the soil was amended with mentha/curry leaf distillates along with *T. harzianum/G. aggregatum* (Pandey & Kalra, 2005b) (Table 3.1).

Table 3.1. Effect of organic materials and bioagents on plant growth of Ashwagandha infected with *Meloidogyne incognita*

Treatment	Dry weight of plant (g)	Root-knot index (RKI)	Total mematodes population (soil + roots)
Untreated – Uninoculated	8.5	—	—
Untreated – Inoculated	5.2 (–38.8)	4.00	6460
Carbofuran	8.0 (–5.9)	1.66	3012
Neem compost + Davana	23.8 (+62.4)	1.66	2820
Neem compost + Curry leaf	7.4 (–12.9)	1.33	2261
Neem compost + Vermicompost	12.0 (+41.2)	1.66	2649
Neem compost + *T. harzianum*	12.8 (50.6)	1.66	2674
Neem compost + *G. aggregatum*	14.3 (+68.2)	3.00	2964
Neem compost + *Mentha* distillate	14.2 (67.1)	1.66	2409
Davana + Curry leaf	10.8 (+27.1)	1.33	2030
Davana + Vermicompost	14.1 (65.9)	3.00	2594
Davana + *T. harzianum*	9.6 (+12.9)	1.66	2110
Davana + *G. aggregatum*	10.3 (+21.2)	1.33	2050
Davana + *Mentha* distillate	13.2 (+55.3)	1.33	2150
Curry leaf + Vermicompost	14.0 (+64.7)	1.66	2440
Curry leaf + *T. harzianum*	15.0 (+76.5)	3.33	3252
Curry leaf + *G. aggregatum*	13.4 (+57.6)	3.00	2635
Curry leaf + *Mentha* distillate	15.3 (+80.0)	1.66	2404
Vermicompost + *T. harzianum*	14.4 (+69.4)	0.66	1400
Vermicompost + *G. aggregatum*	13.6 (+60.0)	1.99	2480
Vermicompost + *Mentha* distillate	13.9 (+63.5)	1.33	2000
G. aggregatum + *Mentha* distillate	15.6 (+83.5)	1.00	1784
CD (P = 0.05)	**0.71**	**0.01**	**465.70**

3.5. COLEUS, *COLEUS FORSKOLII*

Coleus (*Coleus forskohlii*) belongs to the Tulsi family Lamiaceae and is indigenous to India. It is used in Ayurvedic medicines and as a condiment. It shot into prominence in modern medicine with the isolation of the diterpinoid, forskolin, from its tubers. The therapeutical properties of forskolin in the treatment of glaucoma, congestive cardiomyopathy, asthma and certain cancers combined with its use in cosmetics has enhanced importance of coleus in modern medicine. Coleus proved to be the exclusive source of forskolin and therefore, the sudden spurt in the demand for its tubers led to exploitative collection from natural stand. The drug is claimed to improve appetite, facilitate digestion, increase vitality and is useful in anemia and inflammation. Recently it is also being used in the treatment of alcoholic addiction. Forskolin has a multiple biological activities like positive iontropic, antihypertensive, branchospasomolytic, antithrombotic, platelet aggregation inhibiting and adenylate cyclase stimulation. The species is now considered endangered and a prime candidate for cultivation, more especially captive cultivation.

3.5.1. Pests

3.5.1.1. Tinged Bug, *Monanthia globulifera*

(i) Damage: Female bug inserts eggs on leaves and tender foliage. Black, spiny nymphs and lace winged adults cause damage by sucking the sap from the leaves. Attacked leaves develop yellow patches, become brown, shrivel and dry up from margins in case of heavy infestations.

(ii) Management

Removal and destruction of heavily infested leaves reduces the bug population. Acephate or carbaryl or methyl parathion or phenthoate each at 0.05% sprays offer effective control of the bug.

3.5.1.2. Spike Borer, *Helicoverpa armigera*

(i) Damage: Cream coloured, round eggs are seen on tender foliage. Larvae initially feed on tender leaves and later bore into developing spikes by making holes and feeding on buds and flowers.

(ii) Management

Two or three sprays of 0.07% endosulfan or methyl parathion or ethofenprox each at 0.05% at 15 days interval controls the pest.

3.5.1.3. Grasshopper, *Chrotogonus* sp.

(i) Damage: Nymphs and adult grass hoppers attack the crop during the day. They cause damage by making irregular holes on leaves and cutting tender shoots.

(ii) Management

Malathion at 0.1% or carbaryl at 0.2% sprays offer protection to foliage from grasshopper damage. Spraying 2% neem oil checks attack of the pest.

3.5.2. Nematodes

Root-knot nematodes are recognized as the major limiting factors in successful production of coleus crop.

3.5.2.1. Root-knot Nematode, *Meloidogyne incognita*

The root-knot nematode infestation was reported on coleus from Kerala and Orissa. The dry weight of the tubers was reduced by 20% due to root-knot nematodes. The percentage of starch on fresh weight basis showed drastic reduction (16%) in the infested tubers. Senthamarai *et al.* (2006c) reported that *M. incognita* was responsible for 70.2% loss in tuber yield of coleus.

(i) Symptoms: The galls on coleus roots are very big and pronounced. The root-knot nematode damage often leads to crop failure. The infested tubers swell in size with irregular surface and cracking of the skin. When the infestation is severe, rotting sets in even before harvest. Infested tubers rot after harvest and rarely reach market.

(ii) Interaction with Other Pathogens: Simultaneous inoculation of *M. incognita* and *Macrophomina phaseolina* as well as nematode followed by fungus 15 days later, caused 100% root rot disease and significant reduction in plant growth parameters in coleus (Senthamarai *et al.*, 2006a).

(iii) Management

(a) Cultural Methods: Sree Bhadra, a high yielding variety of yam released by Central Tuber Crops Research Institute, Trivandrum, is identified as a resistant trap crop for the root-knot nematode. The nematodes were able to penetrate the root but giant cell formation was not induced and as a result the nematodes die inside the roots. Planting this variety in root-knot nematode infested field helped in clearing the field free of nematodes and giving good tuber yield. Subsequently growing of susceptible crop like coleus escaped nematode damage (Mohandas, 2001).

(b) Chemical Methods: Treatment with DBCP and fensulfothion reduced damage to 4.0 and 5.7%, respectively as against 21% in control. The above treatments also increased the production of healthy tubers to the extent of 53 and 47%, respectively (Pillai, 1976).

(c) Biological Methods: Soil application of *Trichoderma harzianum* and *Pseudomonas fluorescens* each at 2.5 kg/ha recorded increased plant growth and reduced *M. incognita* population (Senthamarai *et al.*, 2006b).

(d) Integrated Methods: Integration of soil solarization in the nursery for 15 days with 150 gauge LDPE film and application of *Paecilomyces lilacinus* + neem cake or *P.lilacinus* + *Bacillus macerans* in the main field are the best treatments in increasing plant height (64.3 and 60.3 cm compared to 40.0 cm in control), number of leaves (593.3 and 583.3

compared to 310.0 in control), weight of tubers/plant (560.0 and 546.6 g compared to 350.0 g in control), yield (11.5 and 11.3 kg/plot compared to 6.9 kg/plot in control) and in reducing root galls (0.3 and 1.0 compared to 50.6 in control), nematode population in soil (25.0 and 30.0/100 ml soil compared to 196.6/100 ml soil in control) and roots (1.0 and 1.6/5 g roots compared to 79.0/5 g roots in control) (Nisha & Sheela, 2006).

3.6. OPIUM POPPY, *PAPAVER SOMNIFERUM*

Opium poppy is grown in Banswara, Bara, Bhilwara, Bundi, Chittorgarh, Jhalawar, Kota and Udaipur districts of Rajasthan; Faizabad, Barabanki, Bareilly and Shahjehanpur in Uttar Pradesh and Neemuch, Mandsaur and Ratlam in Madhya Pradesh. It is a source of raw opium which is used in narcotic, veterinary and pharmaceutical industries. It contains alkaloids like morphine, codeine, narcotine, papaverine and thebain. Poppy seed is used for culinary purpose—breads, sweets and confectioneries. Capsule husk is brewed with tea (Bonda chai).

3.6.1. Pests

3.6.1.1. Cut Worm, *Agrotis ipsilon*

(i) Damage: Female moths lay eggs on lower surface of leaves or on the moist soil. Dark brown larvae cut young plants at ground level and damage shoots of grown up plants after dusk and hide during day time. Attack by the pest leads to retardation in plant growth and yield loss.

(ii) Management

Deep ploughing during summer exposes pupae. Flooding of infested field with water mixed with 0.1% chlorpyriphos and application of 0.1% malathion or 0.2% carbaryl protect newly planted crop from cut worm. Spreading of poison bait made using rice bran, jaggery and chlorpyriphos (in 10 : 1 : 0.5 ratio) in the field kills larvae. Spraying of methyl parathion or phenthoate each at 0.05% also controls the pest effectively.

3.6.1.2. Capsule Borer, *Helicoverpa armigera*

(i) Damage: Spherical, cream coloured eggs are laid on tender leaves and later bore into developing spikes and capsules by making holes.

(ii) Management

Spraying 0.01% fenvalerate or 0.05% methyl parathion or quinalphos or 0.07% endosulfan, when larvae are noticed, controls the borer damage. Spraying of 5% NSKE or 1% neem oil checks feeding by the larvae.

3.6.1.3. Aphid, *Myzus persicae*

(i) Damage: Nymphs and adults colonize on growing shoots and cause damage by sucking the sap. Affected shoots get deformed, wither and retarded in growth.

(ii) Management

Spraying 2% neem/pongamia oil at 10 days interval controls aphids effectively. Dimethoate or phenthoate or profenphos each at 0.05% sprays also reduces aphid population considerably.

3.6.2. Diseases

3.6.2.1. Collar Rot, *Rhizoctonia solani*

(i) Symptoms: Collar rot is one of the serious diseases of poppy. A black lesion develops at the joint of the root and shoot, resulting into falling-off of the plant and rapid loss of crop stand.

(ii) Management

Crop rotation to provide at least 3 years gap for poppy cultivation in the same field is effective. Seed treatment with thiram and soil drench with 0.4% zineb are found effective. Foliar spray of 0.15 to 0.2% ridomil along with micronutrients and urea right from the seedling stage has been found most effective for the control of collar rot.

3.6.2.2. Downy Mildew, *Peronospora arborescens*

Downy mildew has been reported as an important disease from major opium poppy growing countries like India, China, Egypt, France, The Netherlands, Hungary, Greece, Spain, Portugal, Italy, Turkey, Australia, USSR and Yugoslavia. The epiphytotic form of the disease was observed in Argentina, Germany, Iran, Hungary, Austria and India. In India, the occurrence of the disease has been reported from Rajasthan, Madhya Pradesh, Uttar Pradesh, and Delhi.

(i) Symptoms: The plants with systemic infection have two kinds of symptoms *viz.*, lower portion is healthy and upper portion diseased. In the upper diseased portion of the plant, the leaves remain small, chlorotic, curling downward at the edges and are closely placed. Heavy sporulation occurs on the entire lower surface of the diseased leaves. The sporulation on leaves is persistent. Buds generally remain under developed and do not open. The capsule, if formed, shrinks, gets wrinkled and dries gradually after infection. Heavy sporulation occurs on peduncle, buds and capsules which is not persistent. Rudimentary brownish black structures develop in place of normal seeds. Such capsules do not yield opium as well. The infected plants die prematurely.

In the plants with leaf spot infection, the infected area becomes first chlorotic and then necrotic. The leaf spots enlarge by coalescing with each other. Sporulation may occur on the underside of spotted area but is not heavy and not persistent. In spite of excessive leaf spots, plants do not die prematurely.

The nitrogen fixing preceding crops like black gram and groundnut increased the disease incidence. The disease incidence increased with increase in soil moisture and supply of nitrogen. The disease stimulatory effect of nitrogenous fertilizers was mitigated by the application of FYM. The disease severity was more in late sown crops as prime growth period coincided with low temperature and high humidity. As a result, latex, morphine, seed and husk yields were significantly low in the late sown crops.

(ii) Management .

At Udaipur, it was observed that early sown (30th October) crop was less affected by downy mildew incidence. The occurrence and severity of the disease decreased with lower plant density.

Seed treatment with metalaxyl and four sprays of metalaxyl at 0.2% at 35, 55, 75 and 95 days after sowing found most effective in reducing primary and secondary disease incidence. It also increased latex yield (26.7%) and seed yield (24.7%) over control. Apron seed treatment (8–10 g/ha) and sprays of ridomil MZ, Cuman-L, and mancozeb also were highly effective in controlling the disease.

Opium poppy *cv.* Jawahar-540 is having some resistance against the disease has been recommended for Madhya Pradesh.

3.6.2.3. Powdery Mildew, *Erysiphe polygoni*

Powdery mildew has been found to occur in severe form on poppy in Rajasthan, Madhya Pradesh and Uttar Pradesh. Poppy cultivars suffer heavy losses on account of this disease.

(i) Symptoms: powdery mildew is observed in the fields at or after flowering stage *i.e.* nearly 14–16 weeks after sowing. Late sown crop (after 15th November) develops heavy powdery mildew disease. First appearance of the disease is in February or early March. The base of the stem is attacked first where it appears as small, circular patch consisting of radialy arranged mycelium. This enlarges gradually and sporulation occurs soon. Several new colonies continue to appear covering large part of the stem. Seven to ten days after the appearance of the disease, blackening develops on the affected parts. When the powdery mildew develops to its maximum, ½ to ¾ of stem appears black interspersed with green patches. On leaves the mildew spreads all over. Growth of mildew is luxuriant in shade. Late forming branches are severely affected and their growth is checked. The buds on such branches remain rudimentary and wither.

(ii) Epidemiology: The disease is very severe in Rajasthan during February to April when the minimum temperatures were 11.2 to 17.2°C and maximum temperatures were 26.6 to 34.8°C.

(iii) Management

Crop sown before 15th October almost escape the disease. The disease can be effectively and economically controlled by only one spray of 0.5% wettable sulphur at the time of first appearance of the disease in the field or when the crop has reached the age of 80 days or when the maximum daily temperature has reached 27°C and above.

3.6.2.4. Root Rot, *Macrophomina phaseolina*

(i) Symptoms: The disease appeared in the month of December when the plants were about 2-months old. The first visible symptom was in the form of small black spot on the upper part of the tap root near the soil surface. Gradually the leaves started withering and drying. The lesions on the tap root later become deep and the infected plants were broken down at the point of black necrotic lesion. In case of severe infection, about 20% plants died prematurely.

(ii) Management

Soil application of *T. harzianum* reduced the percentage of infected plants (0 to 10% compared to 20 to 40% in control), increased plant height (94 to 100 cm compared to 85 to 88 cm in control) and increased the opium yield (114 to 191 mg compared to 92 to 94 mg in control).

Integrated Disease Management: Seed treatment with metalaxyl 35 SD at 8 g/kg of seed + sprays of 0.2% metalaxyl at 35, 55 and 75 DAS along with recommended agronomic practices was the best to reduce the disease infection by 54.05% at Faizabad, 55.83% at Mandsaur and 82.28% at Udaipur and increased latex yield by 122.2, 72.59 and 53.39% at these places, respectively. Similarly, seed yield at these locations were found to increase by 182.57, 37.32 and 79.72%, respectively.

3.7. SARPAGANDHA, *RAUVOLFIA SERPENTINA*

Sarpagandha (*Rauvolfia serpentina*) is a perennial native Indian herb. Its roots are used for controlling high blood pressure and certain forms of insanity in Ayurvedic system of medicine since ancient times. It has received worldwide recognition after isolation of its bioactive reserpine alkaloid in allopathic medicines. In addition, its sedative property is utilized by Ayurvedic physicians in treatment of insomania, epilepsy and asthama. It is commercially cultivated in Madhya Pradesh, Uttar Pradesh, West Bengal, Assam and Orissa. The root is its economic part, containing 55 alkaloids of which reserpine, deserpidine, ajmalcine, serpentine and yohimbine being pharmacologically active and important. Of these, ajmalicine, reserpine and serpentine are highly potent in medicine. The commercial root crop contains 1.4–3.0% of total alkaloids.

3.7.1. Pests

3.7.1.1. Scale, *Saissetia coffeae*

(i) Damage: Brownish black scales infests lower surface of leaves along midrib and veins. They also infest petioles and tender shoots and deplete the sap from plant. Attacked leaves turn yellow and drop off while plants lose vigour and become stunted in growth.

(ii) Management

Removal and burning of heavily infested plant parts bring down scale population. Spraying of ethion or chlorpyriphos each at 0.05% or acephate at 0.1% effectively controls the scale attack.

3.7.1.2. Grass Hopper, *Orthacric simulans*

(i) Damage: Both young and adult stages damage the crop by making irregular holes on leaves and cutting tender shoots. Attack on young plant result in plant loss.

(ii) Management

Spraying of 4% NSKE or 1% neem oil prevents feeding by the grass hoppers. Foliar sprays of carbaryl or malathion each at 0.1% effectively checks damage by the pest.

3.7.1.3. Leaf Folder, *Glyphodes vertumalis*

(i) Damage: Eggs are deposited on tender foliage. Larvae fold leaves together and feed on green tissue causing drying of leaves and stunted growth of the plant in case of severe attack.

(ii) Management

Collection and destruction of dried and infested plant parts reduce the pest population. Spraying of profenphos or phenthoate or ethofenprox each at 0.05% effectively controls the larval damage.

3.7.2. Diseases

3.7.2.1. Foliar Blight/Spot, *Alternaria tenuis*

(i) Management

Six applications of mancozeb at 0.3% as foliar spray were found to be most effective giving 187.8 g fresh root/plant and 78.8% protection against foliar blight/spot (*Alternaria alternata, Cercospora rauvolfia, Cecospora serpentina, Colletotrichum gloeosporioides, Corynespora cassiicola, Curvularia lunata, Macrophomina phaseolina* and *Rhizoctonia solani*) diseases. Application of two doses of neem cake at 50 g/plant at 30 days interval plus two sprays of *B. subtilis* in September and October along with three foliar sprays of mancozeb at 15 days interval plus two foliar sprays of carbendazim at 0.15% at 21 days interval gave 201.5 g fresh root/plant, and 75.5% protection against foliar blight pathogens.

Seed treatment with contact fungicides is recommended. Spraying of maneb rovral at 0.2% also gives effective control of the disease.

3.7.3. Nematodes

The root-knot nematode is recognized as the major limiting factor in successful cultivation of sarpagandha crop.

3.7.3.1. Root-knot Nematodes, *Meloidogyne* spp.

(i) Management

Soil application of neem cake enriched with *T. harzianum* at 1 MT/ha is recommended.

3.8. KACHOLAM, *KAEMPFERIA GALANGA*

Kacholam (*Kaempferia galanga*), also known as Acangi/Chandramuli, is a glabrous annual medicinal and aromatic herb with very fragrant underground rhizomes. It belongs to the family Zingiberaceae. The plant is generally abundant in China, subtropical himalayas, Nepal and Kumaon.

It is an aromatic herb with tuberous rhizome. Roots are bitter, thermogenic, acrid, carminative, aromatic, depurative, diuretic, expectorant, digestive, antihelmintic, febrifuge and stimulant. Rhizomes are used in dyspepsia, leprosy, skin diseases, rheumatism, asthma, cough, bronchitis, wounds, ulcers, malarial fever, inflammatory tumour and nasal obstruction. The rhizomes also contain resimic acid. Leaves are used for pharyngodynia, ophthalmia, swellings, fever and rheumatism. Oil showed tranquillizing activity of short duration. Essential oil from rhizomes is used in perfumery.

3.8.1. Nematodes

The root-knot nematode is recognized as the major limiting factor in successful production of kacholam crop.

3.8.1.1. Root-knot Nematode, *Meloidogyne incognita*

(i) Symptoms: The root-knot nematode was reported to cause severe damage to the medicinal plant Kacholam (Sheela *et al.*, 1996). *M. incognita* was responsible for 64% loss in yield of kacholam (Sheela & Rajani, 1998).

(ii) Management

(a) Physical Methods: Hot water treatment of rhizomes at 55°C for 5 min was effective in reducing the nematode population (more than 74%) and giving 81% increase in yield (Nisha & Sheela, 2004).

(b) Cultural Methods: Rhizome treatment with neem leaf extract (4%) + garlic extract (1%) and application of neem cake at 200 g/m^2 gave maximum reduction of nematode population in soil (69%), roots (21.5 galls/plant compared to 83.25 galls/plant in control) and root galling (RKI = 1.25 compared to 2.25 in control) (Nisha *et al.*, 2001).

Mulching of kacholam with green leaves of neem and chromolaena at 5 kg/m^2 15 days before planting rhizomes, increased the rhizome yield by 160.7% and 143.3% over control, respectively. The above treatments also reduced root galling (1 as compared to 5 in control), the nematode population both in soil (81.8% and 80.4%, respectively) and roots (89.60% and 86.14%, respectively) (Sreeja *et al.*, 2001) (Table 3.2).

Table 3.2. Effect of mulching of green leaves on root-knot nematodes and yield of kacholam.

Treatment	Root-knot index	Nematodes population (100 g soil)	% redn. over control	Nemtodes population (5 g root)	% rednuc-tion over control	Rhizome yield (kg/4 m^2)	% incre-ase over control
Neem leaf	1	45.44 (6.74)*	81.80	7.00 (2.76)	89.60	5.6	160.7
Glyricidia	2	52.48 (7.24)	79.05	14.67 (2.84)	78.21	3.9	79.6
Mangium	5	178.26 (13.35)	28.50	33.67 (5.87)	49.99	2.6	21.6
Clerodendron	2	95.35 (9.92)	60.60	22.70 (4.82)	6.33	3.4	56.4
Calotropis	2	85.86 (9.27)	65.60	16.00 (4.03)	76.24	3.6	69.8
Cromolaena	1	48.96 (6.99)	80.40	9.33 (3.18)	86.14	5.3	143.3
Control	5	249.37 (15.79)	—	67.33 (8.26)	—	2.2	—
CD (P = 0.05)	—	**(1.02)**	—	**(1.45)**	—	**1.28**	—

*Figures in parentheses are values after square root transformation.

(c) Biological Methods: Rhizome treatment with *Pseudomonas fluorescens*, *Trichoderma viride* and AMF each at 3% w/w significantly reduced the nematode population (more than 74%) and improved plant growth and yield (Nisha & Sheela, 2004). Application of *Glomus fasciculatum* at 200 spores/plant gave maximum reduction of nematode population in soil (43.55%), roots (28.5 galls/plant compared to 83.25 galls/plant in control) and root galling (RKI = 1.75 compared to 2.25 in control) (Rajani *et al.*, 1998).

3.9. CHINESE POTATO, *SOLENOSTEMON ROTUNDIFOLIUS*

Chinese Potato (*Solenostemon rotundifolius*), also known as coleus, is one of the minor tubers used for edible purpose. Its tubers are used as vegetable. It is grown in Kerala, Tamil Nadu and Karnataka. It is also cultivated on a small scale in north-eastern states.

3.9.1. Nematodes

Root-knot nematodes are recognized as the major limiting factors in successful production of Chinese potato crop.

3.9.1.1. Root-knot Nematode, *Meloidogyne incognita*

(i) Management

(a) Physical Methods: Hot water treatment of Chinese Potato tubers at 53°C for 10 min eliminated *Meloidogyne* spp. infection.

(b) Biological Methods: Dipping of plants in 3% solution of *Bacillus macerans* + soil drenching with 2% solution of *B. macerans* 7 days after planting recorded maximum reduction in nematode population and root-knot count (14 galls/g of roots) and gave maximum increase in yield (Sheela *et al.*, 2004).

3.10. HENBANE, *HYOSCYAMUS* SPP.

Henbane (*Hyoscyamus niger, H. muticus, H. albus*) is a popular ancient herb. Henbanes are one of the chief sources of tropane alkaloid *viz.* hyoscine, hyoscyamine and atropine. The hyoscine and its derivatives have been used in pharmaceutical preparations, since they are having anticholingeric, antispasmodic and mydriatic properties. It is used in the treatment of asthama and whooping cough. It also provides relief in gripping pain in intestinal disorders. In India, its small scale cultivation is done in Malwa region in Madhya Pradesh.

3.10.1. Nematodes

The root-knot nematode is recognized as the major limiting factors in successful cultivation of henbane crop.

3.10.1.1. Root-knot Nematode, *Meloidogyne incognita*

Henbane (*H. niger* and *H. muticus*) were severely infested with *M. incognita* and *M. javanica*. Even 3–4 larvae/g of soil cause significant damage to the crop (Pandey, 1990).

(i) Symptoms: Up to 60–70% plants were chlorotic and stunted showing a patchy appearance with fewer smaller leaves and flowers (Haseeb & Pandey, 1989a) (Fig 3.4). Varying sizes of galls were found in the root system.

Figure 3.4. Root-knot nematode on henbane. Left – Healthy; Right – Infected (Courtesy: Haseeb, 1992)

(ii) Management

(a) Cultural Methods: It was observed that essential oils of palmarosa, citronella, basil and menthol mint were quite effective in reducing *M. incognita* population and improving plant growth. The oil of palmarosa at 2 ml/pot was most effective (Pandey *et al.*, 2003) (Table 3.3).

Table 3.3. Effect of essential oils of aromatic crops on growth and root galling of henbane

Treatment	Plant dry weight (g)	No. of leaves/plant	Root-knot index
Control	4.4	39.0	4.00
Menthol mint –0.5 ml	5.3	50.6	2.33
Menthol mint –1.0 ml	8.4	63.6	1.66
Menthol mint –2.0 ml	10.4	72.6	1.33
Basil –0.5 ml	9.6	86.6	2.00
Basil –1.0 ml	11.4	123.3	1.33

Table. 3.3. (Contd.)

Treatment	Plant dry weight (g)	No. of leaves/plant	Root-knot index
Basil –2.0 ml	9.0	116.0	1.33
Citronella – 0.5 ml	10.0	112.0	2.33
Citronella – 1.0 ml	11.3	94.3	1.66
Citronella – 2.0 ml	9.8	88.0	1.33
Palmarosa – 0.5 ml	9.6	79.7	2.00
Palmarosa – 1.0 ml	12.5	83.6	1.66
Palmarosa – 2.0 ml	12.9	106.3	1.00
CD (P = 0.05)	0.555	12.645	0.030

(b) Biological Methods: Negative impact of *Glomus aggregatum, G. fasciculatum* and *G. mosseae* on reproduction of *M. incognita* infecting *H. niger* has been observed (Pandey *et al*., 1999). Significant reduction in *M. incognita* population was observed when henbane plants were inoculated with *Pseudomonas fluorescens* or *G. aggregatum* resulting in higher herbage yield and root colonization with the bioagents.

3.11. BRAHMI, *BACOPA MONNIERI*

Brahmi (*Bacopa monnieri*) is an important medicinal plant and chief source of bacoside A and B which are extensively used in formulation of different drugs useful for improving intellect, asthma and epilepsy.

3.11.1. Pests

3.11.1.1. Grass Hopper and Tobacco Worm, *Spodoptera litura*

(i) Damage: Defoliating insects such as grass hopper and tobacco worm are a menace during summer months. They cause mass defoliation and need to be controlled in time.

(ii) Management

Upon observation of larvae, neem-based pesticides should be applied. Alternatively, spraying of 0.1% dichlorvos is very effective in controlling the pest.

3.11.2. Nematodes

Root-knot nematodes are recognized as the major limiting factors in successful production of brahmi crop.

3.11.2.1. Root-knot Nematode, *Meloidogyne incognita*

(i) Symptoms: Stunting in plant growth, chlorosis and gall formation in the root system give a simple indication of presence of root-knot nematodes. A negative correlation exists between increasing population levels of *M. incognita* and plant growth and yield characteristics.

(ii) Management

(a) Integrated Methods: Neem compound, distillation waste of menthol mint, curry leaf, davana and vermicompost and bioinoculants such as *Glomus aggregatum* and *Trichoderma harzianum* alone or in combination were root-knot nematode suppressive and enhanced the growth and yield of brahmi (Pandey *et al.*, 2002) (Table 3.4).

Table 3.4. Effect of organic materials and bioagents on plant growth, yield and root galling in brahmi infected with *Meloidogyne incognita*

Treatment	Total dry weight (g/pot)	Root-knot index
Control	47.1	3.66
Carbofuran	68.2	2.33
Davana	96.0	0.66
Neem compound	102.6	1.00
Neem cake	100.8	2.33
Linseed cake	64.0	2.66
Sesamum cake	68.3	2.00
Glomus aggregatum	69.5	3.33
G. fasciculatum	69.0	2.66
Trichoderma harzianum	77.3	2.66
CD (P = 0.05)	**3.271**	**0.041**

3.12. BISHOP'S WEED, *AMMI MAJUS*

Ammi majus (Family: Apiaceae) is an important medicinal plant commonly used in the treatment of vitilago leucoderma. It is also used in formulation of suntan lotion.

3.12.1. Nematodes

3.12.1.1. Root-knot Nematode, *Meloidogyne incognita*

The root-knot nematode is a major constraint in the cultivation of *Ammi majus* (Pandey, 2002).

(i) Management

(a) Cultural Methods: Treatment with *Costus speciosus* dried shoot powder at 10 g/kg soil showed the least nematode population (Rf = 0.4) and root-knot index (RKI = 0.6) followed by *Spilanthes acmella* (Rf = 0.6, RKI = 0.6), *C. speciosus* rhizome powder (Rf = 0.5, RKI = 1.0) and neem cake (Rf = 0.5, RKI = 1.0) (Pandey, 2002) (Table 3.5).

Table 3.5. Effect of dried powder of botanicals on growth and root galling in *Ammi majus*

Treatment	Total dry weight (g)	Total nematodes population	Reproduction factor	Root-knot index
Control	13.4	21620	4.3	4.0
Carbofuran	16.9	7000	1.4	2.6
Citronella leaves	21.1	17010	3.4	3.6
Lemon grass leaves	22.2	3890	0.8	2.0
Spilanthes acmella shoot	23.1	3000	0.6	0.6
Costus speciosus shoot	22.7	1816	0.4	0.6
C. speciosus rhizome	21.1	2716	0.5	1.0
Neem cake	22.7	2580	0.5	1.0
CD (P = 0.05)	—	323.113	0.221	0.89

3.13. DIOSCOREA, *DIOSCOREA* SPP.

Various species of *Dioscorea* (*Dioscorea floribunda*) are the source of diosgenin, a steroidal sapogenin, and is the most commonly used precursor for synthesis of certain drugs which include cortisones, sex hormones and oral contraceptives.

3.13.1. Pests

3.13.1.1. Aphids and Red Spider Mites

(i) Damage: Aphids and red spider mites are important pests of medicinal dioscorea.

(ii) Management

Spraying of monocrotophos at 1.5 ml + wettable sulphur at 3 g + teepol at 0.5 ml/litre of water can control these pests.

3.13.2. Diseases

3.13.2.1. Leaf Spot, *Cercospora dioscorea*

(i) Symptoms: The fungus causes many leaf spots scattered all over the leaves. They are irregularly angular and dark brown to black, with indefinite margins. There is heavy defoliation causing considerable reduction in the yield of tubers.

(ii) Management

Spraying of 0.1% benomyl or carbendazim at fortnightly intervals gives effective control of the disease.

3.13.2.2. Leaf Fall/Die-back, *Colletotrichum* sp.

(i) Symptoms: The first symptom observed is the drying of young leaves and the branches from the tip downwards. There is heavy defoliation and die-back. In severe cases, the entire plant may die. The acervuli are formed on the bark of the vines which are dead or dying.

(ii) Management

Spraying of 0.1% carbendazim or 0.2% mancozeb at fortnightly intervals is effective in managing the disease.

3.13.3. Nematodes

Root-knot and dry rot nematodes are recognized as the major limiting factors in successful production of dioscorea crop.

3.13.3.1. Root-knot Nematodes, *Meloidogyne incognita, M. javanica*

Dioscorea floribunda has been reported to be severely parasitized by *M. incognita* and *M. javanica.*

(i) Symptoms: Severe infestation of *D. alata* and *D. spinosa* with *M. incognita* in Kerala was reported (Raveendran & Nadakal, 1975). The economic threshold and economic injury level of *M. incognita* were found to be 250 and 1250 nematodes per plant of *D. rotundata cv.* Igave, respectively. Tubers from pots inoculated with more than 1250 nematodes per plant were so heavily galled that the market value was reduced by 40%. The quality of tubers, an important consideration in growing of this crop, is also affected (Atu *et al.*, 1983).

(ii) Interaction with Other Pathogens: A disease complex involving *Fusarium oxysporum* f. sp. *dioscorea* and *M. incognita*/*P. coffeae* has been reported on *D. rotundata*. A positive interaction between *M. incognita* and *F. oxysporum* f. sp. *dioscorea* occurs in *D. rotundata cv.* Habanero.

3.13.3.2. Dry Rot Nematode, *Scutellonema bradys*

(i) Symptoms: *S. bradys* is responsible for 'dry rot' of yam tubers in Nigeria and Puerto Rico. Hutton (1978) considered *Pratylenchus coffeae* to be the most serious pest of yam in Jamaica, causing 'dry rot' or burning. This condition is characterized by cracking in the skin underlined by a brown, corky rot in the storage tissues. Rot progresses deeper into the yam tissues following harvest and prior to planting or consumption and is more pronounced towards the stem and yam tubers.

(ii) Management of *Dioscorea* Nematodes

A fallow period has been suggested to control *S. bradys* on tubers. Use of nematode-free tubers for propagation will effectively reduce the incidence and damage. Immersion off dioscorea tubers in hot water at 45–51°C for 30 min gave effective control of root-knot nematodes. Treatments at 50–55°C for 40 min has given over a 98% kill of *S. bradys* without damage to tubers.

3.14. SAFED MUSLI, *CHLOROPHYTUM BORIVILIANUM*

Safed musli (*Chlorophytum borivilianum*), known as minor forest produce, is an important medicinal plant belonging to family Liliaceae. This plant is widely distributed throughout India especially in Uttar Pradesh, Bihar, Rajasthan, West Bengal, Madhya Pradesh, Gujarat and Maharashtra. It is a major source of carbohydrates (42%), protein (8.9%), root fibres (3–4%) and saponins (2–17%). Ancient books suggest that various parts of this plant were used in the cure of different human diseases without any side effects.

3.14.1. Diseases

3.14.1.1 Leaf Blight, *Cercospora* sp.

(i) Management

Spraying 1% Bordeaux mixture has been recommended to control the disease.

3.14.2. Nematodes

The root-knot nematode is a major limiting factor in successful production of safed musli.

3.14.2.1. Root-knot Nematode, *Meloidogyne incognita*

Safed musli has been found to be greatly affected with root-knot infection caused by *M. incognita*. A severe loss due to the disease was found under field conditions.

(i) Symptoms: The infested field showed stunting, drying and falling of leaves and severe galling on the root system.

3.15. PYRETHRUM, *CHRYSANTHEMUM* SPP.

The genus *Chrysanthemum* consists of about 300 species of herbs and under-shrubs of which only a few such as *C. cinerariaefolium. C. coccineum* and C. *marschalki*, yield the commercial insecticide pyrethrum. The term pyrethrum is commonly applied to the dried flower heads of the above 3 species which is a source of Pyrethrin I, Pyrethrin II, Cinerin I, Cinerin II, Jasmoline I and Jasmoline II. Pyrethrum is one of the safe insecticides known which has a low mammalian toxicity and is rapidly metabolizes if accidentally swallowed. It acts as a contact poison with rapid knock down effect on insects. It affects the nervous system resulting in muscular convulsions, paralysis and death of insects. Major pyrethrum growing areas include South Indian hills above 1800 m above MSL and Kashmir valley.

3.15.1. Diseases

3.15.1.1. Wilt, *Fusarium* sp.

(i) Symptoms: This fungus is found to cause a severe wilt disease of *Chrysanthemum* spp. The chief symptoms are retarded growth, sudden wilting and death of plants followed by browning and rotting of roots and collar region. Another fungus, *Sclerotinia minor* is also usually found associated with the wilted plants.

(ii) Management

Drenching the basal portion of the infected plants and the soil around with 0.1% benomyl or 0.2% captan is effective in management of the disease.

3.15.1.2. Root Rot, *Fusarium solani*

(i) Symptoms: The fungus causes a root rot disease in pyrethrum. The typical symptoms such as yellowing, wilting and stunting of the aerial parts and reduction in number of tillers, start appearing in June-July. The wilting usually starts with tillers and proceeding inward progressively until most of the tillers and the main plant wilts. Infected root show black discolouration, extensive rotting and disintegration of cortical tissues.

(ii) Management

Seedlings should be grown in disease-free soil.
Raised nursery beds should be prepared.
Nursery beds should be drenched with 0.15% benomyl or 0.1% zineb.

3.16. SENNA, *CASSIA ANGUSTIFOLIA, C. ACUTIFOLIA*

Senna is a well known drug made from leaves and pods. It is grown principally under dry land conditions and usually as a semi-arid irrigated crop. Leaves and pods of senna contain two glycosides *viz.*, Sennoside A and Sennoside B which are used as laxatives. It is also used in medicine as a cathartic. It is used as an ingredient of herbal tea in Europe. The major senna growing areas include Madurai, Tirunelveli and Ramanathapuram in Tamil Nadu; Pune in Maharashtra; Anand and Mehsana in Gujarat and Jodhpur in Rajasthan.

3.16.1. Pests

3.16.1.1. Green Leaf Eating Caterpillar

(i) Damage: The pest feeds on young leaves.

(ii) Management

Spraying endosulfan at 2 ml/litre of water is effective.

3.16.1.2. Pod Borer

(i) Damage: The pod borer bores into pods and feeds on the seed.

(ii) Management

Spraying endosulfan at 2 ml/litre of water is found effective.

3.16.1.3. Aphids

(i) Damage: Aphids suck the sap and devitalize the plant.

(ii) Management

Spraying endosulfan at 2 ml/litre of water gave effective control of aphids.

3.16.2. Diseases

3.16.2.1. Leaf Spot, *Alternaria alternata*

Leaf spot is the most serious among the diseases causing lot of damage to the senna crop. The infested leaves drop off prematurely and affect the yield considerably.

(i) Symptoms: The disease becomes visible first as minute pale yellow spots on the leaf blade and its margins. As the spot grow in size they become circular and irregular in shape with their coalescence and turn dark brown to black in colour. The lesions vary from a minute spot to about 8 mm diameter in size. With the advancement of the disease, leaf tips and margins die and the necrotic tissues increase in size.

The disease is known to adversely affect the sennosides content in leaves and it is inversely proportional to the disease intensity. Sennosides content in leaves of highest disease intensity has almost one-third to that of healthy leaves and about one-half when there is moderate infection. Loss of sennosides in severely affected leaves was over 60% as compared to that in disease-free leaves. Even a mild incidence of the disease had affected the recovery of sennosides considerably (36.63%). As the price of the produce is dependent upon its quality, even a mild incidence of leaf spot affects the revenue of the crop.

(ii) Management

Carbendazim at 0.05% and macuprax at 0.2% gave best control of the disease, but the yield is higher in macuprax treatment. Spraying of 0.15% mancozeb at fortnightly interval for 3 times in a period of 5–6 weeks is also recommended. Spraying of Bordeaux mixture, benomyl, captafol and copper oxychloride is also recommended.

3.16.2.2. Damping-off, *Rhizoctonia bataticola*

(i) Symptoms: In the north-western India, senna is affected by the damping-off disease at seedling stage. The damping-off in seedlings is the most devastating disease.

(ii) Management

Seed treatment with thiram or captafol at 2.5 g/kg of seed gave good protection. The disease incidence under field conditions can be controlled by soil drenching with 0.1% carbendazim solution.

3.16.2.3. Dry Rot, *Macrophomina phaseolina*

(i) Symptoms: The dry rot occurs at late stage of crop growth.

(ii) Management

Soil drenching with thiram or carbendazim gives effective control of the disease.

3.16.2.4. Leaf Spot, *Cercospora* sp., *Phyllosticta* sp.

(i) Symptoms: The leaf spots attack the plants mainly during cloudy weather.

(ii) Management

The pathogens are controlled by spraying 0.15% mancozeb.

3.16.2.5. Leaf Blight, *Colletotrichum* sp.

(i) Symptoms: In Senna, the disease was found to be caused by *Colletotrichum sp.* (new species), which leads to premature defoliation of the leaves.

(ii) Management

The disease has been successfully managed by three applications of propiconazole at 0.1% at 15 days interval.

3.17. LIQUORICE, *GLYCYRRHIZA GLABRA*

Liquorice (*Glycyrrhiza glabra*) is grown in Gujarat, Haryana and Rajasthan. Dried peeled or unpeeled underground stem and roots contain glycyrrhizin, which is 5 times sweeter than sucrose. It is used as flavouring agent in pharmaceutical industry for manufacturing cough syrups.

3.17.1. Pests

3.17.1.1. Termites

(i) Damage: Termites live in above-ground and under-ground mud galleries and colonise the dry parts of plants including roots. Plants are easily uprooted

(ii) Management

Destroy termite mound near the vicinity. Remove above-ground mud galleries. Drench chlorpyriphos at 7 ml/l of water around soil near root zone.

3.17.1.2. Cotton Ash Weevil, *Myllocerus undecimpunctatus*

(i) Damage: The ash weevil causes injury to leaves.

(ii) Management

Foliar spraying of quinalphos at 2 ml/l of water is found effective. Application of neem cake at 0.5 t/ha at planting also gave effective control of the pest.

3.17.2. Diseases

3.17.2.1. Collar Rot, *Macrophomina phaseolina* (= *Rhizoctonia bataticola*)

(i) Symptoms: The increased incidence of the disease was observed with increase in temperature from February to July. During high atmospheric day temperatures above 35°C during summer, the old and new sprouts were attacked near collar region. Leaves on the affected sprouts turned pale yellow and dried up, which subsequently fell down and only deep brown to black branches with sclerotia and pycnidia were seen. In majority of the cases, this disease was confined to collar portion. However, above-ground stem portions were also occasionally affected. High incidence of the disease was observed on young sprouts as well as emerging sprouts during high temperature.

(ii) Management

Reduction in soil temperature during summer by: different crop mulches, increasing the frequency and levels of irrigation and use of soil amendments and organic manures help to reduce the incidence of the disease.

3.17.2.2. Wilt/Root Rot, *Fusarium oxysporum*

(i) Symptoms: Affected plants show discolouration, followed by rotting of under-ground parts (stolons) which results in sudden wilting/drying of the plants.

(ii) Management

Reduction in soil temperature during summer by: different crop mulches, increasing the frequency and levels of irrigation and use of soil amendments and organic manures help to reduce the incidence of the disease.

3.17.2.3. Leaf Spot, *Cercospora cavarae*

(i) Symptoms: This disease is invariably observed in Gujarat starting from late monsoon and prevailing up to onset of summer. The diseased portion on the upper leaf surface turns light in colour which eventually turns brown at later stage. Exactly underneath the upper diseased portion, initially angular water-soaked spots were observed. These spots were restricted to interveinal areas. They were gradually covered with spores and they turned dark brown and became velvety. Under epiphytotic conditions, severe defoliation is also observed. This disease was also reported from north-west himalayas.

(ii) Management

Spraying maneb or zineb at 0.2% gives effective management of the disease.

3.18. QUININE BARK, *CINCHONA* SP.

Cinchona (*Cinchona* sp.) is grown in Nilgiris, Anamalai, Naduvattam, Shevaroy and Palani hills of Tamil Nadu; some parts of West Bengal; South Kanara and Kodagu in Karnataka. Bark is source of quinine and about 20 other alkaloids of which cinchonidine, quinidine and cinchonine are the most important. The other alkaloids of lesser importance are quinamine, quinicine, cinchonicine, hydroquinine, hydroquinidine, hydrocinchonine, quinamine, conquinamine and poricine.

3.18.1. Pests

3.18.1.1. Cockchafer Beetle, *Holotrichia repetita*

(i) Damage: Cockchafer beetle grubs damage seedlings.

(ii) Management

Hardy perennial. Insecticide sprays are not required.

3.18.1.2. Tea Mosquito Bug, *Helopeltis antonii, H. cinchonii*

(i) Damage: Tea mosquito bug infest leaves in nurseries and in the main field. Causes leaf curl by sucking sap from tender leaves.

(ii) Management

Hardy perennial. Insecticide sprays not required.

3.18.2. Diseases

3.18.2.1. Seedling Blight, *Phytophthora parasitica*; Root Rot, *Phytophthora cinnamomi*; Damping-off, *Pythium vexans, Sclerotium rolfsii*

(i) Symptoms: The above pathogens are important which affect seedling stand in the nursery.

(ii) Management

Soil drenching with copper oxychloride at 5 g/litre of water is effective in managing the above diseases.

3.19. BELLADONNA, *ATROPA* SPP.

Indian belladonna (*Atropa accuminata*) and European belladonna (*Atropa belladonna*) are the source for tropane alkaloids which include hyoscine, hyoscyamine and atropine (0.13–0.7%) which are used in the manufacture of tinctures, treatment of whooping cough, epilepsy, Parkinson's disease, renal and bilary colic.

Indian belladonna is grown in the farms of Kashmir State Forest Department which is a remunerative crop for north-western hilly region of India.

3.19.1. Pests

3.19.1.1. Cutworm, *Agrotis flammetra*

(i) Damage: The cut worm attacks tender seedlings during early summer.

(ii) Management

Spraying endosulfan at 2 ml/litre of water is effective.

3.19.1.2. Leaf Defoliator, *Archips imcaceana*

(i) Damage: The pest defoliates the leaves.

(ii) Management

Spraying endosulfan at 2 ml/litre of water is effective.

3.19.2. Diseases

3.19.2.1. Damping-off, *Pythium* spp.

(i) Symptoms: The pathogen is responsible for death of seedlings in nursery.

(ii) Management

Drenching the soil with 3 g/litre of copper oxychloride is effective.

3.19.2.2. Leaf Spot, *Cercospora atropae*

(i) Symptoms: The pathogen is responsible for angular spots on leaf surface.

(ii) Management

Spraying mancozeb at 2 g/litre of water is effective.

3.19.2.3. Mottle Virus

(i) Management

Collection and destruction of affected plants helps to reduce further spread of the disease.

3.20. BACHNAG, *GLORIOSA* SPP.

Gloriosa (*Gloriosa superba*) tubers and seeds contain colchicine at 0.15–0.3% and 0.7–0.9%, respectively. Colchicine is used in treatment of gout.

3.20.1. Pests

3.20.1.1. Lily Caterpillar, *Polytela gloriosae*

(i) Damage: The lily caterpillar is a voracious feeder of leaves and growing point.

(ii) Management

Spraying carbaryl at 3 g/litre of water is effective for the management of lily caterpillar.

3.20.1.2. Green Caterpillar

(i) Damage: The green caterpillar completely eat the flowers.

(ii) Management

Spraying carbaryl at 3 g/litre of water gives effective control of green caterpillar.

3.20.2. Diseases

3.20.2.1. Leaf Blight, *Curvularia lunata*

(i) Symptoms: Leaves turn light yellow and later small black spots are seen which ultimately spread to the entire leaf.

(ii) Management

Spraying mancozeb at 2.5 g/litre of water is effective.

3.20.2.2. Tuber Rot

(i) Symptoms: Tubers become soft and discoloured. Leaves turn yellow and plants die.

(ii) Management

Soil drenching around the base of plants with carbendazim at 2 g/litre of water is found effective.

3.21. GUGGAL, *COMMIPHORA MUKUL*

Guggal (*Commiphora mukul*) is grown in Assam, Gujarat, Karnataka, Maharashtra, Rajasthan and Tamil Nadu. It is a source of Indian Bedellium, an oleo-gum resin obtained by incision of the bark. The oleo-gum resins are mixtures of resins (61%), gum (29.3%), volatile oil (1.45%), occasionally combined with other substances, mainly myrecene, dimyrecene and polymyrecene. Resin is used in incense, perfumery and in indigenous medicine as antiseptic, astringent, carminative and digestant. It is used to treat arthritis, pyorrhoea, tonsillitis, obesity etc.

3.21.1. Pests

3.21.1.1. Leaf Eating Caterpillar, *Euproctis lanata*

(i) Damage: The pest is voracious leaf feeder which causes extensive defoliation.

(ii) Management

Spraying monocrotophos at 2 ml/litre of water is effective in managing the pest.

3.21.1.2. White Fly, *Bemisia tabaci*

(i) Damage: White flies sucks sap from leaves which turn yellow and fall.

(ii) Management

Spraying monocrotophos at 2 ml/litre of water gives effective control of whiteflies.

3.22. LONG PEPPER, *PIPER LONGUM*

Long pepper (*Piper longum*) is grown in Karnataka, Kerala, Madhya Pradesh, Maharashtra, Tamil Nadu, Eastern Uttar Pradesh and lower hills of West Bengal. Its spikes contain the alkaloid piperine (4–5%) and piplatin, piperolactum A, piperolactum B and piporadione. Roots contain alkaloids piper longuminine (0.2–0.25%) and piper longumine (0.02%). Further purification of piper longumine yields alkaloids like cepharadione A, cepharadione B, cepharanone B and aristolactum A11. Dried immature roots and fruits find extensive use in Ayurveda and Unani medicine in treatment of chronic bronchitis, cough and cold.

3.22.1. Pests

3.22.1.1. MealyBugs

(i) Damage: Mealy bugs infest roots of plant and suck the sap.

(ii) Management

Soil drenching with chlorpyriphos at 3 ml/litre of water around the base of plants is effective in managing the disease.

3.22.1.2. Tea Mosquito bug, *Helopeltis theivora*

(i) Damage: Adults and nymphs feed on tender foliage causing necrotic lesions around feeding punctures and leave shot holes on leaf lamina.

(ii) Management

Spray any commonly available systemic insecticide to control the pest.

3.22.2. Diseases

3.22.2.1. *Colletotrichum* Leaf Spot, *Colletotrichum* spp.

(i) Symptoms: The pathogen is responsible for leaf and vine rotting.

(ii) Management

Spraying of 1% Bordeaux mixture or copper oxychloride at 0.2% once during May and another 2–3 sprays during rainy season is effective in controlling the leaf spot disease.

3.22.2.2. *Cercospora* Leaf Spot, *Cercospora* spp.

(i) Symptoms: Necrotic spots and blight appear on leaves during warmer months.

(ii) Management

Spraying 1% Bordeaux mixture or copper oxychloride at 0.2% gave effective control of leaf spot disease.

3.23. TRUE ALOE, *ALOE VERA*, *ALOE INDICA*

Aloe vera is grown in Gujarat, Madhya Pradesh, Maharashtra and Rajasthan. Yellow bitter juice present in the inner parenchymal cells of aloe vera yields 'Drug aloe' (Aloin) and '*Aloe* gel' which are used in medical preparations, cosmetics and ointments.

3.23.1. Diseases

3.23.1.1. Leaf Spot, *Alternaria alternata*

(i) Symptoms: Leaf spot is the main disease which not only affects the growth and yield of the plant but also affects its quality (bad odour of the gel).

(ii) Management

Spraying with 0.2% mancozeb or chlorothalanil gives effective control of leaf spot disease.

3.23.1.2. Leaf Blight, *Colletotrichum* sp.

(i) Symptoms: This is one of the serious diseases of *Aloe vera*, caused by *Colletotrichum* sp.; the disease greatly reduced the yield of gel.

(ii) Management

The disease has been successfully managed by three application of propiconazole at 0.1% at 15 days interval.

3.23.1.3. Black Spot

(i) Management

Among four selected bioinoculants namely, *Bacillus subtilis, Glomus aggregatum, G. intraradices* and *Trichoderma harzianum* evaluated against black spot disease either alone or in different combinations; maximum plant height attained was 56 cm in *G. intraradices* treated plants followed by 53 cm in *B. subtilis* and *B. subtilis + G. aggregatum* treatments. Maximum plantlets production (14) was recorded in the treatments of *B. subtilis + T. harzianum + G. aggregatum* followed by 12 in others. Maximum herb yield obtained was 3.65 kg/plant in *G. aggregatum* treated pots, followed by 3.35 kg/plant in *G. aggregatum + B. subtilis* and *G. intraradices* treated pots.

3.24. BABCHI, *PSORALIA CORYLIFOLIA*

3.24.1. Diseases

3.24.1.1. Collar Rot

(i) Management

Soil application of neem cake at 0.5 MT/ha + seed treatment with *Trichoderma viride* at 4 g/kg of seed + spraying with *T. viride* suspension at 50 g/litre of water + spraying 0.05% carbendazim in December-January significantly reduced the collar rot incidence.

3.25. DUBOISIA, *DUBOISIA MYPOROIDES*

3.25.1.1. Pests

3.25.1. Green Caterpillar

(i) Damage: There was occurrence of green caterpillar which was found feeding on leaves.

(ii) Management

Spraying the crop with 0.03% malathion gave effective management of green caterpillar.

3.26. AMBRETTE, *ABELMOSCHUS MOSCHATUS*

Ambrette, also called as Muskdana and Kasturi Bhendi, grows wild all over the Deccan plateau and Karnataka, in the hilly region and at the foot hills of the Himalayas. It is raised as a kharif crop in Punjab, Nagaland, Arunachal Pradesh, Bihar, Terai area of Kumaon, adjoining areas of Lucknow, Mahmudabad and Sitapur (Uttar Pradesh). Major constituent of oil is ambrettolic acid and its lactone—'ambrettolid'. The volatile oil finds use in perfumes, cosmetics, talcum powder, toiletries and confectionaries. Various parts of the plant find extensive use in Ayurvedic medicine. It is also used in manufacture of 'zarda', a flavoured chewing tobacco.

3.26.1. Pests

Ambrette crop suffers from many pests. The major pests are aphids, thrips, pink bollworm, spotted bollworm and red spider mites.

3.26.1.1. Aphids, Thrips

(i) Damage: The early season sucking pests *i.e.* aphids and thrips cause considerable damage to the crop in early stage of its growth, particularly in the early sown summer crop.

(ii) Management

The foliar spray of dimethoate, endosulfan, malathion, phosphamidon and vamidothion reduces the risk of aphids and thrips.

3.26.1.2. Spotted Bollworm

(i) Damage: The spotted bollworm attacks succulent shoot, vegetative buds, flowers and pods during vegetative growth and fruiting stage. The infested shoots above the point of infestation become brown, bend down and die. The fruits are also infested which causes damage to seeds in pods.

(ii) Management

Endosulfan at 0.07% can be sprayed at 10–15 days interval right from the stage of one month old crop to the last picking to control the pest. To eradicate spotted bollworm, methyl parathion, quinalphos, toxaphene, dimethoate and phosphamidon can be applied.

3.26.1.3. Pink Bollworm

(i) Damage: The deep pink coloured caterpillar of pink bollworm also infests the crop. The caterpillar bore into the pods and seeds.

(ii) Management

Healthy seeds free from pink bollworm infestation should be sown as precautionary measure. Chlorpyriphos + cypermethrin each at 0.05% can be sprayed for the control of bollworms.

3.26.1.4. Red Spider Mite

(i) Damage: The web forming red spider mite feeds on the under surface of leaves and produces white specks. The infested plants become yellow, weak and shed leaves.

(ii) Management

Dimethoate, phosphamidon, folidol each at 0.05% can be sprayed once on the infested plants to ward off red spider mites.

3.26.1.5. Red Cotton Bug

(i) Damage: The red cotton bug feeds on the surface of hairy leaves.

(ii) Management

Spray cypermethrin or fenvalerate each at 0.01% to control the red cotton bug.

3.26.1.6. Leaf Hopper

(i) Damage: Nymphs and adults of leaf hopper are prevalent throughout the cropping season and maximum damage is caused during July-September in kharif season and during November-January in winter season. Severe infestation results in the retarded growth of plants and premature defoliation.

3.26.1.7. Leaf Roller

(i) Damage: Caterpillars of leaf roller roll the leaves from edge to midrib forming a cover around it and feed inside which results in defoliation.

3.26.2. Diseases

3.26.2.1. Anthracnose, *Colletotrichum hibiscicum*

(i) Symptoms: The fungal pathogen infects almost all parts of the plant. Infection at seedling stage has been observed.

(ii) Management

The seed should be treated with agrosan GN or ceresan to save the crop from anthracnose before sowing. Seed treatment with contact fungicides and spray of Bordeaux mixture are recommended.

3.26.2.2. Fusarium Wilt

(i) Symptoms: Fusarium wilt occurs in seedling stage which also precedes its death. The disease prolongs over 2–3 weeks.

(ii) Management

The crop may be sprayed with 1% Bordeaux mixture to control the infection.

3.26.2.3. Leaf Blight

(i) Symptoms: Leaf blight disease causes damage to ambrette by way of premature leaf drying which results in death of leaves and pods. During high humidity, a white colony growth appears on diseased surface of leaves. Dark brown patch covers the whole leaf causing defoliation and death of plants.

(ii) Management

Seed treatment with ceresan makes plants less susceptible to the disease. Repeated spray of 0.5% zineb at monthly interval controls the leaf blight effectively.

3.27. ROSEMARY, *ROSMARINUS OFFICINALIS*

Major growing areas of rosemary include Nilgiris and Bangalore. Leaves and flowers yield "Rosemary oil" used in perfumery, cosmetics and culinary purpose.

3.27.1. Diseases

3.27.1.1. Leaf Blight, *Rhizoctonia solani*

(i) Symptoms: Rosemary plantation is adversely affected by this soil-borne fungus causing severe leaf blight disease. The incidence of the disease is more after rains.

(ii) Management

A fortnightly spray of 0.1% benomyl or 0.3% mancozeb is effective for management of the disease.

3.28. SOUTH AMERICAN MARIGOLD, *TAGETES MINUTA*

3.28.1. Diseases

3.28.1.1. Wilt and Rot, *Sclerotinia scletiorum*

(i) Management

Deep ploughing and seed treatment with thiram at 2 g/kg of seed gives effective control of the disease.

3.28.1.2. Collar Rot, *Sclerotium rolfsii*

(i) Management

Deep ploughing and seed treatment with thiram at 2 g/kg of seed gives effective control of the disease.

3.28.1.3. Little Leaf (Phytoplasma)

(i) Management

The disease can be controlled by spraying streptomycin at 5 g/litre of water.

3.28.1.4. Mosaic Virus

(i) Management

Collection and destruction of infected plants reduces the incidence of the disease.

3.29. KALMEGH, *ANDROGRAPHIS PANICULATA*

3.29.1. Diseases

3.29.1.1. Leaf Blight, *Colletotrichum* sp., *Rhizoctonia* sp.

(i) Symptoms: Two distinct leaf blight diseases of kalmegh have been investigated. One type is caused by *Colletotrichum* sp. and another by *Rhizoctonia* sp. *Rhizoctonia* sp. was found more destructive than the *Colletotrichum* sp. However, both caused premature defoliation of leaves.

(ii) Management

The disease has been successfully managed by three applications of propiconazole at 0.1% at 15 days interval.

Table 3.6. Pest and disease management in medicinal plants

Crop	Disease	Etiology	Control measures
Opium poppy *(Papaver somniferum)*	Downy mildew	*Peronospora arborescens*	Seed treatment with metalaxyl and four sprays of metalaxyl at 0.2% at 35, 55, 75 and 95 days after sowing found most effective in reducing primary and secondary disease incidence. It also increased latex yield (26.7%) and seed yield (24.7%) over control. Opium poppy *cv.* Jawahar-540 is having some resistance against the disease and recommended for Madhya Pradesh.
Senna (*Cassia angustifolia*)	Leaf spot	*Alternaria alternata*	Spraying of Bordeaux mixture, benomyl, captafol and copper oxy chloride is recommended.
Isabgol (*Plantago ovata*)	Downy mildew	*Peronospora alta, P. plantaginis, Pseudoperonospora plantaginis*	Seed treatment with metalaxyl at 4 g/kg seed and 3 sprays of metalaxyl (0.2%) at 15 days interval starting from first appearance is best for disease control. December sown crops are mostly disease free.
Sarpagandha (*Rauvolfia serpentina*)	Leaf blight and bud rot	*Alternaria tenuis*	Spraying of maneb (0.3%) or rovral (0.2%); seed treatment with contact fungicides.
Liquorice (*Glycyrrhiza glabra*)	Leaf spot	*Cercospora cavarae*	Spraying maneb or zineb at 0.2%.
Long pepper (*Piper longum*)	Leaf spot rotting	*Colletotrichum* spp., *Cercospora* spp.	Spraying of Bordeaux mixture-once during May and 2–3 sprays during rainy season.
Khasi kateri (*Solanum viarum*)	Wilt	*Fusarium oxysporum*	Direct sowing of seeds instead of transplanting of seedlings can reduce the disease incidence
Solanum (*Solanum viarum*)	Fruit borer	*Leucinodes orbonalis, Helicoverpa armigera*	Spray monocrotophos at 0.5 ml/l
	Leaf eating caterpillars	*Spodoptera litura, Spilosoma obliqua*	Spray quinalphos at 2.0 ml/l

4 AROMATIC CROPS

In recent years, there has been an increased interest in the cultivation of aromatic plants to meet the requirements of cosmetic, flavouring and perfumery industries. The aromatic plants provide raw material for the production of flavours, herbal cosmetics, perfumery etc. Aromatic plants and their products, particularly the essential oils, are now becoming one of the more important export items from many developing countries of Asia. Chemically essential oils are terpenes, which act as carriers of the aromatic substances. Most essential oils also contain camphors, and the more odiferous compounds present in them consist of oxygen derivatives of terpenes, alcohols, esters, aldehydes and ketones. India has enjoyed a pre-eminent position in the manufacture of superior perfumes and aromatics by using essential oils. The important aromatic plants are lemon grass, vetiver, patchouli, palmarosa, citronella, mints, geranium, lavender, basil, jasmine etc.

In the world market, India stands in second position. Most of the essential oils produced though marketed within the country, but a sizeable amount has also been exported. Among the essential oils exported from India are Japanese mint oil, ambrette seed essence, sandal wood oil, citronella oil, lemon grass oil, palmarosa oil, ginger oil, clove oil, eucalyptus oil, vetiver oil, pepper oil, etc. The annual exports of the derivatives of aromatic plants are to the tune of Rs. 600–700 million.

4.1. JASMINE

Jasmine (*Jasminum* spp.) is one of the leading traditional flowers of India. Its flowers are used for making garlands, adorning hairs of women, in religious and ceremonial functions, and for producing perfumery oil. It is commercially cultivated in Karnataka, Tamil Nadu, Andhra Pradesh, Maharashtra, Gujarat, Haryana, Punjab, West Bengal, Rajasthan, Assam and Uttar Pradesh. Apart from internal trade, fresh flowers of jasmine are exported to Malaysia, Singapore and Sri Lanka.

4.1.1. Pests

4.1.1.1. Bud-worms, *Hendecasis duplifascialis*

(i) Damage: This is a serious pest of *Jasminum sambac* and *J. pubescens*. The pest appears in severe form during post-monsoon months. The eggs are laid on flower clusters. The caterpillars are green with a black or brown head and feed on fresh leaves and floral parts and leave behind faecal pellets entangled in the silk strands (Fig 4.1). Webbing is seen in severe infestation and pupation takes place in soil or in the webs. Severely affected bushes show burnt appearance. Pupation takes place in the soil or inside galleries. The moth is small and white in colour (Prasad *et al.*, 1983).

Figure 4.1. Bud worm on jasmine

(ii) Management

- Collection and destruction of infested plant parts.
- Hand picking of larvae and pupae to reduce the insect population.
- Spray acephate 75 SP at 1.5 g/l or cypermethrin 25 EC at 1 ml/l or quinalphos 25 EC at 2 ml/l or profenphos 50 EC at 1 ml/l.
- Botanicals like neem seed kernel extract 5% or 1% neem oil are also effective against the pest (Nelson *et al.*, 1993).
- This pest is highly susceptible to *B. thuringiensis* (Sundarababu, 1969).

4.1.1.2. Web Worm, *Nausinoe geometralis*

(i) Damage: This is a serious pest of *J. sambac*. Eggs are seen on matured leaves. Hatched larvae scrape leaf tissue initially and web leaves together with silken threads. Green coloured caterpillar with dark warts and hairs on the body feed gregariously on leaves resulting in skeletonization and drying.

(ii) Management

Collection and destruction of webs having larvae and pupae reduce the damage. Spraying of 0.1% carbaryl, 0.05% quinalphos or 0.01% fevalerate was observed to be effective against the web worm (Sandhu & Shukla, 1984). Methyl parathion or phenthoate each at 0.05% also provide good control of the pest.

4.1.1.3. Blossom Midge, *Cantarinia maculipennis*

(i) Damage: The maggot attacks *J. sambac* and *J. auriculatum* petals and the flower shows the signs of purple discolouration 4–5 days after oviposition. These maggots are dull white and feed on inner petals, anthers and stigma. Mature yellowish orange maggots fall down to the soil and pupate. Pupation occurs in a thin papery white case in the top layer of the soil. The life cycle is completed in 13–18 days.

(ii) Management

- Soil application of carbofuran 3G at 40 g per bush or foliar application of cypermethrin 25 EC at 1 ml/l (David & Natarajan, 1988) or triazophos 40 EC at 1.5 ml/l or profenphos 50 EC at 1 ml/l at fortnightly interval.
- Apply neem cake at 250kg/ha as soil treatment.
- Spraying 2% neem oil mixed with teepol at 0.05% effectively reduces the damage (David *et al.*, 1992).

4.1.1.4. Gallery Worm, *Elasmopalus jasminophagus*

(i) Damage: The pest attacks all *Jasminum* species. The larvae are green in colour with red head and brown streaks on the body. It builds galleries of fecal matter and silk on the terminal leaves, shoots and flower heads and feeds inside. Pupation takes place inside the galleries and the adult is dark grey in colour.

(ii) Management

Collection and destruction of infested plant parts reduce the pest population. Cypermethrin at 0.05% or acephate at 0.01% or methyl parathion at 0.05% sprays offer very effective control of gallery worm.

4.1.1.5. Shoot Borer, *Sycophila* sp.

(i) Damage: The pest is observed on *J. grandiflorum*. Larvae tunnel shoots, feed inside by making grooves and pupate inside the shoots. Series of holes are seen on damaged shoots.

(ii) Management

Application of carbofuran at 1.0 kg *a.i.*/ha or spraying of 0.06% triazophos or 0.05% chlorpyriphos at fortnightly intervals check shoot borer attack.

4.1.1.6. Gall Mite, *Aceria jasmini;* Carmine Mite, *Tetranychus cinnabarinus*

(i) Damage: Gall mite is a major pest of *Jasminum auriculatum* and *J. pubescens*. Colonization and feeding by the pest results in the formation of felty outgrowths resulting in malformation of vegetative and floral parts (Fig 4.2) and consequently reduces crop yield.

Carmine mite attacks jasmine during summer on leaves resulting in blotching, browning and falling of leaves.

(ii) Management

- Spraying of 0.3% wettable sulphur 80 WP at 3 g/l or 0.06% triazophos (40 EC at 1.5 ml/l) with the onset of monsoon at 14 days intervals, repeated 2–3 times, provides effective control of gall mite (Shivaramu & Jhansi Rani, 1996).

Figure 4.2. Gall midge infestation on jasmine

- Sprays of 0.05% ethion or 0.01% fenproparthin and 0.02% dicofol are effective in reducing galls by the mite (Mallikarjunappa *et al.*, 1991).
- Karuppuchamy *et al.* (1986) reported successful control of adults and egg hatching of spider mite by spraying 0.05% dicofol or 0.3% wettable sulphur.
- If necessary spray neem or pongamia oil at 1% (10 ml/l).

4.1.2. Diseases

4.1.2.1. Wilt, *Sclerotium rolfsii*

(i) Symptoms: The young leaves turn yellow and the twigs start drying from tip downwards. The plants wilt in patches.

(ii) Management

- Soil drenching with 1% Bordeaux mixture or 0.3% copper oxychloride reduced damage due to wilt.

4.1.2.2. Leaf Spot, *Cercospora jasminicola*

(i) Symptoms: Many species of *Jasminum* are susceptible to this disease. The symptoms are noticed mainly on leaves as circular to oval necrotic spots which show dark brown to reddish borders. The under side of the spots show dark grey coloured growth of the pathogen. Severe infection often leads to heavy defoliation. Symptoms are sometimes observed on stems.

(ii) Management

- The disease can be controlled by spraying 1% Bordeaux mixture or 0.1% benomyl or carbendazim at 15 to 20 days interval. The frequency of application may be increased during rainy season.

4.1.2.3. Rust, *Uromyces hobsoni*

(i) Symptoms: This disease is prevalent in some species of *Jasminum* and is more common in south India. Symptoms are noticed on all the aerial parts of the plant including flowers. The earliest symptoms are yellowish to orange coloured spots on leaves, young twigs and floral buds. These spots enlarge and swell considerably. The lower side of the swollen region shows several orange yellow cup-like projections which contain orange yellow spores of the fungus. The infected leaves and young stems become greatly hypertrophied and distorted. Towards the end of the season, dark brown to black swollen structures (called telial galls) develop on areas which earlier showed orange yellow cup-like projections.

(ii) Management

- The infected leaves and other plant parts should be collected and destroyed to reduce the inoculum for the next season.
- The disease can be controlled to some extent by spraying 0.2% mancozeb or zineb.

4.1.3. Nematodes

The root-knot and the burrowing nematodes are major problems in jasmine cultivation. Bajaj (1989) recorded *Tylenchulus semipenetrans* on *J. sambac* from Haryana.

4.1.3.1. Root-knot Nematode, *Meloidogyne incognita*

M. incognita was reported on *J. sambac* and *J. flexile* from Tamil Nadu by Rajendran and Rajendran (1979).

(i) Symptoms: The roots exhibited very small swellings and enlarged rootlets. The minute galls were more in *J. sambac.* Pale coloured leaves and dieback symptoms were associated (Rajendran & Rajendran, 1979).

(ii) Management

(a) Cultural Methods: The planting material should be raised in nematode-free or solarized nursery beds. Treatment of nursery beds with neem or pongamia cakes at 2 kg/sq. m. helps in reducing the nematode population.

(b) Chemical Methods: Application of phorate at 4 g *a.i.* per plant during May and September months and incorporation of 20 kg FYM increased flower yield of jasmine by 50% and reduced the root-knot nematode population by 70% (Sundarababu, 1992).

Nursery beds should be treated with carbofuran 3G or phorate 10G at 25 g/sq. m. and main field with carbofuran 3G or phorate 10 G at 2.5 kg *a.i.*/ha before planting.

(c) Integrated Methods: Cuttings or planting material should be raised in nematode-free or treated nursery beds. Application of neem cake enriched with *T. harzianum* at 1 t/ha is recommended.

Nursery beds should be treated with neem or pongamia cakes enriched with mycorrhizal spores at 1 kg/sq. m.

4.1.3.2. Burrowing nematode, *Radopholus similis*

Symptoms: Khan and Parvatha Reddy (1989) recorded a high population of *R. similis* from stunted *J. pubescens* plants in Bangalore. Roots of infected plants exhibited yellow to brown lesions. In some cases blackening also occurs. Infected plants become severely stunted. Branches become dry with chlorotic leaves which ultimately drop.

4.2. MINTS

Mints (*Mentha* spp.) are a group of aromatic herbs belonging to the family Lamiaceae, which are considered to be the most important cash crops in Indo-Gangetic plains. Cultivation of mints has been done in a large scale in many tropical and sub-tropical countries of the world including India, China, Brazil, Japan and USA. Indian farmer grow it as a bonus crop on an area of more than 1,50,000 ha in central and northern parts of Indo-Gangetic plains. The crop fit well in the traditional food based cropping system with other crops like paddy, wheat, potato, mustard, maize, okra, carrot, onion, spinach, pigeonpea, cowpea, etc. Similarly it can be grown as an intercrop with sugarcane and some legumes. In India, the highest area under commercial cultivation of mint has been in Uttar Pradesh (Terai region), but during last few years it has spread through out the country, especially in north and north-western states including Madhya Pradesh.

Among different oil yielding species of mint, Japanese mint (*M. arvensis*) occupies the highest area under mint cultivation followed by Peppermint (*Mentha spicata*), Spear mint (*M. piperata*), Bergamot mint (*M. citrata*) and Scotch spearmint (*M. cardiaca*).

Japanese mint essential oil is used as a source of menthol, menthyl acetate, menthone and terpenes which finds wide use in medicine, perfumes, food and cosmetics. The oil of peppermint, which contains less menthol but has a sweeter aroma and taste, is used in tooth paste, chewing gums, candies, high grade liquors, medicines and other pharmaceutical preparations. Spearmint oil having carvone and limonene, is mainly used for flavouring; whereas, Bergamot mint oil is mainly used in perfumery and flavouring due to higher content of linayl acetate and linalool.

4.2.1. Pests

4.2.1.1. Semilooper, *Thysanoplusia orichalcea*

(i) Damage: The pest infests the crop round the year. It has overlapping generations and the life cycle is completed in 27 to 38 days (Prem Sagar & Ramji, 1991). Eggs are found on lower surface of leaves. Larvae devour the leaves and cause severe defoliation.

(ii) Management

Spraying of methyl parathion or quinalphos each at 0.05% or fenvalerate at 0.01% provides god control. NSKE 4% or 1% neem oil reduces damage by semilooper to foliage.

4.2.1.2. Leaf Webber, *Syngamia abruptalis*

(i) Damage: Infestation of leaf webber is severe during May-June and August-October. Larvae web leaves together and feed on them resulting in skeletonization and drying of damaged leaves. Pupation takes place inside the web. The life cycle is completed in 31 to 43 days in different months.

(ii) Management

Clipping and destruction of webs reduce webber population. Spraying of 0.005% cypermethrin followed by 0.05% quinalphos or 0.1% dichlorvos or 0.07% endosulfan gives successful control of leaf webber.

4.2.1.3. Aphid, *Aphis affinis*

(i) Damage: This pest is active during March-April. About 15 generations are reported on this crop (Prem Sagar & Singh, 1981). Colonies of the aphid are found on tender shoots and lower side of leaves. They suck the sap resulting in discolouration, deformation and falling of leaves. Affected shoots become stunted in growth and wither.

(ii) Management

Spraying of 4% PSKE or 1% neem or pongamia oil reduces aphid infestation. Dimethoate or oxydemeton methyl each at 0.05% also provides good control of the aphids.

4.2.1.4. Whitefly

(i) Damage: The initial adult population of whitefly was found building up from March and was maximum in the month of May.

(ii) Management

Among genotypes evaluated for their resistance potential to whitefly population, significantly low number of flies was observed in *cv.* 'Kushal'.

4.2.1.5. Leaf Roller, *Syngamia abruptalis*

(i) Damage: The caterpillar folds the leaf in the form of a roll and feed inside the leaf tissue. The edges of the leaves are held together with silk-like filaments. Leaves get rolled in August-September.

(ii) Management

Spraying of 0.1% phosphamidon or 0.07% endosulfan or thiodan at 1.5 ml/litre of water gives effective control of leaf roller. Two-three sprayings are sufficient to control the pest.

4.2.1.6. Red Pumpkin Beetle, *Aulocophora favicollis*

(i) Damage: Leaves and buds are eaten away.

(ii) Management

Spraying 0.05% malathion twice at 20 days interval is effective for the management of beetles.

4.2.1.7. Hairy Caterpillar, *Diacrisia obliqua*

(i) Damage: The caterpillar starts eating the under-surface of the leaves and the leaves are damaged.

(ii) Management

Spraying of folidol or malathion each at 0.05% twice at 20 days interval is effective for the control of the pest.

4.2.1.8. Cutworms, *Agrotis flammatra*

(i) Damage: Pest damages the young plants at the collar region during spring season.

(ii) Management

Soil treatment with Phorate 10G before planting the crop is recommended.

4.2.2. Diseases

4.2.2.1. Powdery Mildew, *Erysiphe cichoracearum*

This disease was first reported by Ganguli and Pandotra (1962) on Japanese mint. It causes heavy toll in April-May leading to reduction in yield by 20% due to defoliation.

(i) Symptoms: The important symptoms of the disease are; small chlorotic spots appearing on the surface and brownish discolouration on the corresponding lower surface prior to the appearance of powdery phase. Later, circular white powdery patches appear mostly on the upper portions of the surviving leaves (Fig 4.3). Heavy defoliation occurs in severe cases having a banded appearance on stem and branches. Later, these white patches turn to purplish black and produce large number of perithecia.

Figure 4.3. Powdery mildew on mint leaf

(ii) Management

To control powdery mildew, remove infected plants and do not water from above.

4.2.2.2. Wilt, *Verticillium albo-atrum*

(i) Symptoms: Another fungal wilt disease *Verticillium dahliae* is a major problem on mint crops. The typical symptoms of the disease are dwarfing, unilateral development of the branches, etiolation of leaves, wilting and ultimate death of the plants (Fig 4.4). The pathogen is soil-borne and spread through soil, roots and stolons used as seed material. None of the commercial varieties are resistant.

Figure 4.4. Verticillium wilt of mentha

(ii) Management

The disease can be controlled using crop rotation and fungicides.

4.2.2.3. Stolon Rot, *Macrophomina phaseoli* (*Rhizoctonia bataticola*)

The disease was reported on *Mentha cordiaca* by Green (1961) and on *M. arvensis* and *M. spicata* by Hussain and Janardhan (1965).

(i) Symptoms: The important primary symptoms are yellowing of leaves and stunted growth of the plants. The wilting of the plants occurs in advance stages leading to the death of above-ground portion. Under-ground stolons exhibit pinkish brown lesions in the earlier stages and gradually turn into dark brown to black patches. These patches increase in size and finally result in the soft decay of a portion or the entire affected stolon.

(ii) Management

Avoid excess irrigation. Uproot and burn the affected plants. Apply *Trichoderma* sp. to soil.

4.2.2.4. *Sclorotium* Rot, *Sclorotium rolfsii*

Ganguly and Pandotra (1964) reported this disease from India.

(i) Symptoms: The characteristic symptoms of the disease are appearance of white mycelial strands around infected regions of the stem, which gradually become pale and the plant starts wilting. In later stages, sclerotial bodies develop and start girdling the stem which results in toppling of the plant at the collar region and finally the death of the infected plant.

(ii) Management

Pandotra and Shastri (1968) reported that *Mentha spicata* var. *luciniata* and *M. arvensis* var. *piperascens* were able to tolerate the disease even after infection.

4.2.2.5. Rust, *Puccinia menthae*

The most significant disease problem encountered with mint is a rust fungus which, if left unchecked, will totally defoliate the plants. Scotch spearmint appears to be more sensitive to rust attack than the native spearmint, but both are prone to significant oil loss if the rust is left unchecked. Ganguly and Pandotra (1962) reported this disease from India. Although this disease is known to cause severe losses in other countries where this crop is grown extensively, but under Jammu conditions, this disease appears almost at the end of the crop season and causes comparatively very little or no damage to the standing crop. This disease mainly causes defoliation which ultimately reduces the yield of mint oil.

(i) Symptoms: Brown, circular pustules appear on the undersides of leaves followed by leaf yellowing and defoliation. A yellow spore stage (uredospores) is in reddish blisters on young shoots in spring, and a black spore stage (teliospores) is on stems and regrowth in fall (Fig 4.5). Telial stage is characterized by swelling on the upper portion of the stem. Swollen shoots with elongated, chlorotic internodes are associated with systemic infection.

Figure 4.5. Lesions of rust on the underside of mint leaf

Puccinia menthae overwinters on mint stubble and on wild and escaped mint. It completes its complex life cycle on one host (autoecious) and is macrocyclic (five spore stages), of which the aecial stage can be systemic in plants.

(ii) Management

Rabbing (flaming) in April 1 to May 15 is absolutely necessary for rust control.

Early harvest (before leaves drop due to rust) may limit rust damage.

Spray chlorothalonil or pycroclostrobin or propiconazole. Begin applications when plants are 5 to 10 cm high or when conditions become favourable for disease. Make second application 10 to 14 days after the first application. Do not apply within 90 days of harvest.

The disease can be controlled using crop rotation.

Removal and destruction of disease affected plants.

4.2.2.6. Leaf Blight, *Alternaria* sp.

(i) Symptoms: The main symptoms of the disease are appearance of the round to oval or slightly irregular dark brown spots on the upper surface of the leaves with concentric zones surrounding the pale yellow margin. Later, these spots coalesce, enlarge and occupy the entire leaf or a portion of the leaf which leads to the dropping of leaves.

(ii) Management

To control leaf blight, remove infected plants and try reducing soil moisture.

4.2.2.7. Leaf Spot, *Curvularia lunata*

Thakur *et al.* (1974) reported this disease on Japanese mint.

(i) Symptoms: The typical symptoms appear in the form of small dirty brownish spots, scattered all over the leaves in initial stages. These spots are very minute, which increase in their size later on and form big spherical or irregular patches. In the advance stage, the spots coalesce and form large dark brown patches of irregular size. The symptoms of the disease appear at all the stages of the crop growth of leaves, but the bigger spots generally appear on the older and lower leaves. The remaining portion of the infected leaves become chlorotic and finally wither away.

(ii) Management

To control leaf spot, remove infected plants and try reducing soil moisture.

4.2.2.8. Begomo Virus

(i) Symptoms: Typical symptoms were foliar rugosity, light yellow mosaic, upward curling, crinkling and retarded growth leading to drastic reduction in the herb's yield (Fig 4.6). In 2007, the disease incidence was recorded to be in the range of 50 to 60%. The disease also appeared in the Terai region of Uttaranchal province. From the naturally infected plants, the disease was transmitted to healthy seedlings of *M. spicata cv.* Viridis by whiteflies (*Bemisia tabaci*), but not by aphids or mechanical inoculation. Viruliferous whiteflies did not transmit the infection to other mint species (*M. arvensis; M. citrata* or *M. piperita*) in glasshouse tests.

Figure 4.6. Symptoms of Begomo virus on mentha leaves

(ii) Management

Out of four samplings in field from March to May, the Begomo virus infection was also found minimum in variety Kushal as compared to others. The maximum infestation of Begomo virus was found in cultivar Shivalik.

4.2.3. Nematodes

Among the recognized species of plant parasitic nematodes, root-knot and root lesion nematodes have been considered most important.

4.2.3.1. Root-knot Nematodes, *Meloidogyne incognita, M. javanica, M. hapla*

Root-knot disease of Menthol/Japanese mint, spearmint, scotch spearmint, peppermint and Bergamot mint caused by *M. incognita* and *M. javanica* was observed for the first time in Lucknow and Terai region of Uttar Pradesh which reduce the herbage yield and oil content (Haseeb & Pandey, 1989b). *M. incognita* dominates over *M. javanica* in mixed infection. The quality of mint oil was also adversely affected due to nematode infection. The root-knot nematode, *M. incognita* caused 40.2% loss in herbage yield and 46.6% loss in oil yield in menthol mint (Pandey, 2001).

(i) Symptoms: Initial damage occurs at 1 juvenile/g of soil with marked reduction in chlorophyll content, rate of photosynthesis and oil yield (Pandey, 1988). Symptoms of the disease include occasional yellowing of leaves which in a month time spread to a large portion of the foliage. Growth ceases soon after yellowing. Leaves turn yellow and thin, scorching easily and eventually turning brown. Initially symptoms appear in patches as the reduced plant growth with smaller leaf size and temporary wilting under slight stress of water during hot sun. As the crop grow, especially after first harvest, symptoms become more severe and appears as stunted growth with yellowing of leaves, while veins remain green.

The below-ground symptoms are the large numbers of small sized galls with large egg masses on the roots. Under severe infection, initiation of lateral roots and rootlets on suckers is checked. As a result, uptake of nutrients is inhibited, which in turn produces deficiency symptoms on the aerial portion of the plants (Haseeb & Pandey, 1989b).

(ii) Interaction with Other Pathogens: The severity of root-knot disease of Japanese mint was increased under field conditions in the presence of *Rhizoctonia solani.* The *M. incognita-R. solani* disease complex caused severe reduction in growth and oil yield at lower inoculum densities, and at higher inoculum levels, young plants could not survive.

(ii) Management

(a) Physical Methods: Hot water treatment of *M. arvensis* suckers at 48°C for 30 min gave complete eradication of *M. incognita* without any adverse effect to suckers and roots.

(b) Cultural Methods

Nematode-free Planting Material: The nematodes which damage mints are generally endoparasitic (*M. incognita* and *P. thornei*) in nature and they can easily spread through suckers (propagating materials) from infested planting sites to uninfested planting sites. Therefore, it is essential and utmost important to use nematode-free planting material to check the spread of nematodes (Pandey *et al.*, 2003).

Crop Rotation: Cultivation of some non-host crops in rotation with Japanese mint, water logging, rice cultivation during rainy season caused reduction in the population of root-knot nematodes. Crop rotation sequences of Menthol mint with maize, potato, mustard, pigeonpea, paddy and wheat were effective in reducing the root-knot nematode population and in increasing the yield (Table 4.1).

Table 4.1. Effective crop rotation sequences for the management of root-knot nematode in menthol mint

Crop rotation sequence	Root-knot index	Net benefit (in rupees)
Maize–potato–menthol mint	Mild	83,000
Paddy–potato–menthol mint	Mild	80,000
Maize–mustard–menthol mint	Mild	76,000
Pigeon pea–menthol mint	Mild	72,000
Paddy–menthol mint	Mild	75,000
Paddy–wheat–menthol mint	Mild	68,000

Organic Amendments: The severity of the nematode was reduced to a considerable extent in clayey soils. The high organic matter content in the soil suppressed the nematodes and enhanced the growth and oil yield. Soil application of neem/castor/mahua oil cakes at 1.0 g N/kg of soil was found effective against *M. incognita* and resulted in increase of oil yield and also plant growth parameters (Pandey *et al.*, 1992). Incorporation of neem cake 15 days before transplanting of *M. spicata* has been found the best approach for managing the population of *M. hapla* (Khan & Khanna, 1992).

The application of hydro-distillation wastes of *M. arvensis, M. piperita, Cymbopogon martini* (palmarosa) and *C. winterianus* (citronella) in the soil (1 month before transplantation) was found effective in reducing the nematode population under fie d conditions.

Application of different oil seed-cakes improved the growth of Japanese mint under field conditions coupled with increased oil yield and reduced root-knot and lesion nematode population. However, best results were achieved when the soil was treated with neem cake. The use of vermicompost and different distillation wastes were found to enhance the growth and yield of different mint species and reduced nematode population significantly.

Maximum reduction in *M. incognita* population in Menthol mint was recorded in neem cake treated soil followed by mustard cake and *Trichoderma harzianum*. Higher herbage yield (21 t/ha) was observed in the soil treated with oil cakes in comparison to control (14.66 t/ha). The crop produced had significantly higher oil content (2.7%) and yield (151 kg/ha) in neem cake treatment followed by mustard cake (2.7% oil and 143.6 kg/ha yield), *T. harzianum* (2.5% oil and 116.4 kg/ha yield) and *Glomus aggregatum* (2.4% oil and 112 kg/ha yield). The untreated crop gave 1.8% oil and 70.4 kg/ha yield (Pandey *et al.*, 1998) (Table 4.2).

Table 4.2. Effect of bio-organics and a pesticide on herb and oil yield of menthol mint and root-knot indices of *Meloidogyne incognita*

Treatment	Herbage yield (t/ha)	% oil content	Oil yield (kg/ha)	Root-knot indices
Untreated control	1.466	1.8	70.4	3.66
Carbofuran	1.880 (+28.4)	2.3	114.2 (+62.2)	2.33 (–34.7)
Mustard cake	1.998 (+36.3)	2.7	143.6 (+104.0)	2.00 (–45.4)
Neem cake	2.100 (+43.2)	2.7	151.0 (+114.5)	2.00 (–45.4)
Glomus aggregatum	1.758 (+19.9)	2.4	112.2 (+59.4)	2.66 (–27.3)
Trichoderma harzianum	1.750 (+19.4)	2.5	116.4 (+65.3)	2.66 (–27.3)
CD (P = 0.05)	**0.284**	**0.5**	**40.4**	**0.1**

(c) Chemical Methods: Application of carbofuran at 3 kg *a.i.*/ha increased plant growth and oil yield of *M. arvensis* and reduced the root-knot nematode population.

(d) Biological Methods: Application of arbuscular mycorrhizal fungi such as *G. aggregatum, G. fasciculatum* and *G. mosseae* improved plant growth, enhanced herbage and oil yield and effectively inhibited root-knot nematode infection (Pandey *et al.*, 1997) (Table 4.3).

(e) Host Resistance: The lowest root-knot (*M. incognita*) infection level was found on Kalka. Spear mint *cvs*. Neera and Arka were resistant to *M. incognita* (Pandey, 2003). Peppermint *cv*. Kukrail; Spearmint *cvs*. Arka, Neera and Neer-Kalka; and Bergamot mint *cv*. Kiran were found resistant to *M. incognita* (Pandey & Patra, 2001). The genotypes; MAS-35, MAS-216 and cultivar *Shivalik* were found moderately resistant to *M. incognita*.

Table 4.3. Effect of arbuscular mycorrhizal fungi on the productivity of mint infected with *Meloidogyne incognita*

Treatment	Fresh herbage weight (g)	% Oil yield	% Mycorrhizal colonization	Root-knot index
Untreated - Inoculated	285.0	0.38	—	3.6
Glomus aggregatum (Ga)	322.0	0.40	42.3	2.3
G. fasciculatum (Gf)	333.0	0.41	58.2	2.0
G. mosseae (Gm)	388.0	0.48	63.5	1.6
Ga + Gf + Gm	360.0	0.46	78.5	1.3

(f) Integrated Methods: Application of carbofuran and neem cake in combination improved the growth of Japanese mint and oil yield and reduced root-knot nematode population very effectively.

The maximum reduction in *M. incognita* population was recorded in beds treated with mustard cake along with *T. harzianum* and produced significantly higher herbage and oil yields (Pandey, 2005a).

4.2.3.2. Lesion Nematodes, *Pratylenchus minyus, P. thornei, P. penetrans*

Pathogenicity of *P. penetrans* on *M. piperita* has been demonstrated and reported to cause yield losses up to 34%. *M. arvensis* favours the high population build up of the nematode followed by *M. cardiaca, M. piperita* and *M. citrata.*

(i) Symptoms: High population of *P. thornei* was found to be associated with different mints and caused wilting and chlorosis of the plants (Haseeb, 1992). The lesion nematode causes light brown to dark coloured lesions on the suckers and roots. In severe cases whole underground portion becomes black and rotting of cortical portion takes place, first due to the nematode infection and later because of the invasion of secondary pathogens on necrotic cells. The other symptoms include stunting of the plant, reduction in growth of root/suckers and herbage yield within 4–5 months after nematode infection (Bergeson & Green, 1979; Rhoades, 1984).

Reduction in foliage and root/sucker growth occurs up to 39 and 66%, respectively (Bergeson, 1963).

(ii) Interaction with Other Pathogens: *Pratylenchus penetrans* and *P. minyus* increased the incidence and severity of *Verticillium albo-atrum* and *V. dahliae* f. sp. *menthae* wilt on mentha.

Bergeson (1963) reported that when *P. penetrans* infected *M. piperita* plants were inoculated with *V. albo-atrum,* diagnostic symptoms of wilt appeared approximately two weeks earlier than the plants inoculated with the fungus in the absence of nematode.

Presence of *P. miniyus* increased both incidence and severity of wilt (*V. dahliae* f. sp. *menthae*) disease, reduced the period for the development of wilt by 2 to 3 weeks, increased the reproduction of nematode and reduced the dry weight of peppermint plants up to 68% when both the pathogens were present (Faulkner & Skotland, 1965).

(ii) Management

(a) Cultural Methods: Application of neem and mustard cakes at 1.5 t/ha reduced the lesion nematode population (*P. thornei*), improved the growth and oil yield of *M. piperita, M. citrata* and *M. spicata* (Shukla & Haseeb, 1996).

(b) Chemical Methods: Mint plants treated with phenamiphos at 5.6 kg per ha gave maximum reduction of *P. scribneri* population (Rhoades, 1984). Ingham *et al.* (1988) found that ethoprophos (Mocap) at 6 kg/ha was highly effective in reducing the population of *P. penetrans* both in soil and root tissue of *M. piperita*. However, oxamyl (Vydate) at 1 kg/ha reduced the population of *Pratylenchus* spp. on the crop significantly.

(c) Host Resistance: Menthol mint *cvs*. Kosi and Kalka were resistant to *P. thornei* and had least *M. incognita* infection. Bergamot mint *cv*. Kiran and spear mint *cv*. Neer-Kalka were having least *P. thornei* infection.

4.2.3.3. Reniform Nematode, *Rotylenchulus reniformis*

Haseeb (1992) reported *R. reniformis* on *M. arvensis, M. cardiaca, M. citrata, M. piperita* and *M. spicata. M. piperita* and *M. spicata* are good hosts of *R. reniformis*. The reniform nematode was found to reproduce well on *M. arvensis* and to cause significant reduction in herb and oil yields.

(i) Symptoms: *R. reniformis* was observed to cause stunted growth of mint plants, withering of branches with chlorotic leaves at Hessaraghatta near Bangalore (Khan & Parvatha Reddy, 1992).

4.3. BASIL, *OCIMUM* SPP.

Among several species of basils, sweet basil (*Ocimum basilicum*) is considered to be the most important crop for its high quality of essential oil. Major components of its oil are linalool (43–50%), menthyl chavicol (18–33%), eugenol and isoeugenol (5–6%) and the minor constituents are alpha and beta pinene, camphor, geraniol, etc. The oil of sweet basil owes its importance to its extensive use in condimentary products, cosmetics, toiletry, perfumery and confectionary industries, particularly in European countries.

4.3.1. Pests

4.3.1.1. Aphid, *Aphis gossypii*

(i) Damage: Dark brown adults and green coloured nymphs colonize tender shoots, leaves and inflorescence. They damage plants by sucking sap resulting in yellowing and deformation of leaves (Fig 4.7). Attacked buds and flowers drop-off.

(ii) Management

Spraying 2% neem/pongamia oil causes considerable mortality of aphids. Dimethoate or oxydemeton methyl each at 0.05% also provides good control.

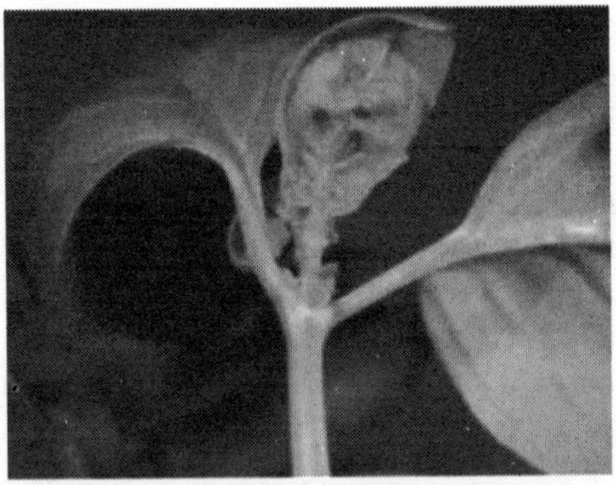

Figure 4.7. Aphid infestation on basil leaves

4.3.1.2. Whitefly, *Dialeurodes* sp.

(i) Damage: White, elongated eggs are laid singly on lower side of leaves. Nymphs and adults suck sap from leaves causing curling, withering and drying of leaves. Sooty mold develops on heavily infested leaves resulting in leaf drop.

(ii) Management

Spraying 0.05% dichlorvos or 0.0015% deltamethrin controls adults. Triazophos or methomyl each at 0.05% sprays cause mortality of nymphs and eggs. Butani (1982) recommended spray of endosulfan or malathion each at 0.05% and application of carbofuran granules at 2–3 g/plant in case of severe infestation.

4.3.1.3. Tinged Bug, *Monanthia globulifera*

(i) Damage: Spiny, black coloured nymphs and lace winged adults suck sap from ventral side of leaves resulting in yellowing, curling and drying of damaged parts.

(ii) Management

Spraying of methyl parathion or fenthion or phenthoate each at 0.05% provides effective control. Sprays of 1% neem/pongamia oil cause considerable mortality of bugs.

4.3.1.4. Leaf Folder, *Syngarina abruptalis*

(i) Management

Spraying 0.05% dimecron gives effective control of leaf folder.

4.3.2. Diseases

4.3.2.1. Fusarium Wilt, *Fusarium oxysporum* f. sp. *basilicum*

(i) Symptoms: A devastating basil wilt disease is caused by a soil-borne pathogenic fungus, *Fusarium oxysporum* f. sp. *basilicum*. *Fusarium* sp. causes a major production-limiting disease in commercial farms. Plants infected with this disease usually grow normally until they are 15 to 30 cm tall, then they become stunted and suddenly wilt (Fig 4.8). Early signs of this fungal disease include slow growth and yellowing of the young shoots. Advanced symptoms include wilting, die-back, and discoloured stems. Infected plants may show a darkened discolouration of the stem beneath the bark. Longitudinal sections of the stem will reveal this symptom. Under moist conditions, a light white to pink, cottony, growth can be observed along the infected stem areas. The pathogen is seed-borne and survives in the soil for many years. Once a field has become infested with the *Fusarium* wilt pathogen, infective propagules may persist in the soil for 8–12 years. Interestingly, only sweet basil is affected.

Figure 4.8. Fusarial wilt of basil

(ii) Management

Soil fumigation with methyl bromide before planting is effective in controlling the disease.

Control can be achieved by using clean seed and fresh potting soil. Seed can be disinfested by soaking in 10% household bleach (1 part bleach to 9 parts water) for 1 min, thoroughly rinsing with tap water, and planting immediately. The seed should be soaked in cold water for four hours followed by a heat treatment of 20 min in 133–136°F.

Outdoor plantings should be rotated out of basil. Growers should avoid growing sweet basil or members of the mint family. Mints will not exhibit symptoms of the disease but may carry over the inoculum from year to year.

The spread of *Fusarium* can be reduced by quick removal of all diseased plants and leaves. Avoid planting in wet, poorly drained areas.

The basil cultivar "Nufar" has shown some resistance. Some of the specialty basils, such as lemon basil and purple basil, show some resistance to the disease.

4.3.2.2. Leaf Spot, *Colletotrichum sp.*

(i) Symptoms: Dark spots form on leaves and the dead tissue within leaf spots may drop out, causing a shot-hole symptom. The disease can cause defoliation, tip dieback, stem lesions, and sometimes loss of entire plants. Spores are water-splashed from diseased tissue.

(ii) Management

Sow seeds in sterile containers in sterile soil. Since wet conditions favour disease development, reduce leaf wetness periods by reducing humidity and increase plant spacing to increase air movement. Avoid overhead irrigation. Remove diseased plants to reduce inoculum levels.

4.3.2.3. Damping-off, *Rhizoctonia solani*, *Pythium* spp.

(i) Symptoms: Seedlings either rot and do not germinate or germinate and collapse. The disease frequently attacks the base of the seedling and leaves a shriveled dried-up stem at the base of the seedling.

Symptoms appear as wilting and a slow or rapid collapse of the plant. The roots can appear brown and water-soaked instead of white. A water-soaked lesion can often appear at the base of the stem.

(ii) Management

Control can be achieved by using sterile potting media, clean pots, and fungicide-treated seed. If seeds in a seed tray become symptomatic, discard all of the seeds since asymptomatic seeds may be infected. Affected plants should be rouged out and discarded.

4.3.2.4. *Botrytis* Blight, *Botrytis cinerea*

(i) Symptoms: *Botrytis* blight occurs on many different kinds of herbs. The disease appears when plants are grown under conditions of high humidity, such as in a terrarium. The disease appears as tan to brown spots on leaves that can be associated with a grayish mold colonizing the damaged area. Senescing flowers are particularly susceptible and frequently have papery spots on them.

(ii) Management

Botrytis blight can be suppressed by avoiding overhead irrigation and watering in the morning to allow time for leaves and flowers to dry. Proper plant spacing can improve air drainage and rapid leaf drying. Control can also be achieved with the use of fungicide sprays (neem oil and thiophanate-methyl) applied as soon as symptoms are visible.

4.3.2.5. Leaf Spot, *Cercospora canescens*

(i) Symptoms: The disease causes characteristic leaf spot on both surfaces which are brownish black on the upper surface and light grey on the lower surface. A large number of spots may form on a single leaf which often coalesce and become necrotic. The infected leaves dry up and there is heavy defoliation.

(ii) Management

Spraying of 0.1% benomyl or carbendazim at fortnightly intervals is effective in controlling the disease.

4.3.2.6. Leaf Spot, *Glomerella cingulata*

(i) Symptoms: The infection starts from the upper surface of leaves usually from the leaf margins and tips. At first, the spots are small and irregular. Gradually infection spreads to cover the entire leaf surface. The leaf spots coalesce and become irregular large patches of dark brown colour. This results in heavy defoliation.

(ii) Management

Spraying of 0.1% benomyl or carbendazim at fortnightly intervals is effective in controlling the disease.

4.3.2.7. Bacterial Leaf Spot, *Pseudomonas cichorii, P. syringae*

(i) Symptoms: Spots on leaves are water-soaked, dark, and may be both angular and delineated by small veins in the leaves or show as irregular leaf spots. A wet stem rot may occur. Bacterium is reported to be seed-borne. The disease is favoured by wet, humid conditions and is disseminated by splashing water or by handling and spreading infected tissue.

(ii) Management

Decrease moisture on plants with low humidity and sufficient plant spacing for adequate movement to reduce leaf wetness periods. Keep the foliage dry. Use disease-free seed and transplants. Remove diseased leaves and plants to reduce inoculum levels. Avoid overhead irrigation. Use clean, sterile equipment and do not move between infected and healthy plants. Spray ridomil or mancozeb for the control of the disease.

4.3.3. Nematodes

Root-knot nematodes are recognized as the major limiting factors in successful production of basil crop.

4.3.3.1. Root-knot Nematode, *Meloidogyne incognita*

Recently, *M. incognita* and *M. javanica* have been found to be the major problems in the cultivation of *Ocimum basilicum, O. canum, O. sanctum, O. gratissimum* and *O. kilmandescharicum* (Haseeb *et al.*, 1988; Haseeb & Pandey, 1987). Balasubramanian and Rangaswami (1964) reported *O. sanctum* as a host of *M. javanica*. Krishnamurthy and Elias (1967) reported that *O. basilicum* was severely affected by *M. incognita*.

(i) Symptoms: Basil is very susceptible to root-knot nematodes, *Meloidogyne* spp. These microscopic round worms damage the roots and impede the plant's ability to take up water and nutrients from the soil. As a result, affected plants may show symptoms of nutrient deficiency, wilting, and yield decline. Galling and root rot occur on plants that are heavily infected. The nematode is responsible for significant reduction in foliage and root growth over a period of 10 months (Rhoades, 1988).

(ii) Management

The addition of organic matter such as chicken manure (500 g per hill) may improve soil conditions and reduce the effect of nematodes and other pathogens. Other cultural practices for disease control include rotation with non-susceptible plants and soil solarization. Soil solarization techniques employ a clear plastic sheet placed over tilled soil for six or more weeks. The plastic cover retains heat in the soil and reduces nematode numbers. Soil solarization is effective only if sufficient sunlight is received to accumulate heat beneath the plastic cover.

Haseeb *et al.* (1988) evaluated neem oil cake, aldicarb, bavistin, mentha oil cake and carbofuran against *M. incognita* on *O. basilicum* and observed neem oil cake as the best for plant growth and oil yield. Application of mahua cake also increased the oil yield.

4.3.3.2. Foliar Nematode, *Aphelenchoides* spp.

(i) Symptoms: Patches can appear on the leaf and appear light or dark brown. The disease is caused by microscopic worms called nematodes that tunnel in the leaf.

(ii) Management

Affected leaves should be cut off and discarded. Avoid splashing water from leaf to leaf. The nematodes can live as saprophytes in soil or potting mix for a short time so replanting the plant into fresh soil mix at the time the leaves are picked off can increase the control.

4.4. LEMON GRASS

Lemon grass (*Cymbopogon flexuosus*), West Indian lemon grass (*Cymbopogon citratus*) and Jammu lemon grass (*Cymbopogon pendulus*) are grown in Assam, Kerala, Maharashtra, Tamil Nadu and parts of Uttar Pradesh. The chief constituent of palmarosa oil is Citral. Citral is the starting material for preparation of α-Ionone (used in flavours, cosmetics and perfumes) and β-Ionone (used in the manufacture of synthetic vitamin A).

4.4.1. Pests

4.4.1.1. Stem Boring Caterpillar, *Chilotrea* sp.

(i) Damage: The most important pest species infecting lemongrass is *Chilotrea* sp. The caterpillar is white in colour with a black head and black spots on the body. It bores into the stem and remains there, feeding on the shoot. It is usually found at the bottom of the stem. As the grass is cut a few centimeters above the ground, the caterpillar, which is lodged lower down, is not affected. The first symptom of the attack is the drying up of the central leaf. Subsequently, the whole shoot dies, resulting in a significant reduction in the yield of the grass.

(ii) Management

The dry stubbles are set on fire during the off-season in summer. The caterpillars lurking inside the stubbles are thus destroyed.

The affected shoots are pulled out and destroyed.

When attacks are serious, a spray of Folidol E 605 is used.

4.4.1.2. Shoot Borer, *Chilo infuscatellus*

(i) Damage: Female moth lay eggs in batches on lower side of leaves. Larvae cut a hole on stem near ground level and feeds inside the shoot. Attacked shoots wilt and dry consequently.

(ii) Management

Cutting and burning of dried shoots and earthing up to a height of 25 cm around stem are effective in preventing further attack. Application of carbofuran granules at 1.5 kg *a.i.*/ha and spraying of chlorpyriphos or phenthoate each at 0.05% or ethofenprox at 0.01% provide good control of the borer.

4.4.1.3. Spittle Bug, *Clovia bipunctata*

(i) Damage: This is a commonly found pest of lemon grass. Eggs are deposited on leaves. Nymphs secrete frothy spittle mass and remain inside. They cause damage by sucking sap resulting in yellowing of leaves.

(ii) Management

Collection and destruction of leaves with spittle bugs. Spraying fenthion or methyl parathion each at 0.05% or acephate at 0.1% checks the bug infestation. Saradamma and Dale (1983) recommended spray of 0.05% quinalphos for the control of this pest.

4.4.1.4. Aphid, *Macrosiphum muscanthii*

(i) Damage: Infestation of aphid is mainly found on tender growing shoots. They suck the sap and cause deformation and stunted growth of the attacked plant.

(ii) Management

Two sprays of 1% neem/pongamia oil at 10 days interval cause considerable mortality of aphids. Dimethoate or oxydemeton methyl or profenphos each at 0.05% checks aphid damage.

4.4.1.5. Termites, *Microtermes obesi*

(i) Damage: Attack of termites to the crop is severe during summer when dry soil conditions prevail. Termites damage roots and stem portion by building earthen galleries and feeding inside. Infested plants show wilting and drying of leaves followed by death of the plant.

(ii) Management

Avoid dry soil condition by irrigating frequently. Scrape earthen galleries on stems and spray 2% neem oil or 0.1% malathion or 0.07% endosulfan. Drench the soil with 0.1% chlorpyriphos after raking up to kill soil inhabiting stages. Killing of the queen by destroying the colony provides permanent control.

4.4.1.6. Scale Insect, *Duplachionapsus divergens*

(i) Management

The pest can be controlled by spraying 0.05% quinalphos.

4.4.1.7. Whitefly, *Tetralauerodes semilunaria*

(i) Management

Spraying 0.05% dimethoate is effective for the management of the pest.

4.4.2. Diseases

4.4.2.1. Leaf Blotch, *Curvularia dreschslera*

Barua and Bordoloi (1983) reported the presence of *Curvularia verruciformis* to occur on lemon grass. *Curvularia andropogonis* causing severe leaf blight disease has been reported on lemon grass in different parts of India during rainy season (Alam & Hussain, 1976; Alam *et al.*, 1986). A leaf spot caused by *Curvularia trifolii* was found to be severe on some lemon grass varieties at Lucknow. The pathogen causing leaf blotch (*Curvularia* sp.) can also infect *Cymbopogan martini* var. *motia, C. pendulus* and *C. winterianus* (Gupta *et al.*, 1977).

(i) Symptoms: The disease is characterized by the appearance of minute, circular, reddish-brown spots, mostly on the leaf margin and tips. These spots later enlarge and coalesce to form longitudinal, elongated, reddish-brown, necrotic lesions resulting in premature drying of leaves. Older leaves are more susceptible for infection. Lenticular spots have a distinct dark brown border. The size of the spots varied from 1–8 mm in length and 0.5 to 3.0 mm in width. The spots often coalesce and form blotches.

(ii) Management

Spraying of 0.2–0.3% mancozeb at 15 days interval is effective in managing the disease.

4.4.2.2. *Helminthosporium* Leaf Spot, *Drechslera sacchari* (*Helminthosporium sacchari*)

(i) Symptoms: The disease is characterized by the appearance of oval-shaped brown spots with pale centres and dark purple borders which in advanced stages, coalesce to form necrotic blight areas.

(ii) Management

Spraying of 0.2–0.3% mancozeb is effective in managing the disease.

4.4.2.3. Red Leaf Spot, *Colletotrichum graminicola*

(i) Symptoms: The disease appears on lower leaves in the rainy season and forms small brownish spots, scattered all over the ventral surface of the leaves. Under favourable conditions, the disease also infects the leaf sheaths and midrib. As the disease advances, many such spots may coalesce to form bigger patch leading to drying of the leaves. Later, the spots show concentric rings in the centre. Under warm and humid conditions, enormous black fruiting bodies of the pathogen are present on older lesions.

The pathogen perpetuates in the remaining portions of the plants (stumps) on the field from one season to another.

(ii) Management

Two sprays of carbendazim at 0.1% or 3 sprays of maneb at 0.3% starting from first appearance of disease at 20 days or 10–12 days intervals, respectively gives effective control of the disease.

4.4.2.4. Rust, *Puccinia nakanshikii*

(i) Symptoms: Rust disease on lemon grass was reported from Sri Lanka (Bandare, 1981) and Hawaii (Gardener, 1985). Mildly and moderately infected leaves show discrete, linear, cinnamon brown uredinia on the abaxial surfaces, associated with chlorotic streaks. Severe infection causes extensive necrosis corresponding with coalesced uredinia. Telia also occur on severely infected leaves. No aecial host is known.

Varma (1987) reported the presence of rust caused by *Puccinia purpurea* from Kerala.

(ii) Management

Uproot and destroy the disease affected plants.

4.4.2.5. Little Leaf/Grassy Shoot, *Balansia sclerotica*

This disease is now becoming more serious on lemon grass and palmarosa as reported from Madhya Pradesh (Indore), Kerala and Uttar Pradesh (Lucknow). It is being a floral disease; the oil yield is adversely affected because the inflorescence consists of higher percentage of oil. This disease also affects seed production.

(i) Symptoms: The characteristic symptoms are little leaf formation and stunting of the plants. In place of normal inflorescence, proliferated, small, leafy structures are produced. The symptoms appear like viral/mycoplasma infection.

(ii) Management

Uproot and destroy the disease affected plants.

4.4.2.6. Smut, *Tolyposporium christensenii*

The disease was reported on lemon grass from India by Raghunath (1965). It is occurring in severe form in Kerala on lemon grass and on palmarosa in Madhya Pradesh (Indore). This is also an inflorescence disease which adversely affects the oil yield and seed production.

(i) Symptoms: The symptoms are characterized by the transformation of seed into the smut sori which are slightly long and covered by dull yellow membrane.

(ii) Management

Uproot and destroy the disease affected plants.

4.4.2.7. Clump Rot, *Fusarium equisetti*

(i) Symptoms: A clump rot disease of lemon grass was recorded at Odakali, Kerala (Thomas, 1987).

Table 4.4. Disease Management

Name of disease	Causal organism	Symptoms	Control
Red Leaf Spot	*Colletotrichum graminicola*	Brown spots with concentric rings in the centre appearing on the lower surface of the leaves The spots may be formed on leaf sheaths and midrib; Later the spots merge to form bigger patches and the affected leaves dry away.	Two sprays of bavistin 0.1% just after the appearance of the disease at an interval of 20 days. Three sprays of dithane M-45 (0.2%) at an interval of 10–12 days.
Leaf Blight	*Curvularia andropogonis*	Minute, circular, reddish brown spots mostly on the margins and tip of the leaves which merge to form elongated reddish brown necrotic lesions resulting in premature drying of leaves, older leaves are more suscepti-ble to infection.	Spraying dithane Z-78 (0.2%) or 0.3% copper oxychloride at an interval of 15 days.
Little Leaf or Grassy Shoot	*Balansia sclerotica*	Stunted growth; little leaf formation in place of normal inflorescence.	Spraying dithane Z-78 (0.3%) just before flower-ing stage at an interval of 10–12 days; use of fresh seedlings for plantation and crop rotation.

4.5. PATCHOULI

Patchouli (*Pogostemon patchouli*) is a highly aromatic bushy under-shrub. It is cultivated for its highly fragrant leaves which contain a very sweet smelling oil of lasting sticky odour. The crop is grown in small pockets in Karnataka, Kerala and Tamil Nadu. It has a very characteristic aroma and blends well with other essential oils. The oil is used in very low concentration (2 ppm) in scenting soaps, cosmetics, after-shave lotions, detergents and many fancy products. The oil is also used to flavour foods, beverages, candy and baked products. In combination with sandal wood oil, it is used in blending tobacco and making incense sticks.

4.5.1. Pests

4.5.1.1. Leaf Caterpillars, *Pronomis profusalis, Psara stultalis*

(i) Damage: *Pronomis profusalis* larvae feed extensively on tender leaves leaving only veins and cause foliage loss. Caterpillars of *Psara stultalis* characteristically rolls the leaves and feed inside causing browning and drying of attacked leaves. Damage by these caterpillars affects growth of the plant.

(ii) Management

Clipping and burning of webs and rolled leaves bring down the pest population. Two to three sprays of 0.01% fenvalerate followed by quinalphos or methyl parathion each at 0.05% are recommended for the control of these pests (Regupathy *et al.*, 1979).

4.5.1.2. Pentatomid Bug, *Nezera viridula*

(i) Damage: Green coloured nymphs and adults suck the sap from tender leaves and shoots. Deformed leaves and stunted growth of the plant are the symptoms of attack.

(ii) Management

Collection and destruction of egg masses prevent damage. Spraying of carbaryl or malathion each at 0.1% or fenthion at 0.05% protects foliage from the bug damage.

4.5.2. Diseases

4.5.2.1. Wilt, *Rhizoctonia* spp.

(i) Symptoms: The disease is characterized by sudden wilting of plants. Infected plants are found to have a brown discolouration at the collar region and the main roots below. The disease is very severe under wet or moist conditions of soil.

(ii) Management

Soil drenching with 0.2% PCNB (Brassicol) at the base of the infected plants helps to reduce the disease incidence.

4.5.3. Nematodes

Root-knot, lesion and spiral nematodes are recognized as the major limiting factors in successful production of patchouli crop.

4.5.3.1. Root-knot Nematodes, *Meloidogyne* spp.

(i) Economic Importance: It was responsible for 47.0 and 86.7 per cent loss in shoot weight and shade dried leaf yield, respectively (Prasad & Reddy, 1984).

(ii) Symptoms: In patchouli, the root-knot nematode causes chlorotic leaves, premature drying and shedding of leaves (Prasad & Reddy, 1984).

(iii) Management

(a) Cultural Methods: Neem cake at 4 t/ha controlled the root-knot nematodes (Sarwar *et al.*, 1982).

(b) Chemical Methods: Aldicarb at 1.5 kg *a.i.* per ha was effective in controlling phytonematodes infecting patchouli.

(c) Biological Methods: Application of 150 kg N and 50 kg *Glomus fasciculatum*/ha enhanced fresh herbage (297.2 g/plant compared to 212.7 in control), dry herbage (92.0 g/plant compared to 64.0 in control) and oil yield (110.4 kg/ha compared to 28.0 kg/ha in control); increased mycorrhizal colonization (90.5% compared to 65.5% in control); reduced *M. incognita* population (145/200 ml soil compared to 395/200 ml soil in control) and recorded the highest cost: benefit ratio (1: 3.67) (Sumathi *et al.*, 2006).

(d) Host Resistance: Planting of cultivated patchouli scion grafted on wild patchouli rootstock was effective for the management of root-knot nematodes (Krishna Prasad, 1978).

(e) Integrated Methods: The following practices may be adopted to manage root-knot nematodes on patchouli:

- Raise cuttings or planting material in nematode-free or treated nursery.
- Treat nursery beds with carbofuran 3G at 25 g/sq. m. or phorate 10G at 10 g/sq. m.
- Treat nursery beds with neem/pongamia cake at 2 kg/sq. m.
- Treat nursery beds with neem/pongamia cake at 1 kg/sq. m. along with *Glomus fasciculatum* at 50 g/sq. m.
- Treat the main field with carbofuran/phorate at 2.5 kg *a.i.*/ha before planting.

4.5.3.2. Lesion Nematode, *Pratylenchus brachyurus*

P. brachyurus was a major problem on patchouli in Annamalai hills of Tamil Nadu and Hessaraghatta in Karnataka (Khan & Parvatha Reddy, 1992).

(i) Symptoms: The infected plants invariably exhibited wilting symptoms.

4.5.3.3. Spiral Nematode, *Helicotylenchus dihystera*

(i) Symptoms: *H. dihystera* causes heavy damage to patchouli roots and is responsible for decline.

(ii) Management

(a) Chemical Methods: Aldicarb, carbofuran and phorate each at 2 to 3 kg *a.i.* per ha significantly reduced *H. dihystera* and increased leaf yield of patchouli.

4.5.3.4. Root-knot Nematode and Root Rot Disease Complex

(i) Management

(a) Integrated Methods: Combined mortality due to root rot and root-knot nematode could be minimized by the application of *Trichoderma harzianum* + karanj cake at 5 t/ha.

4.6. DAVANA

Davana (*Artemisia pallens*) is the most important aromatic herb for its high value essential oil used in the perfumery and cosmetic industries. Its foliage and floral tops produce a viscous essential oil, emitting a delicate, persistently fruity fragrance which is used in floral decorations, bouquets and cosmetics. It is also used in flavouring of cakes, pastries, beverages and tobacco products. India is one of the major exporters of davana oil to the rest of the world (mostly to USA). It is grown in 2,000 ha mainly in Karnataka and to a lesser extent in Tamil Nadu and Maharashtra.

4.6.1. Pests

4.6.1.1. Bug, *Nysius ericae*

(i) Damage: Small bugs are found mainly at flowering stage of the crop. They suck the sap from leaves resulting in yellowing and drying of leaves.

(ii) Management

Collection and destruction of leaves infested with bugs reduces the pest incidence. Spraying malathion or dichlorvos each at 0.1% or quinalphos at 0.05% at the commencement of flowering is effective in managing the pest.

4.6.1.2. Semilooper, *Thysanoplusia orichalcea*

(i) Damage: Larvae feed voraciously on leaves and cause defoliation of the plants. Eggs are laid on lower surface of leaves. This pest infests the crop throughout the year. It has four overlapping generations and the life cycle is completed in 27 to 38 days.

(ii) Management

Collection and destruction of larvae reduces the damage. Spraying of 0.01% fevalerate or 0.1% malathion or methyl parathion or chlorpyriphos each at 0.05% check the damage by semilooper.

4.6.2. Nematodes

Root-knot nematodes are recognized as the major limiting factors in successful production of davana crop.

4.6.2.1. Root-knot Nematode, *Meloidogyne incognita*

The root-knot nematode is a major problem in the cultivation of davana. *M. incognita* is responsible for more than 30% reduction in oil yield of davana (Haseeb & Pandey, 1990).

(i) Symptoms: The main symptoms were chlorotic and stunted plants with less number of flower buds (which are the major source of essential oil) showing patchy appearance in the field. Their roots were severely galled by root-knot nematodes (Fig 4.9). One larva/2 g soil has been found as economic threshold level of *M. incognita* on this crop.

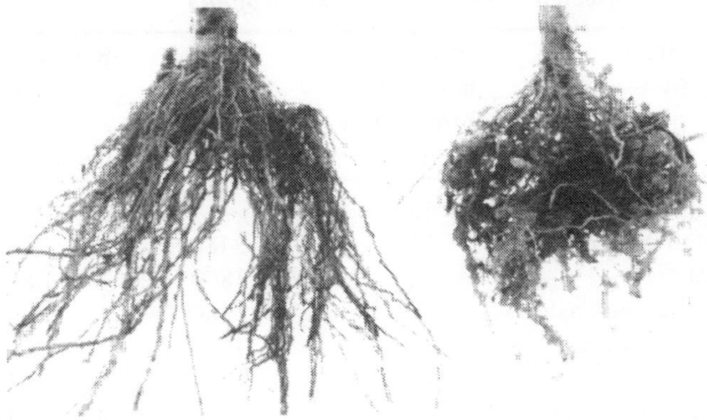

Figure 4.9. Root-knot nematode on davana. Left–Healthy roots; Right–Infected roots (Courtesy: Haseeb, 1992)

(ii) Management

(a) Cultural Methods: Soil application of neem cake at 1 g N/kg soil significantly suppressed root-knot nematode population and increased oil yield (Haseeb & Butool, 1991). Better efficacy of neem cake over the other organic materials *viz.*, mahua cake, castor cake and nematicides like aldicarb and carbofuran has been demonstrated on *M. incognita* infecting *A. pallens* (Pandey, 1994b) (Table 4.5). Increase in nematicidal efficacy of neem cake and FYM for 90 to 120 days in comparison to 60 days for fenamiphos has also been noticed (Anita & Vadivelu, 1997).

Table 4.5. Effect of oil cakes and pesticides on plant growth and yield of davana infected with *Meloidogyne incognita*

Treatment	Fresh weight (g)	% Oil content	Root-knot index
Control	52.5	0.25	3.6
Mahua cake	54.9	0.27	1.4
Castor cake	73.4	0.33	2.4
Neem cake	87.5	0.36	1.4
Carbofuran	75.0	0.33	2.4
Aldicarb	83.3	0.35	1.4
Bavistin	53.0	0.27	2.6

A considerable enhancement of herbage biomass, flower buds and oil yields were noticed when davana plants were treated with the distillation waste of palmarosa, citronella, curry leaf, menthol mint and vermicompost of pyrethrum, marigold and menthol mint. Least nematode infections were recorded in distillation waste of curry leaf, palmarosa, lemon grass and vermicompost of pyrethrum, marigold and menthol mint (Pandey, 2005b) (Table 4.6).

Table 4.6. Effect of distillation waste and vermicompost on growth and root-knot development of davana infected with *Meloidogyne incognita*

Distillation waste/ vermicompost	Fresh wt. of herbage & flower buds (g)	Total nematode population (soil + root)	Reproduction factor	Root-knot index
Untreated-Uninoculated control	40.0	—	—	—
Untreated-Inoculated control	24.7	3480	3.48	4.00
Carbofuran control	36.8	1800	1.80	2.66
Lemon grass (Distillation waste)	41.8	1260	1.26	2.00
Palmarosa (Distillation waste)	48.8	1100	1.10	1.66
Citronella (Distillation waste)	44.2	1040	1.04	1.66
Menthol mint (Distillation waste)	42.7	1640	1.64	2.33
Curry leaf (Distillation waste)	58.5	1020	1.02	1.00
Geranium (Distillation waste)	32.3	1680	1.68	2.66
Patchouli (Distillation waste)	31.3	1800	1.80	2.66
Marigold (Distillation waste)	41.6	1600	1.60	2.33
Menthol mint (Vermicompost)	51.3	1260	1.26	2.00
Citronella (Vermicompost)	51.0	1380	1.38	2.00
Chrysanthemum (Vermicompost)	44.0	1240	1.24	1.66
Marigold (Vermicompost)	37.9	1040	1.04	1.66
CD (P = 0.05)	**2.176**	—	—	**0.118**

(b) Biological Methods: A considerable enhancement of herbage biomass, flower bud and oil yields were noticed when davana plants were treated with the bioagent, *Trichoderma harzianum* (Pandey, 2005b) (Table 4.7).

Table 4.7. Effect of vermicompost and bio-agents on growth and root-knot development of davana infected with *Meloidogyne incognita*

Distillation waste/ vermicompost	Fresh wt. of herbage & flower buds (g)	Total nematode population (soil + root)	Reproduction factor	Root-knot index
Untreated-Uninoculated control	40.5	—	—	—
Untreated-Inoculated control	27.3	3480	3.48	4.00
Carbofuran control	39.0	1760	1.76	2.66
Trichoderma harzianum	56.7	1480	1.48	2.00
Glomus aggregatum	29.5	2080	2.08	2.66
G. fasciculatum	30.4	1880	1.88	2.66
CD (P = 0.05)	**2.174**	—	—	**0.137**

4.7. SCENTED GERANIUM

Scented geranium (*Pelargonium graveolens*) is an aromatic perennial herb. It is grown in cooler, sub-tropical climate of Mysore and Bangalore (Karnataka), and extended to Hyderabad (Andhra Pradesh). Its oil has refreshingly delicate long-lasting rosy odour with fruity under-note, containing 66–78% primary alcohols (rhodinol). It is used in manufacturing of perfumes, creams, talcum powder and body lotion. It is stable in alkaline medium and is therefore used for scenting soaps.

4.7.1. Pests

4.7.1.1. Whitefly

(i) Damage: They are commonly found on geraniums. The adults look like tiny moths, white in colour about 1–2 mm long and can be found on the underside of leaves.

If not acted upon, small scales or eggs which will be of next generation are also observed. These tiny insects excrete honeydew which they obtain from the plant sap. A black sooty mould starts to grow on these deposits. The adults have a waxy coating which repels water making it hard to use any insecticides.

(ii) Management

One way to remove this waxy covering is to spray with a weak solution of a detergent (a few drops in 1 pint of water). This will remove the coating and weakens the insect and then kills it.

The parastic wasp, *Encaria* sp. which lays its eggs within the eggs of the whitefly can be used as a biological control agent.

4.7.1.2. Aphids

(i) Damage: Aphid is a small winged insect like the whitefly but is easier to eradicate. The indication of crimped leaves will be the first signs and on the underside of the leaves a small colony of aphids may be observed. These tiny, soft-bodied, pear-shaped insects suck plant juices and can spread viral diseases.

(ii) Management

They can be controlled by knocking them off the plants with a strong stream of water, application of insecticidal soap (pay special attention to treating the undersides of leaves) or malathion, spraying with horticultural oil, and releasing beneficial insects, such as parasitic wasps, ladybugs, green lacewings, and aphid midges.

4.7.1.3. Caterpillars

(i) Damage: Several species of caterpillars chew geranium leaves. Caterpillars in summer can be a nuisance. They are small, green in colour and about 6–8 mm long. The first signs are holes in the leaves where they have been feeding.

(ii) Management

Caterpillars can be handpicked and destroyed or they can be controlled with an application of Bt (*Bacillus thuringeinsis*) which kills caterpillars, but does not harm beneficial insects. Cultivating the soil in winter will kill overwintering pupae. Attracting or releasing beneficial insects such as ladybugs and parasitic wasps will also help to control caterpillars.

4.7.1.4. Thrips

(i) Damage: Thrips can be a problem in late summer. Small grayish marks on the top side of the leaves is an indication of this pest. They are grey-green in colour, 3 mm long and jump when disturbed.

(ii) Management

Remove by hand or spray insecticide

4.7.1.5. Vine Weevil

(i) Damage: The main damage is done by the larva which lives in the soil and chews on the root system of the plant. First indication will be a sudden wilting of the plant. By uprooting the plant, white fat maggots with a brown head and curled within the compost can be seen.

(ii) Management

Treat with a soil sterilising liquid and change the type of compost that is used.

4.7.1.6. Bronze Butterfly

(i) Damage: This butterfly has an insatiable appetite for geraniums and its caterpillars cause the damage. The butterfly eggs are laid on the flower buds and the caterpillars then burrow into them, eating there way into the stem.

(ii) Management

Spray with a strong insecticide every two weeks to keep this pest under control.

4.7.1.7. Bud Worm, *Heliothis (Helicoverpa) virescens*

(i) Damage: The adult budworm moth has light green wings with a series of lighter colour bands that span a distance of 2.5–3.75 mm. Adults lay eggs singly on leaves of plants during early evening. Eggs are laid on the flower buds of geranium. The larval budworms are striped and quite variable in colour from dark brown to red or even green depending in part on the plant they are feeding on. Young larvae tunnel primarily through small buds and stems while larger caterpillars chew holes in flower buds, petals and leaves. Larvae take one month to eat their way to maturity. It completes two generations in a year.

Damage by the larvae includes feeding on developing ovaries and flower buds, tunneling in plant stems, and feeding on plant leaves (Fig 4.10) and flower petals. Damage to the developing flower buds seriously reduces the show of blooms with some budworm infested plantings producing no blooms during most or all of the growing season. In the early stages of development, the larvae are small enough to tunnel into flower buds and plant stems. The damage by the small larvae as they tunnel through plant stems results in the introduction of various disease organisms into the plant and results in stems that break readily.

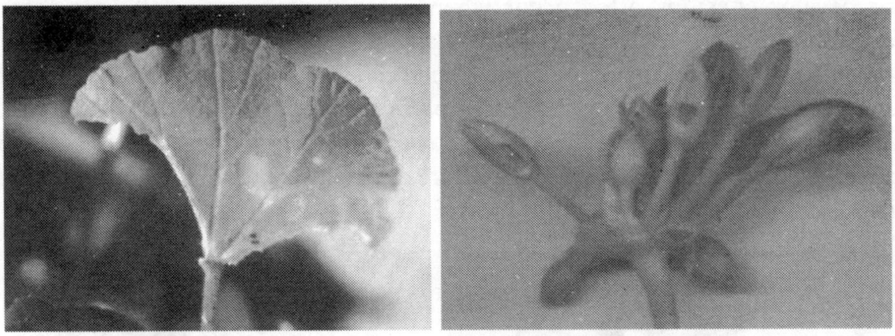

Figure 4.10. Bud worm infestation on leaf and buds of geranium

The budworm overwinters as pupa in earthen cells several inches beneath the soil surface. In areas where winter temperatures dip below 20°F, pupa do not successfully survive the winter. Adults normally emerge in late spring.

(ii) Management

Hand picking of the caterpillars is strongly recommended. A braconid wasp *(Cardiochiles nigriceps)* and a sarcophagid fly parasitize caterpillars. *Polistes* wasps also destroy many larvae.

Insecticides which have shown some degree of success are Orthene, Sevin, the pyrethroids Talstar and Scimitar and the biological control treatment *Bacillus thuringiensis*. Insecticidal treatments should be repeated often to disrupt the life cycle. Alternating insecticides used in these control efforts will help reduce the chances of this pest developing additional insecticidal resistance.

4.7.1.8. White ants (Termites), *Microtermes obesi, Odontotermes obesus*

(i) Damage: The adults are brownish dirty white in colour and damage root and stem through out the year.

(ii) Management

Application of chloropyriphos at 600–800 ml/ha with irrigation water gives effective control of termites.

4.7.2. Diseases

4.7.2.1. Botrytis Blight, *Botrytis cinerea*

(i) Symptoms: Botrytis gray mold is caused by the fungus *Botrytis cinerea*. Symptoms can develop at any stage and on any plant part (Fig 4.11). Under conditions of high moisture and humidity, infected plant tissues and debris develop the diagnostic gray, fuzzy growth characteristic of the fungus *Botrytis cinerea*. The symptoms on leaves vary from distinct spots or lesions to large dead areas, often with concentric rings. V-shaped lesions can also develop. Lesions often occur when spent flower petals fall onto the leaves. Symptoms on stems appear at the base of plants as light- to dark-brown lesions. These lesions grow until they rot the entire stem. Plants can develop brown lesions after cuttings are taken. Flowers can turn brown and drop prematurely during periods of high moisture and relative humidity. Senescing flowers are covered with a gray, fuzzy mass of the fungus. Botrytis on the flower of a Geranium will greatly reduce the appearance of the flower and will result in lower sales for the grower.

Figure 4.11. *Botrytis* blight symptoms on geranium leaves and flowers

The fungus can survive in the field year round as conidia, mycelium, or sclerotia on living or dead plant material. The fungus is very active and rapidly reproduces when relative humidity is at or above 85%, with little or no air circulation. The optimum air temperature for germination is between 72 and 77°F. The fungus penetrates the living tissue through wounds in the plants from broken stems, damaged leaves from transport, or from cuttings that were harvested.

(ii) Management

- Practice good sanitation by pulling weeds and removing plant debris.
- Use soil amendments to develop a well-drained soil.
- Use disease-free plants.
- Avoid overhead irrigation and water early in the day and not later than 11 a.m.
- Use adequate plant spacing to promote good air circulation.
- Remove diseased plants and plant parts. Remove old flowers and senescing leaves.
- Fungicides are available for chemical control.

The easiest way to control this disease is through cultural practices. These practices include cleaning up waste around the growing areas including leaves, soil, and removal of excess water that adds to relative humidity. By improving air circulation with Horizontal Air Flow (HAF) fans, the leaves of the plants will dry faster and not be as susceptible to infestation. Chemical treatments of Daconil are effective but should be used only when needed. Over use of these chemicals can possibly lead to resistance by the fungus.

4.7.2.2. Black Leg/Root Rot, *Pythium* sp.

The Black Leg disease is caused by the fungal pathogen *Pythium* sp. *Pythium* is very commonly found in greenhouses where bedding plants are grown. *Pythium* is normally associated with various root rots of poinsettias, petunias, impatiens, and geraniums. In geraniums, *Pythium* can infect the roots when the media conditions are extremely saturated. The fungus initially attacks the root tips and works its way up to the crown of the plant through the vascular system. Black Leg can attack at any stage of development. When the symptoms appear it is usually too late and any healthy roots are dead by the time the problem is noticed.

(i) Symptoms: Root and stem rots cause yellowing, stunting and wilting of the above-ground portions of the plant. Plants often have a noticeable brown to black lesion at the soil line, since the fungi attack and girdle the plant. Under wet, humid conditions, a gray or off-white webby growth can be seen on the infected plants (Fig 4.12).

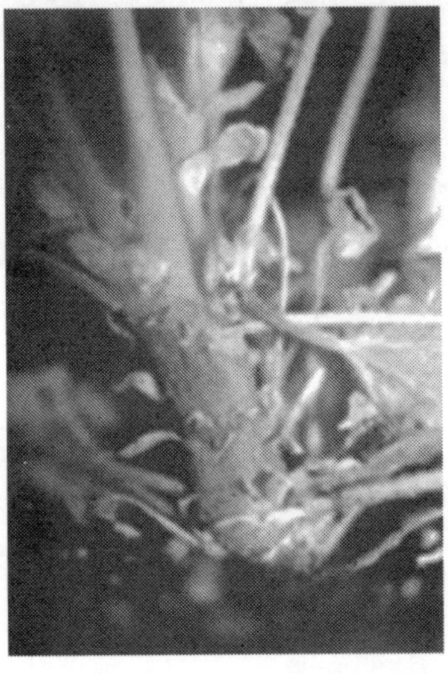

Figure 4.12. Black leg on geranium

(ii) Management

Improve soil drainage, use soil amendments and avoid standing water. Water management is essential to control root rots.

- Avoid heavy use of nitrogen fertilizer.
- Use disease-free plants.

- Remove diseased plants.
- Apply fungicides as soil drenches.

4.7.2.3. Leaf Rust, *Puccinia pelargonii-zonalis*

(i) Symptoms: Leaf rust, caused by the fungal pathogen *Puccinia pelargonii-zonalis,* is characterized by small yellow spots on the upper surface of the leaf and rust-colored pustules that develop in the spots on the underside of the leaf. The pustules break open and release spores. Spores are often formed in a circular fashion around the original pustule. When infection is severe, leaves become yellow and drop prematurely. Pustules containing the rust-colored spores can be seen on stems, petioles and stipules. The disease is more common in greenhouse production, but it is a potentially important disease in the landscape (Fig 4.13).

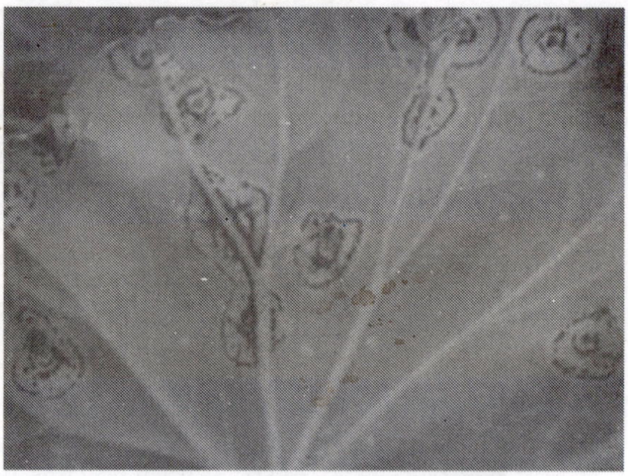

Figure 4.13. Leaf rust symptoms on Geranium

(ii) Management

- Practice good sanitation by pulling weeds and removing plant debris; use amendments to create a well-drained soil.
- Use disease-free plants.
- Avoid overhead irrigation.
- Remove diseased plants.
- Apply fungicides as soil drenches.

4.7.2.4. Alternaria and Cercospora Leaf Spots

(i) Symptoms: *Alternaria* leaf spots are small and blister-like, and appear on the undersides of older leaves. Mature spots become sunken and brown with diffuse yellow halos. If conditions remain suitable for disease development, the spots are apparent on upper leaf surfaces as well, and lesions show concentric rings. Chlorosis and defoliation occur when infection is severe.

Cercospora leaf spots appear as small, pale, sunken lesions that later turn gray and produce many spores. This gives an appearance of dark, raised centres. Cholorosis may be seen close to the lesions, and defoliation also occurs when infection is severe.

(ii) Management

- Remove and discard infected leaves and plant debris.
- Do not let the foliage stay wet very long.
- Water before 11 a.m.
- Thin annuals and prune perennials selectively to improve air circulation and increase light penetration.
- Fungicides are available for chemical control.
- Three to four sprays with 0.2% captafol or 0.3% chlorothalanil or 0.1% ridomil at 12–15 days interval gives effective control of the disease.

4.7.2.5. Wilt, *Fusarium oxysporum* f. sp. *redolens*

(i) Symptoms: A slow wilting caused by *Fusarium* is a very severe disease of scented geranium. The infection seems to take place through the cracks on the lower part of the stem or branches at or near the soil level. The disease is characterized by wilting of side branches of the plant which slowly proceed inwards finally resulting in the death of the entire plant. The leaves and stem of the infected plants become brown and there is rotting of plant at collar region which proceeds downwards involving the roots. The disease is found to be more severe during rainy months.

(ii) Management

Avoid water logging and planting in areas where this disease already exists. Treat the rooted cuttings with 0.3% mancozeb before planting. Drenching of the soil at the base of the affected plant with 0.1% benomyl, carbendazim or captan helps in reducing the disease incidence.

4.7.2.6. Bacterial Wilt/Blight, *Xanthomonas campestris* pv. *pelargonii*

(i) Symptoms: Bacterial wilt or blight of geranium is caused by *Xanthomonas campestris* pv. *pelargonii*. This is the most destructive disease associated with greenhouse geranium production today. The bacterium enter the geranium through wounds and natural openings. Once inside the geranium, the bacteria multiplies and spread systemically through the plants vascular system. As the bacteria multiplies it literally plugs up the vascular system and causes the plant to wilt due to lack of water transported up to the leaves of the plant. If the infected plant goes undetected for just a short time, the bacteria can spread from plant to plant and infect an entire crop. The V-shaped lesions on the leaves of infected plants are one way of determining if you have a possible infection of *Xanthomonas*. Symptoms on leaves include small distinct, water-soaked or brown and V-shaped spots (Fig 4.14) first visible on the under-surfaces. Some leaves develop distinctly darkened veins and wilt at the leaf margins. When infection is systemic, the entire plant becomes flaccid, and branches wilt and die back. Stems blacken and shrivel into a dry rot. Infected young plants fail to root properly and slowly rot from the base upward. Stems become dull black-brown. Plants with systemic infection collapse and die.

Figure 4.14. Bacterial wilt symptoms on geranium.

(ii) Management

- Sanitation: Pull weeds and remove plant debris.
- Use soil amendments to develop well-drained soil.
- Use disease-free plants.
- Avoid overhead irrigation and water early in the day and not later than 11 a.m.
- Avoid growing perennial geraniums (*Geranium* spp.) near *Pelargonium* spp.
- Remove diseased plants as soon as possible.
- Space plants to promote good air circulation.
- Avoid unnecessary handling of plant material and wash hands frequently.
- Chemical control is available but is rarely needed in landscape situations.

4.7.2.7. Tomato Spotted Wilt and Impatiens Necrotic Spot Viruses

(i) Symptoms: Symptoms of these viruses are highly variable and include stunting, ring spots, and sunken purple-brown lesions on leaves, stems and petioles. The symptoms vary with plant age, physiological condition, level of infestation and timing.

(ii) Management

- Use disease-free, virus-indexed cuttings.
- Destroy diseased plants.
- Avoid unnecessary handling of plant material.
- Pull and control weeds.
- Pay attention to insect control to avoid spreading of the disease. Monitoring and managing thrips is very important.

Table 4.8. Disease Management

Disease/Pathogen	Symptoms	Management
Alternaria Leaf Spot, *Alternaria tenuis*	Water-soaked spots on underside of leaf enlarge to 1/4–1/2 inch; concentric rings of dark tissue form in spots.	Avoid overhead watering. Remove and destroy crop debris from benches, pots, and walkways. Apply azoxystrobin, copper hydroxide, chlorothalonil, fludioxonil, or mancozeb.
Bacterial Blight, *Xanthomonas campestris* pv. *pelargonii*	Two types of symptoms may develop: (i) Small spots, less than 1/4 inch, develop on underside of the leaf and become sunken and well defined; the leaf wilts and dies as the bacterium spreads through water-conducting vessels of vein and petiole. V-shaped areas form with wide part of V on leaf margins and point of V on veins; (ii) Lower leaves wilt at margins while the blade and petiole remain turgid; the vascular tissue of the main stem on the side of the affected portion of the plant is discolored gray-brown and may ooze bacteria when cut; lower leaves die and fall.	Purchse culture-indexed cuttings or grow plants from seed; immediately discard infected plants after a positive diagnosis is made; avoid overhead irrigation; discard all unwanted geraniums at season's end; do not take cuttings with knives; wash hands frequently when breaking out cuttings; or disinfest the cutting knife by dipping in alcohol and flaming or leaving in ZeroTol or 1 Clorox to 9 parts water for 10 minutes.
Bacterial Fasciation, *Rhodococcus fascians*	Short, thick, fleshy, aborted stems that form at base of main stem at or below soil level are pale green or green-yellow; the rest of the plant appears healthy.	Purchase culture-indexed plants; discard infected stock plants and cuttings and infested media.
Blackleg, *Pythium*	Cuttings first develop a brown water-soaked rot at the base; the coal black rot proceeds 3–4 inches up stem and kills the cutting rapidly.	Use sterile propagation media; discard infected cuttings since affected rooted cuttings later develop root rot (see *Pythium* Root Rot); keep hose ends off the ground.
Botrytis Blight, *Botrytis cinerea*	Flowers turn dark and fall prematurely; where infected petals land on leaves, an irregular leaf spot forms; cutting stubs have a brown rot that may extend down stem; cuttings develop a dark	Avoid damaging plants in any way; do not leave large cutting stubs; remove and destroy fading flowers and leaves; heat, ventilate, and space plants to ensure good air circulation and

Table 4.8. (Contd.)

Disease/Pathogen	Symptoms	Management
	brown rot near the base.	low humidity; apply a mixture of chlorothalonil & mancozeb; use copper hydroxide, fenhexamid, fludioxonil, or iprodione alone only once as a final clean up of the crop and only if disease pressure is low.
Pseudomonas Leaf Spot, *Pseudomonas cichorii*	Elliptical, water-soaked 1/4- to 1/2-inch spots form on leaves; spots become dark brown to black and irregularly shaped; a yellow halo may or may not surround each spot.	Purchase disease-free cuttings; do not irrigate overhead; keep leaf surfaces dry at all time.
Pythium Root Rot, *Pythium*	Root tips appear translucent and water-soaked and contain microscopic spores; the outer layers of root tissue strip off when pulled from soil leaving the central core of vascular tissue bare; the fungus may invade the lower stem.	Pot in sterile media; keep hose ends off the ground; discard blackleg-infected cuttings even if well rooted. Apply fosetyl-AL, mefenoxam, etridiazole, etridiazole + thiophanate methyl, metalaxyl, or propamocarb.
Rust, *Puccinia pelargonii-zonalis*	Chlorotic specks on upper leaf surface appear directly opposite pustules of rust-colored spores on underside of leaf; spores erupt in concentric rings forming a "target" spot.	Purchase rust-free cutings; do not irrigate overhead; discard unwanted geraniums at season's end; apply chlorothalonil, myclobutanil, flutolanil, oxycarboxin, or triadimefon to protect plants.
Southern Bacterial Wilt, *Pseudomonas solanacearum*	Lower leaves wilt, yellow, and fall; vascular tissue of affected stems turns brown or black.	Purchase culture-indexed plants; do not bring ground-planted geraniums into the production area or propagate from them; destroy infected plants.
Verticillium Wilt, *Verticillium albo-atrum* or *V. dahliae*	Middle and upper leves collapse, dry, and fall; vascular tissue of affected stems is browned; symptoms are readily confused with those of bacterial blight.	Purchase culture-indexed cuttings; use sterile potting mix; destroy infected plants.
Viruses	Symptoms vary with virus, cultivar, and growing conditions; symptoms include: mottling, ring spotting, leaf cupping, yellowing of veins, leaf malformation, leaf spotting, and loss of zonation.	Purchase virus-indexed cuttings; maintain strict insect and mite control.

4.7.3. Nematodes

The root-knot nematode is recognized as the major limiting factor in the successful production of scented geranium crop.

4.7.3.1. Root-knot Nematodes, *Meloidogyne* spp.

(i) Distribution: *M. hapla* was found to be widely distributed on geranium in Nilgiri Hills of Tamil Nadu (Khan & Parvatha Reddy, 1992).

(ii) Management

(a) Chemical Methods: Aldicarb at 1 g *a.i.* per plant gave best control of *M. incognita* and increased shoot height, number of branches and leaf yield in gernanium.

4.8. CHAMOMILE

Chamomile (*Matricaria recutita*), an important medicinal and aromatic herb, is widely cultivated for its flowers which are used as medicinal tea (which acts as mild sedative and digestive and provides relief in cough and cold) in Europe and the USA. Their extracts are being used in antiflogistic creams and ointments. The oil of chamomile is utilized in alcoholic and nonalcoholic beverages, ice creams, candies, baked foods as well as in high-grade perfumery throughout the world. It is traditionally employed in manufacturing of pain-relieving balms. It is used in scenting of shampoos, pomades and face creams. It is grown in small area in North-Central parts of Uttar Pradesh and Himachal Pradesh.

4.8.1. Pests

4.8.1.1. Aphid, *Aphis faba*

(i) Damage: The nymphs and adults suck cell sap from the inflorescence of plants.

(ii) Management

Spraying dimethoate at 2 ml/litre of water is effective for the management of aphids.

4.8.1.2. Lygied Bug, *Antographa chryson*

(i) Damage: Lygied bug is one of the serious pests of chamomile. The pest causes defoliation of the plant. Both nymphs and adults cause damage by sucking cell sap from the plants. The adults are 6–7 mm long, black with white patches on the wing, which extend slightly beyond the abdomen. They are mostly prevalent in February and March.

(ii) Management

Spraying 0.05% malathion or dimethoate at 2 ml/litre of water gives effective control of *A. chryson*.

4.8.1.3. *Nysius minor*

(i) Damage: The pest is responsible for flower shedding.

(ii) Management

Spraying carbaryl 50 WP at 3 g/litre of water is effective for the management of *N. minor*.

4.8.2. Nematodes

The introduction of chamomile as a commercial crop brought into the limelight the threat posed by root-knot nematode, *M. incognita* for its cultivation (Pandey *et al.*, 1999).

4.8.2.1. Root-knot Nematode, *Meloidogyne incognita*

(i) Symptoms: *M. incognita* causes considerable reduction in growth, flower buds and essential oil yield of chamomile (Pandey *et al.*, 1999).

(ii) Management

(a) Chemical Methods: Maximum reduction in root-knot severity and nematode population occurred with 4-chlorosalycylic acid, O-acetylsalicylic acid and 2-chloronicotinic acid. The result envisages that these chemical activators may be used in integrated management system for protection against root-knot nematodes. (Pandey & Kalra, 2005) (Table 4.9).

Table 4.9. Effect of chemical activators on root-knot nematodes, plant growth and flower yield

Treatment	Total nematode population (soil + roots)	Reproduction factor (Rf = Pf/Pi)	Root-knot index (RKI)	Dry weight of plant (g)	Flower yield (g)
Control	3018	3.01	4.00	4.1	3.3
Isonicotinamide	2498	2.49	3.00	3.0	3.3
2-Chloronicotinic acid	1710	1.71	2.00	6.3	6.1
5-Nitrosalicylic acid	1818	1.81	2.00	7.7	6.3
4-Chlorosalicylic acid	1412	1.41	1.33	9.6	6.7
DL-2 Aminobutyric acid	1777	1.77	2.00	6.7	6.1
2 Aminobutyric acid	2020	2.02	2.33	4.3	3.7
0-Acetyl salicylic acid	1640	1.64	1.66	10.6	7.4
4-Amino salicylic acid	2176	2.17	2.66	6.7	6.2
Salicylic acid	2010	2.01	2.00	8.6	6.5
Carbofuran	1920	1.92	2.00	4.9	6.0
CD (P = 0.05)	**321.83**	**—**	**0.021**	**—**	**0.01**

(b) Integrated Methods: Integration of *G. mosseae* and neem cake reduced the severity of root-knot disease and enhanced the growth, biomass and flower yield by 12 to 37% of chamomile.

4.9. CITRONELLA

Citronella (*Cymbopogon winterianus*) is a large, perennial, stemless, aromatic grass. Its oil from leaves has a strong characteristic lemonic odour extensively used in scenting soaps, cosmetics, producing deodorants and mosquito repellent products. The oil is used as starting material to produce pure citronellol, geraniol and related upstream high-value aroma chemicals used in perfumery industry. It is cultivated in 6,000–10,000 ha land in high rainfall tracts of Assam, Arunachal Pradesh, Meghalaya, Nagaland and Manipur. It is also cultivated on small scale in Karnataka, coastal Andhra Pradesh and Orissa.

4.9.1. Pests

4.9.1.1. Aphid

(i) Damage: The adults which are yellowish green in colour damages leaves from vertical sides during January-February.

(ii) Management

Spraying dimethoate at 0.3–0.7 kg *a.i.*/ha gives effective control of aphids.

4.9.1.2. Army Worm, *Mythimna separata*

(i) Damage: The larvae which are green in colour cause defoliation during March-April.

(ii) Management

Spraying malathion or quinalphos each at 0.05% gives effective control of army worm.

4.9.1.3. Grasshopper, *Macrosiphum muscanthii*

(i) Damage: Adult and nymphs with stout body, grayish in colour cause defoliation during August.

(ii) Management

Spraying malathion at 0.05% gives effective control of army worm.

4.9.1.4. Shoot Borer, *Chilo infuscatellus*

(i) Damage: Larvae are dirty white in colour with fine longitudinal strip on the body, bore the stem of young crop and causes dead heart during January.

(ii) Management

Spraying chlorpyrifos at 600–800 ml/ha with irrigation is effective in controlling shoot borer.

4.9.1.5. Termite, *Microtermes obesi*

(i) Damage: Adult workers are whitish in colour destroy roots of plants throughout the year.

(ii) Management

Spraying chlorpyrifos at 600–800 ml/ha with irrigation is effective in controlling termites.

4.9.1.6. Thrips, *Anaphothrips sudanensis, Chloethrips holorphnus*

(i) Damage: Nymphs and adults of *Anaphothrips sudanensis* which are yellowish brown in colour and *Chloethrips holorphnus* are completely brown in colour bore into the stem of young crop and cause dead heart.

(ii) Management

Spraying 0.05% dimethoate or phosphamidon at 0.3–0.7 kg *a.i.*/ha is effective in controlling thrips.

4.9.1.7. White grub, *Holotrichia consanguinea*

(i) Damage: Grub/larva is dirty brown in colour with dark brown colour head feed on roots during July.

(ii) Management

Spraying chlorpyrifos at 600–800 ml/ha with irrigation is effective in controlling termites.

4.9.2. Diseases

4.9.2.1. Leaf Blight/Blotch, *Curvularia andropogonis*

The pathogen attacks citronella at Arunachal Pradesh, Manipur, Meghalaya, Nagaland and other citronella growing areas (Gupta *et al.*, 1977).

The pathogen which causes severe leaf blight on Jawa citronella was also found on lemon grass in different parts of India during rainy season (Alam & Husain, 1976; Alam *et al.*, 1986).

(i) Symptoms: The fungus affects the crop in the beginning of monsoon. The disease appears as minute, circular, reddish-brown spots, mostly on the leaf margin and tips during the month of July. These spots later enlarge and coalesce to form longitudinal, elongated, reddish-brown, necrotic lesions resulting in premature drying of leaves. Older leaves are more susceptible for infection. In case of serious infections, the entire leaves dry up resulting in considerable decrease in leaf and oil production. Black dots in the from of sporulation of the fungus are observed on dead spots of the leaves.

Alam *et al.* (1986) reported that two strains of *Curvularia andropogonis* probably appear in nature, one causing leaf spot (strain CLS) and another causing leaf blight (strain CLB). Strain CLS infects all the four species of *Cymbopogan* severely, whereas strain CLB was severe on *C. winterianus* and *C. flexuosus* and do not infect *C. citratus* and *C. martini* var. *motia*.

(ii) Management

The incidence of the disease can be reduced significantly in the field by prophylactic spraying of 0.3% mancozeb or zineb or 0.1% ridomil at an interval of 15 days.

4.9.2.2. Lethal Yellowing, *Pythium aphanidermatum*

(i) Management

The disease could be easily managed under field conditions by slip treatment and three foliar sprays either with 0.1% ridomil-mancozeb or previcur-N.

4.9.2.3. Collar Rot and Wilt, *Fusarium moniliforme*

(i) Management

Foliar spraying with carbendazim or Topsin-M each at 0.2% gives highly effective control of the disease.

4.9.3. Nematodes

Root-knot and lesion nematodes are recognized as the major limiting factors in successful production of citronella crop.

4.9.3.1. Lesion Nematodes, *Pratylenchus* sp.

(i) Symptoms: Very high populations of *Pratylenchus* sp. were encountered from the rhizosphere of apparently diseased plants of citronella which were stunted, chlorotic with less number of tillers (Haseeb, 1992).

4.9.3.2. Root-knot Nematode, *Meloidogyne incognita*

(i) Symptoms: The root-knot nematodes were consistently isolated from the rhizosphere of citronella. Pandey (1994) reported that *M. incognita* was responsible for 50% loss in oil yield of citronella.

4.10. SAGE, *SALVIA OFFICINALIS*

4.10.1. Diseases

4.10.1.1. Root Rot, *Fusarium solani, Rhizoctonia solani*

(i) Management

T. viride was more effective in reducing the disease incidence (38.88%) over control followed by *T. harzianum* (36.11%).

4.11. PALMAROSA

Palmarosa (*Cymbopogon martinni* var. *motia*), also called as Rusa, Rusha and Rosha, is grown in Andhra Pradesh, Madhya Pradesh, Maharashtra, Karnataka, Tamil Nadu and Uttar Pradesh. The major trading centres include Amravati, Akola and Elichpur in Maharashtra; Khandwa and Betul in Madhya Pradesh; Zaheerabad and Mehboobnagar in Andhra Pradesh. Oil of palmarosa is used for fine perfumery and valued for its geraniol content. Oil imparts rose like aroma to wide variety of soaps and tobacco products. Oil contains 70–80% geraniol.

4.11.1. Pests

4.11.1.1. White Grub, *Holotrichea consanguinea*

(i) Damage: White grubs attack on roots and leaves.

(ii) Management

Spraying chloropyriphos 20 EC at 600–800 ml/ha with irrigation gives effective control of white grubs.

4.11.1.2. Aphid, *Aphis gossypii*

(i) Damage: Nymphs and adults suck the cell sap from the inflorescence of plants.

(ii) Management

Spraying dimethoate or folidol each at 0.1% is effective in managing aphids.

4.11.1.3. Termite, *Microtermes* sp.

(i) Damage: Newly planted woody seedlings are more vulnerable and the seedlings die after stem from the base is damaged.

(ii) Management

Application of chloropyriphos 20 EC at 600–800 ml/ha with irrigation is effective for the management of termites.

4.11.1.4. Thrips, *Haplothrips* sp.

(i) Damage: Nymphs and adults suck the sap from young shoot tips and leaves, also feed on floral parts, thus affecting seed set.

(ii) Management

Spraying dimethoate or folidol each at 0.1% gives effective control of thrips.

4.11.2. Diseases

4.11.2.1. Leaf Spot, *Colletotrichum graminicola*

This disease was first recorded on palmarosa in India by Patwardhan (1963) from Pune.

(i) Symptoms: The disease appears on lower leaves in the rainy season and forms small brownish spots, scattered all over the ventral surface of the leaves. Under favourable conditions, the disease also infests the leaf sheaths and midrib. As the disease advances, many such spots may coalesce to form bigger patch leading to drying of the leaves. Later, the spots show concentric rings in the centre. Under warm and humid conditions, enormous black fruiting bodies of the pathogen are present on older lesions.

The pathogen perpetuates in the remaining portions of the plants (stumps) on the field from one season to another.

(ii) Management

Two sprays of carbendazim at 0.1% or 3 sprays of maneb at 0.3% starting from first appearance of disease at 20 days or 10–12 days intervals, respectively gives effective control of the disease.

4.11.2.2. Leaf Blotch, *Curvularia andropogonis, C. trifolii*

This disease was first reported by Upadhyaya *et al.* (1975) from Jorhat, Assam. It was reported to occur very severely under favourable climatic conditions of north-eastern region of India. This pathogen not only infects palmarosa, but also *Cymbopogan winterianus* (citronella), *C. flexuosus* (lemon grass) and *C. pendulus* in Arunachal Pradesh, Manipur, Meghalaya, Nagaland, Uttar Pradesh, Madhya Pradesh, Bihar, Karnataka and Jammu and Kashmir (Gupta *et al.*, 1977). Janardhan *et al.* (1980) reported that the disease causes more than 31% total oil and 17.8% geraniol loss in palmarosa. The disease occurs in epiphytotic form during August and October.

(i) Symptoms: The disease appears as minute, circular, reddish-brown spots, mostly on the leaf margin and tips. These spots later enlarge and coalesce to form longitudinal, elongated, reddish-brown, necrotic lesions resulting in premature drying of leaves. Older leaves are more susceptible for infection.

(ii) Management

The disease can be effectively controlled by foliar spraying of 0.3% mancozeb at 15 days interval before the onset of the disease or even during initial stages of infection.

4.11.2.3. Leaf Blight, *Ellisiella caudata*

(i) Symptoms: Leaf blight is one of the serious diseases of palmarosa. The disease appears in epiphytotic form during rainy season and causes considerable loss in the production of herbage and essential oil yield. Initial symptoms appeared in the form of small grey coloured necrotic spots all over the surface of infected leaves. In severe stage of infection, the necrotic lesions increase in size and coalesce to form big patches leading to premature drying of infected leaves. The fungus produces masses of spores on the dried necrotic lesions.

(ii) Management

The disease can be effectively controlled by foliar spraying with captafol or chlorothalanil each at 0.3% at 15 days interval.

4.12. VETIVER

Vetiver (*Vetiveria zizanoides, V. lawsoni*) is grown in Madurai, Thanjavur, Tirunelveli and Trichy districts of Tamil Nadu; Kurnool and East Godavari districts of Andhra Pradesh; Palghat, Kozikode, Thrissur and Wayanad districts of Kerala. Cultivation is picking up in Madhya Pradesh, Uttar Pradesh and Haryana. Oil of vetiver is used in soap making. It is also used for making aromatic mats, brooms, screens, pillows and mattresses. Vetiver is used in soil conservation. It has deep, dense and strong fibrous root system. Grass is unpalatable to cattle. It is planted on the edges of field bunds and banks of major irrigation canals to prevent soil erosion.

4.12.1. Pests

Termites sometimes attack vetiver, but seemingly only in arid regions. Serious threat from stem borer has also been reported from China. Sometimes, rats also attack vetiver roots.

4.12.1.1 Grub, *Phyllophaga serrata*

(i) Damage: Grubs of a beetle have been reported infesting the vetiver roots.

(ii) Management

The pest can be controlled by broadcasting 65% lindane at 1.25 kg/ha before the final ploughing.

4.12.2. Diseases

Vetiver is subjected to a few diseases only due to its sturdiness.

4.12.2.1. Leaf Blight, *Curvularia trifolii*

(i) Symptoms: A leaf blight disease was reported from India (Sarwar & Shrinath, 1965). Tan to dark, oval spots appear on the leaves later turning black as the leaves turn pale yellow. The spots become necrotic and coalesce to form blighted appearance. Leaves become yellow and gradually dry out.

(ii) Management

Spray copper oxy chloride at 3 g/litre of water twice or thrice depending on the intensity of disease.

4.12.2.2. *Gloeocercospora* Leaf Spot, *Gloeocercospora sorghi*

(i) Symptoms: Puranik and Suryanarayana (1966) reported a new leaf spot disease on vetiver caused by *Gloeocercospora sorghi*. The disease appears as small, diffused brown spots with irregular margins on leaves. The isolate readily infected seedlings of some sorghum varieties.

4.12.2.3. *Helminthosporium* Leaf Spot, *Helminthosporium* sp.

(i) Symptoms: The occurrence of this disease was reported by Thomas (1987) from Odakali, Kerala. The spots were found scattered throughout leaf and sheath. They are oblong or oval and have a dark purplish brown colour.

4.12.2.4. Smut, *Ustilago vetiveriae*

(i) Symptoms: This pathogen infects the inflorescence of vetiver in India (Padwick, 1946). In place of seed, the smut sori are present.

Table 4.10. Pest and disease management in Aromatic plants

Crop	Disease	Etiology	Findings & Recommendations
Palmarosa, Lemongrass (*Cymbopogon* spp.)	Red leaf spot	*Collectorichum graminicola*	Two sprays of carbendazim (0.1%) or 3 sprays of maneb (0.3%) starting from first appearance of disease at 20 days or 10–12 days intervals, respectively.
Rose (*Rosa damascena*)	Die back	*Diplodia rosarum*	Pruned ends of the shoots should be treated with copper fungicides.
Ambrette seeds (*Abelmoschus moschatus*)	Anthracnose	*Collectotrichum hibiscicum*	Seed treatment with contact fungicides and spray of Bordeaux mixture are recommended.
Jasmine (*Jasminum* spp.)	Bud worm	*Hendecasis duplifascialis*	Spray cypermethrin at 0.5 ml/l or quinalphos at 2 ml/l.

5 TUBER CROPS

Tuber crops form an important part of the diet in many areas of the tropics and subtropics. They are most important after cereals and grain legumes. They are mostly grown for the carbohydrates stored in underground storage organs *viz.*, roots or rhizomes. In addition, the leaves or leaf stalks of some of the tuber crops are rich in protein, vitamins and minerals and are used for either human consumption or cattle feed. Tuber crops are a major source of food and calories in many tropical countries. The annual tropical root crop production (potato, sweet potato and cassava) is in the range of 11.835 million metric tonnes in 1.909 million ha with productivity of 19.7 tonnes/ha (Table 5.1). Root and tuber crops are predominantly grown in the upland or rain fed areas, mainly by resource poor small and marginal farmers. The major tuber crops include cassava (tapioca), sweet potato, colocasia, amorphophyllus, yams and coleus. Cassava and sweet potato are grown to a larger extent in southern and eastern India. Their cultivation has spread to Maharashtra, Gujarat and north-eastern states as well. The Salem belt in Tamil Nadu and the Samalkot belt in Andhra Pradesh are known for cassava as an industrial crop. Yams and Colocasia are popular in Andhra Pradesh, Tamil Nadu, West Bengal, Uttar Pradesh and Orissa states.

Table 5.1. Area, production and productivity of tuber crops in India (2005–06)

Tuber crop	Area ('000 ha)	Production ('000 tonnes)	Productivity (tonnes/ha)
Potato	1542.0	29118.0	18.8
Sweet potato	122.8	1066.5	8.9
Cassava (Tapioca)	244.6	7854.9	31.4
Total	1909.4	11835.4	19.7

(*Source:* Indian Horticulture Database 2006)

Tuber crops, especially cassava can be successfully substituted as alternative source of feed for livestock in both tropical and subtropical countries. Cassava has also been used as a substrate for the production of ethyl alcohol. The tropical root crops, in general, have a great potential in meeting basic food and energy needs of the developing world. Tuber crops are very efficient in solar energy transfer, having a clear superiority over cereals in biological efficiency and are comparatively free from severe pests and diseases. Cassava and sweet potatoes account for about 30% of the total production of root crops from developing countries. The potential of root crops as nutritionally rich sources of β-carotene, anti-oxidants, dietary fibre and minerals like calcium have begun to be recognized as a result of the multifarious research programmes worldwide.

5.1. POTATO

Potato (*Solanum tuberosum*) is one of the important vegetable crops grown in India. It is being cultivated in 1.550 million hectares producing 29.094 million tonnes of potatoes with an average yield of 18.8 tonnes per hectare (Tables 5.2 and 5.3). It thrives best in cool climate. Therefore it is a summer crop in the hills and a winter crop in the plains. The major potato growing states include Uttar Pradesh, West Bengal, Bihar, Punjab, Gujarat, Madhya Pradesh, Assam, Uttarakhand, Haryana and Karnataka.

Table 5.2. Major potato producing countries of the world

Country	Area (in '000 ha)	Production (in '000 MT)	Productivity (in MT/ha)	% Share of world production
China	4884	73460	15.0	22.4
Russia	3071	37280	12.1	11.4
India	1550	29094	18.8	8.9
Ukraine	1516	19462	12.8	5.9
USA	439	19091	43.5	5.8
Germany	277	11624	42.0	3.6
Poland	588	10369	17.6	3.2
Belarus	462	8185	17.7	2.5
Netherlands	156	6777	43.4	2.1
France	156	6605	42.2	2.0
Others	6102	105368	17.3	32.2
Total	**19200**	**327314**		

[Source: FAO (except Indian data) (Indian data source: Indian Horticulture Database 2006)]

Table 5.3. Area, production and productivity of leading potato growing states in India (2005–06)

State	Area (in '000 ha)	Production (in '000 MT)	Productivity (MT/ha)
Uttar Pradesh	443.2	9986.8	22.5
West Bengal	320.6	7106.6	22.2
Bihar	308.9	5702.5	18.5
Punjab	75.0	1223.5	16.3
Gujarat	41.0	1088.5	26.6
Madhya Pradesh	53.3	799.7	15.0
Assam	90.1	629.1	7.0
Uttarakhand	22.2	450.6	20.3
Haryana	19.0	374.4	19.7
Karnataka	57.5	368.4	6.4
Others	119.5	1363.6	11.4
Total	**1550.2**	**29093.7**	

(Source: Indian Horticulture Database 2006)

5.1.1. Pests

5.1.1.1. Aphid, *Myzus persicae*

(i) Damage: Wingless green peach aphids tend to be pear shaped. They may have variable colors but usually are straw-colored or light-green (Fig 5.1). Hind cornicles slant so they converge. A key characteristic is tubercles at the base of the antennae. Potato aphids are more elongate and larger, about 1/6 to 1/8 inch, than the green peach aphid. They often are bright green, but pinkish forms occur. They are similar in general appearance to the English grain aphid. Both potato and green peach aphids produce winged forms. Winged aphids are most important in spread of viruses to plants since they move readily within and between fields. Wingless green peach aphids tend to occur on lower, older potato leaves. Potato aphids more commonly colonize the upper part of the potato plant. During the summer, aphids reproduce asexually (no males) and give live birth to young. Under optimal conditions, a generation may be completed in 10 days. Green peach aphids overwinter in the egg stage on various *Prunus* species, particularly plum and apricot, in and around towns. It may also be introduced on infested greenhouse plants.

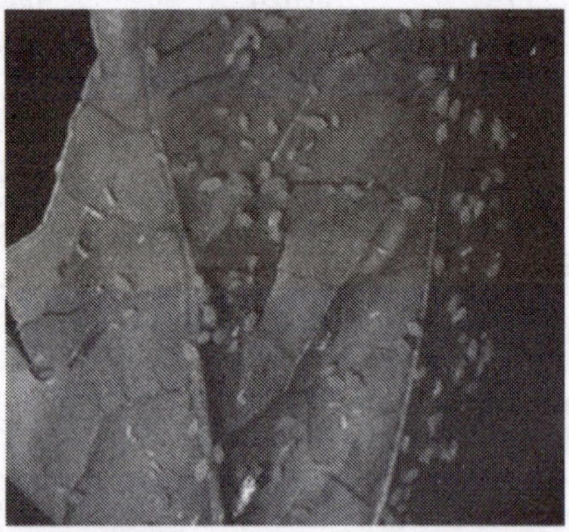

Figure 5.1. Potato leaf infested with aphids

(ii) Management

Cropping in aphid-free periods or when their population is low by adjusting the planting dates in the Indo-Gangetic Plains, where about 90% of seed crop is grown *i.e.*, up to 15th October in North Western Plains, up to 5th November in North Eastern Plains. Remove all weed hosts susceptible to viruses and for aphids (especially those having yellow flowers) from within and around the vicinity of seed potato plots. The haulm cutting of seed crop should be done as soon as the aphid number crosses the critical level *i.e.*, 20 aphids/100 leaves. Care must be taken to prevent any regrowth of stumps before harvest.

Monitor flights of winged aphids by using yellow traps (yellow dishpans with water and a bit of bleach). Sweep fields at least weekly to monitor aphids. Make sure to get the sweep net deep in the canopy since a lot of aphids are found on lower leaves. Pull up yellowish plants to check for aphid colonies. Apply pesticides only when necessary. Use of some insecticides can exacerbate aphid problems, because the control has more effect on natural enemies than on aphids. Insecticides can only provide limited control of aphid transmitted virus diseases. The green peach aphid can be highly resistant to insecticides registered on potatoes. Therefore, choose insecticides carefully for effective control. Protect natural predator populations that are important in aphid management. Trap crops around the edges of fields can help aphid control. Direct damage from aphid feeding in a strictly commercial field needs high populations. If aphid levels are approaching 50 aphids per plant, treat the field.

Several predators and parasites help control aphids. Among the more important insects are lady beetles, damsel bugs, green lacewings, minute pirate, and parasitic wasps.

5.1.1.2. Cut Worm, *Agrotis ipsilon, A. flammatra*

(i) Damage: Cutworms feed at night. During the day they can be found under clods of soil or in cracks in the ground near injured plants (Fig 5.2). Cutworms have a smooth appearance, three pairs of legs and five pairs of prolegs. Damage early in the season includes stems cut off at or below ground level. Later, chewed foliage is the most common symptom. Tubers that are exposed on the surface or in cracks in the soil can be chewed by some cutworms. Cutworms overwinter as immature larvae or eggs in the soil.

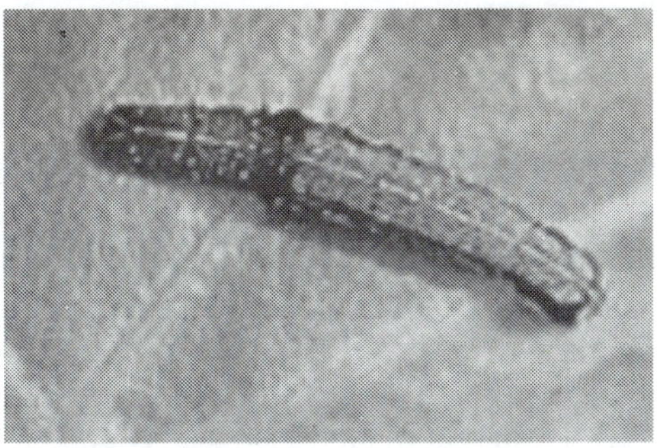

Figure 5.2. Cut worm of potato

(ii) Management

Hot weather ploughing in plains and autumn ploughing in hills reduce the population of immature stages. A number of birds such as crow, mynah, starling, etc. feed on insects that get exposed upon ploughing. Planting of potato on 25th November recorded least foliage damage (1.25%).

Some of the natural enemies which play important role in the management of cut worms are *Broscus punctatus, Liogryllus bimaculatus, Anpopus hypsipylae, Ichneumon* sp., and *Turanogonia chinensis*.

Bacillus thurigiensis is quite effective against cut worms. The percentage foliage damage in *Bacillus thurigiensis* treated plots ranged between 0.0 to 2.0% compared to 1.6 to 6.4%

in control. Similarly, tuber damage recorded at harvest was also less in *B. thurigiensis* treated plots. Further, significantly higher yield of potatoes was recorded from *B. thurigiensis* treated plots.

5.1.1.3. White Grub, *Holotrichia consanguinea*

(i) Damage: White grubs can severely damage seed pieces and tubers (Fig 5.3). These insects feed on grass roots and should be considered as a serious threat when potatoes are to be grown in ground immediately following sod.

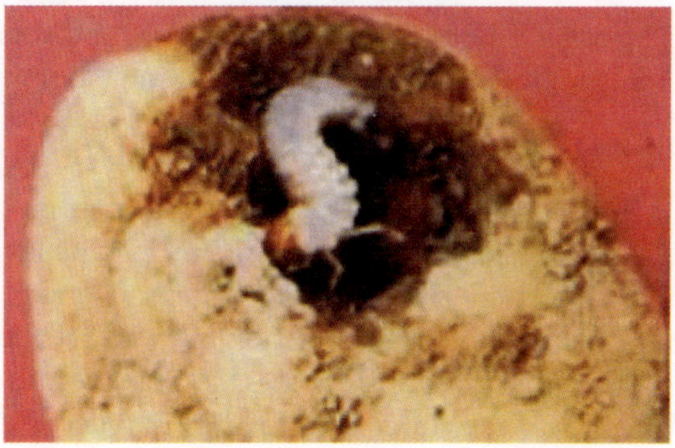

Figure 5.3. White grub damage to potato tuber

(ii) Management

Mechanical control by collection of beetles should be organized on community basis and started in late May or early June after the first shower by vigorous shaking of host trees/shrubs and then killed. Repeated ploughing before monsoon (April–May) exposes the grubs and pupae for predation by natural enemies (birds) or may be hand collected and destroyed. Flooding of the fields for 7–10 days, wherever possible may be undertaken. Adopt suitable crop rotations. Apply only well rotten FYM/compost in the fields.

A majority of beetles are attracted to the light source. Hence, light traps may be operated for mass collection.

5.1.1.4. Tuber Moth, *Phthorimaea operculella*

Tuber moth is one of the most serious pests of potato in the field as well as in storage. On an average 105 eggs are laid by each female at an average temperature of 28°C with an incubation period of 4.93 days. The larval, pupal and adult stages last for 8.87, 5.10 and 24 days, respectively. The pest is observed during December to March in the field and April onwards in stored potatoes.

(i) Damage: The tuber worm is a small caterpillar, the larva of an inconspicuous gray and black moth called the potato tuber moth. The larva is usually dull white with a brown head. The tuber worm can have several generations a year in warm areas. Tuber worms also feed on other solanaceous species like nightshade, tomato and eggplant. On potato tubers, the larva feed just below the surface and sometimes deep into the flesh. The tubers can appear black since the tuber worm tunnels are filled with larval feces (Fig 5.4). In any case, tuber worm feeding leaves the tuber un-marketable. Tuber worms will feed on the tubers. Varieties that have tubers well protected by soil can tolerate higher populations of tuber worms.

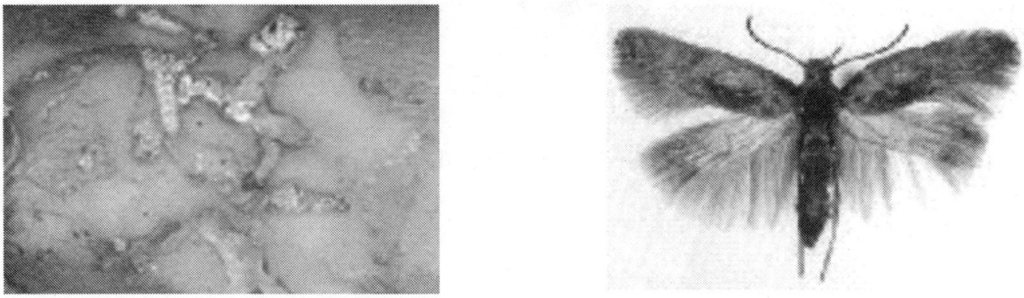

Figure 5.4. Tuber moth damage on potato

(ii) Management

The pest is controlled through deep planting, high ridging and cold storage of seed. Larvae failed to enter the potato tubers through wet soil layers. Keep the tubers in storage before sun set covered with 5 cm sand layer to prevent the adult moths from laying eggs on the eyes of the tubers (Patel, 1980). Use 2.5 cm thick layers of dried lantana or eucalyptus leaves below and on top of the potato heaps which checks tuber moth infestation. Planting varieties that set tubers deep in the soil and hilling the plants up well to protect the tubers from tuber worm can help reduce damage.

In country stores, use of sex pheromones on water traps at 4 traps/100 m^3 store space are effective for mass trapping of males. Pheromones are available to monitor the tuber moth populations. The trap consists of a dishpan of water with a little soap to break the surface tension of the water. Place a tent over the trap to help catch adult moths and place the pheromone inside to attract them. If you catch 10–20 per day you should treat the field.

Foliar application of carbaryl at 0.15% and chlordane 5% dust at 20 kg/ha is found effective in controlling the pest under field conditions. Pesticide treatment must occur before vine kill.

Use of biocontrol agents like *Bacillus thuringiensis* and Granulosis Virus are also advocated.

5.1.1.5. Epilachna Beetle, *Epilachna vigintioctopunctata*

(i) Damage: Grubs scrape the chlorophyll from the leaves leaving only veins (Fig 5.5).

(ii) Management

Apply 5% carbaryl dust at 25 kg/ha.

Figure 5.5. Epilachna beetles on potato leaves

5.1.2. Diseases

5.1.2.1. Early Blight, *Alternaria solani*

Early blight of potato, caused by the fungus *Alternaria solani,* poses a significant risk to crop productivity in the field and to tuber quality in storage. This disease can occur over a wide range of climatic conditions and can be very destructive, sometimes resulting in complete defoliation of plants. Early blight is found in all potato growing regions and is a major disease in hot climates, especially where crops are irrigated. This disease is most severe on maturing or nutritionally deficient potato vines.

(i) Symptoms: Initial infection is most frequent on older leaflets on lower portions of plants. Lesions first appear as small, (1–2 mm) dark brown to black spots. The spots enlarge and develop concentric rings of raised and depressed necrotic tissue which gives the lesion a characteristic "target spot" appearance. Advanced lesions often have angular margins because of limitation by leaf veins and a narrow chlorotic halo frequently surrounds the spot (Fig 5.6). As the disease progresses, affected leaves turn yellow and senescent and either dry up or fall off.

Figure 5.6. Early blight on potato leaves and tuber

Tuber infection occurs less frequently than leaf infection and is usually evident only after several months of storage. Tuber lesions are dark, sunken, irregularly shaped and often surrounded by a raised violet border (Fig 5.6). The underlying flesh is dry, leathery, and usually brown. In advanced decay, the tissue is often water-soaked and yellow to greenish yellow. Lesions can enlarge during storage, and tubers can become shriveled in advanced cases. Early blight tuber lesions are not as prone to invasion by secondary organisms as compared to other tuber rots.

(ii) Disease Cycle: *Alternaria solani,* can persist in crop debris, soil, infected tubers, or other solanaceous hosts. Spores are carried primarily by wind and infect potato leaves under ideal conditions of warm temperatures (68–86°F) and high moisture levels. Disease development is also favored by high relative humidity (about 95 percent). Spores are frequently produced after heavy dews and frequent rains and are blown in from mycelium growing on debris or on infected cultivated plants. The germinating spores penetrate susceptible tissue directly or through wounds and produce new conidia that are further spread by wind, splashing rain, etc. Periods of alternating wet and dry weather favor spore dispersal and disease spread. Early blight is often more prevalent on older, senescing tissues and particularly when plants have been predisposed by injury, poor nutrition, insect damage, or other type of stress. Because of the relative resistance of young to intermediate-aged potato plants, major outbreaks of disease do not occur until late in the season.

If sufficient moisture is present, spores can germinate and infect tubers. Infection of potato tubers occurs mostly through wounds inflicted during harvest. Immature tubers are particularly susceptible to infection by *A. solani.*

(iii) Management

Use disease-free seed tubers for planting. Avoid cultivation of solanaceous crops nearby potato fields. Spray 1% Bordeaux mixture to control the disease. Use disease resistant *cvs.* like Kufri Sindhuri, Kufri Jyoti, Kufri Lalima for planting.

The disease is controlled primarily through the use of resistant cultivars, cultural practices, and foliar fungicides such as chlorothalonil, maneb, triphenyl tin hydroxide and mancozeb. Cultivars with levels of field resistance are available, however no immunity to disease exists in commercial potato cultivars or in their wild parents. Highly susceptible cultivars such as Red Norland, Norchip, and Superior should be avoided in locations where early blight is prevalent and disease pressure is high. Field resistance to foliage infection is associated with plant maturity. Late maturing cultivars are usually more resistant than early maturing cultivars; therefore, one should avoid planting early and late varieties in the same or adjacent fields.

Cultural practices, such as crop rotation, removal and burning of infected plant debris, and eradication of weed hosts helps reduce the inoculum level for subsequent plantings and susceptible crops. Because *A. solani* persists in plant debris in the field from one season to the next, rotation with non-host crops reduces the amount of initial inoculum. Other cultural control measures include avoiding irrigation in cool, cloudy periods, harvesting fully mature tubers and avoiding bruising or other mechanical damage and adjacent potato and tomato fields planted the previous season should be plowed immediately after harvest. Current recommendations include a combination of cultural and chemical measures to retard the development of early blight and reduce the impact of this disease on yield and tuber quality.

To minimize tuber infection after harvest, tubers should be stored under conditions that promote rapid suberization. Suberization of the wounded surface provides a significant degree of protection against the penetration of *A. solani.* The storage of tubers for three weeks at approximately 61°F with more than 90% relative humidity will facilitate rapid wound healing.

The most common and effective control method of early blight is the application of foliar fungicides. The protectant fungicides recommended for late blight control (*i.e.*, maneb, mancozeb, chlorothalonil, and triphenyl tin hydroxide) are also effective against early blight when applied at approximately 7–10 days intervals. The application of foliar fungicides is not needed in plants at the vegetative stage, when they are relatively resistant. Accordingly, spraying should commence at the first sign of disease or immediately after bloom. Frequency of subsequent sprays should be determined according to the genotype and age-related resistance of the cultivar. Protectant fungicides should be applied initially at relatively long intervals and subsequently at shorter intervals as the crop ages.

Early season applications of fungicides before secondary inoculum is produced often have minimal or no effect on the spread of the disease. Early blight can be adequately controlled by relatively few fungicide applications if the initial application is properly timed. Regular inspection of fields after plants reach 30 cm in height is recommended in order to detect early infections.

Disease forecasting models can be used to predict the onset of early blight and specify when the initial fungicide application of a season should be applied. For example, the computerized forecasting system, modified FAST, was developed to identify periods when environmental conditions are favorable for the development of *A. solani* on potato and to provide an efficient fungicide application schedule. This forecasting model calculates the increased susceptibility of potato plants according to their physiological age. It determines the number of days (P-days) after potato emergence at which plants become susceptible to the disease. In general, early maturing cultivars should be sprayed at approximately 250P-days, whereas fungicide applications for late maturing cultivars should commence at approximately 300P-days.

5.1.2.2. Late Blight, *Phytophthora infestans*

The late blight of potato shot into prominence during Irish potato famine in 1844–45 which resulted in one million deaths due to starvation and another 1.5 million fled the country when entire potato crops rotted in the field or in storage because of infection by *Phytophthora infestans.* The losses are caused by the foliage infection which leads to premature death of the plant and consequently the reduction in tuber yield. The losses are also caused directly by tuber infection which leads to rottage of infected tubers in the field and stores (Bhattacharya *et al.*, 1990). The disease is capable of wiping out entire potato commercial fields within a week if the conditions are wet. Farmers who grow potatoes are at serious risk of losing their entire income for the season. Losses are more in hilly regions where crop is grown under rain-fed conditions as compared to plains. In India losses range between 21–74% in Shimla hills, 11–38% in Mukteshwar hills, 33–65% in Darjeeling hills and 32–39% in Ooty hills.

(i) Symptoms: Late blight affects the foliage and tubers of potato (Fig 5.7). The characteristic foliar symptoms on any susceptible plant are irregular to circular lesions. Initially, in wet weather, the lesions may appear water-soaked; in dry weather, the centres tend to dry out. The dark centre is often surrounded by yellow, chlorotic tissue or brown, collapsed tissue. These spots often occur at the tips or edges of leaves. If moist weather persists, the leaf spots enlarge rapidly to blight larger areas of the foliage. Dark brown discolouration of leaf stalks and stems is also commonly encountered. The symptoms frequently develop first on lower leaves but can develop at any level in the canopy. The symptoms are usually most dramatic inside the canopy. During humid weather, a white fuzz or mildew-like growth appears on the leaves (upper or under sides) and on stems. This indicates that the fungus is producing spores.

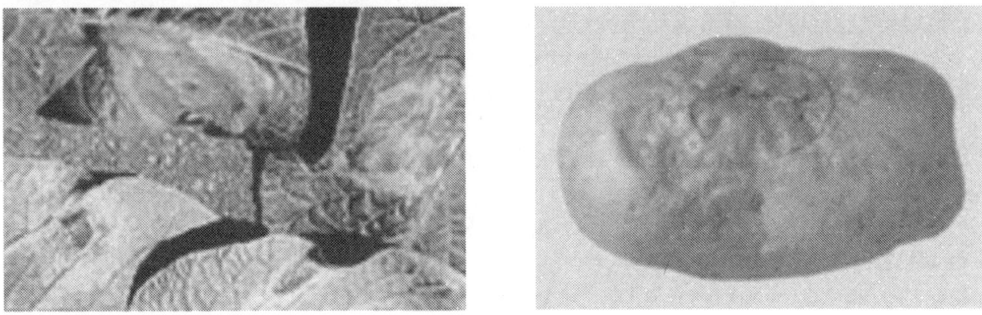

Figure 5.7. Late blight on potato leaves and tuber

Infected potato tubers are discolored by a reddish-brown, granular rot. The rot can be superficial or penetrate more deeply into the tubers (2.5 cm or more). Tubers must be cut open to observe this symptom. The rot continues to develop after the tubers are harvested. Further decay of the tubers is often hastened by secondary fungi and bacteria. The soft rot bacteria often turn tubers into a foul smelling liquid.

Tubers become infected when spores of the fungus wash down through the soil and come into contact with the fleshy organs.

(ii) Epidemiology: The infected seed tubers and refuse piles in the temperate regions and cold stored infected tubers in the sub-tropical regions serve as the primary source of the inoculum.

The fungal growth and spores produced could only survive in living tissues. Infected potato tubers used for seed, infected tubers left in cull piles, and unharvested, infected tubers left in the ground were the sources of inoculum, or spores, for the disease from one growing season to another.

The development of the disease is favored by day time temperatures of 21–27°C (70–80°F) and night time temperatures of 10–16°C (50–60°F). The presence of free moisture (such as dew, rain or fog) on the foliage also favors the disease. Free moisture is a prerequisite for spore germination and infection and thus is essential for disease development. As the foliage becomes further blighted, more spores are produced and spread of the disease to adjacent plants becomes more likely. As the season progresses, potato tubers can become infected as spores from the blighted foliage are washed down through the soil and contact developing tubers.

(iii) Management

The effective management of late blight of potato can be achieved by reduction of both foliar and tuber infection by adopting three pronged attack *i.e.*, by employing chemicals, cultural practices and host resistance in an integrated manner.

(a) Cultural Methods: The most effective strategy for managing late blight is to avoid sources of inoculum (spores). Do not save potato seed from year to year. Use healthy certified potato seed. Many states that produce potato seed have seed certification programs to ensure that the seed meets certain standards for disease levels. Using certified seed also reduces the carry over of a number of potato viruses. Certified seed, however, is not a guarantee that late blight or other diseases will not be present. Examine your seed carefully before you plant, and plant only sound, blemish-free tubers. Examine tubers for late blight symptoms by cutting them open with a knife. Immediately sterilize or thoroughly wash the knife if late blight symptoms are found.

Use disease-free seed tubers for planting. Ridges should be made up high enough to cover all daughter tubers and reduce chances of their infection upon exposure.

Destroy any rejected seed tubers that you do not plant (bury them deeply in an area away from the field, burn them, or discard them in a plastic bag in the trash). Make sure any tubers put into compost piles are completely decomposed. Pull up and destroy potato plants that come up from tubers left in the field from the previous season.

If late blight becomes severe, remove diseased plants by digging them up. Destroy these plants immediately by one of the following: burying them deeply in an area away from the field, burning them, or by bagging them in a plastic bag and discarding the bag. These steps will help avoid production of a larger number of spores. Harvest all potato tubers in the field. If late blight occurs when the tubers have already 'sized up', harvest the crop as soon as possible to avoid post-harvest tuber rot. Again, destroy diseased foliage and stems.

It is best to avoid wetting the leaves when you irrigate; soak the ground around the plants instead of spraying the plants with water. If this is not possible, water your garden mid-day so that the foliage dries off quickly. Eliminate weeds around the field so as to maximize air circulation around the plants. Additionally, earthing up the soil around the plants to provide a more substantial barrier against tuber infection from spores washing down through the soil.

The following cultural practices may be adopted to reduce the disease incidence:

- Use of disease-free seed.
- Removing/rouging of ground keepers.
- Adjustment in planting time.
- High ridges and proper earthing up to avoid exposure of seed tubers.
- Water management–light or no irrigation when the blight is likely to appear.
- Killing of haulms at 75% crop damage.

(b) Host Resistance: There are several potato varieties that show resistance to late blight. These varieties will slow down, but do not prevent, the development of the disease. Elba is currently the most resistant potato variety available. Potato varieties with moderate levels of resistance include: Kennebec, Sebago, Allegany, and Rosa.

In north Indian plains, use disease resistant *cvs.* like Kufri Sindhuri, Kufri Jyoti, Kufri Badshah, Kufri Chipsona 1 and 2, Kufri Jawahar and Kufri Sutlej. In hills, use disease resistant *cvs.* like Kufri Jyoti, Kufri Giriraj in Himachal Pradesh hills, Kufri Kanchan and Kufri Jyoti in Darjeeling hills, Kufri Swarna and Kufri Thenamalai in Nilgiri hills.

(c) Chemical Methods: Potatoes are susceptible to late blight at any time during the growing season. If the growing season is wet, late blight occurrence is very likely. Fungicides may be necessary to protect your plants from infection. Continuous fungicide coverage is necessary to protect plants from infection. Spray 1% Bordeaux mixture to control the disease.

When disease first appears, begin treatment with chlorothalonil, maneb or copper oxychloride. Additional pesticides that may be used later in rotation include Heritage or products containing the active ingredients *Bacillus subtilis* or basic cupric sulfate.

The following strategies can be adopted for chemical control:

- Application of prophylactic spray of contact fungicide and subsequently need based fungicide sprays.
- Application of 1 or 2 sprays of systemic fungicides mixtures (metalaxyl, ridomil, etc.) in areas where blight comes every year in epiphytotic form. First and last spray should be of contact fungicides (mancozeb, captafol, copper oxychloride, Bordeaux mixture) to avoid development of resistant strains of the fungus.

(d) Biological Methods: Biocontrol agents like *Pencillium* sp. and *Trichoderma* sp. have been found effective.

5.1.2.3. Dry Rot or Black Scurf, *Rhizoctonia solani*

Rhizoctonia stem and stolen canker, commonly called black scurf because of its tuber phase, is an occasionally serious disease of potato found in all production areas of the world. This disease affects potato development from emergence to harvest. Under cold, moist soil conditions following planting, *Rhizoctonia* can affect emergence and the developing plant. The development of resting structures (sclerotia) causing the black scurf or "dirt that won't wash off" phase of this disease occurs after vine desiccation. Most reports of this disease claim it will limit total yield of a potato crop only under extreme conditions. *Rhizoctonia* can seriously limit marketable yield, however, by causing infected plants to produce an abnormally high percentage of tubers that are misshapen, knobby and cracked while also adversely affecting size distribution. *Rhizoctonia* will invariably decrease the amount of bruise-free potatoes during harvest due to the increased percentages of these misshapen and knobby tubers.

(i) Symptoms: Perhaps the most readily apparent phase of *Rhizoctonia* disease is the black scurf or sclerotia present on tuber surfaces. These sclerotia can vary in size from very small, flat, superficial black specks to large, raised, irregularly shaped masses that can cover a major portion of the tuber. While black scurf can, under extreme conditions, affect the marketability of table stock/fresh pack potatoes, it is an extremely important seed-borne phase of the pathogen that deserves attention by all potato producers

Lesions characteristic of *Rhizoctonia* on stems and stolons are brown to black and sunken. These cankers can continue to expand and are capable of girdling stems and stolons of young developing plants (Fig 5.8). *Rhizoctonia* infection of older plants very seldom leads to girdled stems that die. However, the health of these plants can be severely compromised and they can frequently become more susceptible to other diseases, particularly early blight.

Figure 5.8. Black scurf on collar region of potato

Potato plants are most severely affected in the spring when underground sprouts can be killed prior to emergence. The secondary sprouts that develop are generally less vigorous and emerge much later causing irregular, uneven stands. Sprouts of severely affected plants that do not die are frequently stunted and remain so for the rest of the growing season. Infections of stolons that occur early in the growing season frequently result in pruning to tuber formation or abortion of tubers early in their development.

Mid-season *Rhizoctonia* infections of potato plants will result in deep, sunken cankers on the main stem. The above ground portion of the plant will appear yellow with some purpling and upward curling of the foliage. These symptoms are frequently confused with leaf roll or purple top. A key in differentiating *Rhizoctonia* from these two diseases is the frequently formation of aerial tubers in the lower leaf axils and the brown, sunken lesions just below the soil surface.

(ii) Epidemiology: *Rhizoctonia solani* can be either a soil-borne or seed-borne pathogen. The fungus survives in soil as mycelium in decomposing plant tissues. It also survives as sclerotia on tuber surfaces (seed-borne) or in the soil for extended periods. Populations of *R. solani* decline in the absence of a susceptible host although the rate of decline is affected by soil type, rotational crops and possibly the amount of organic matter present in the soil.

Disease development on emerging sprouts is favored by cold, wet soil conditions. Although these conditions may not increase disease incidence, disease severity is generally much greater. These conditions slow sprout development but favor germination of sclerotia and infection causing cankers to develop on young, under-developed tissues. Tuber-borne inoculum is very important in this phase of the disease while soil-borne inoculum is believed to be generally more important in stem and stolon infections. However, tuber-borne inoculum is also important in the pruning of the first stolons developed by the plant. When stolons become pruned, the secondary stolons are generally weaker and much shorter in length. Tubers produced on these shorter stolons are generally the ones that become misshapen because they tend to grow around the plant stem. Cool, moist conditions, with moisture being the most critical factor, also favor disease development on stems and stolons. Disease development is optimum at around 65°F and decreases as soil temperatures increase.

(iii) Management

Management of *Rhizoctonia* canker must be integrated since no single tactic is totally effective. First and foremost is crop rotation. In this regard it is important to consider both the number of years between potato crops and the rotational crops used in the system. Most research has shown that one potato crop in three or four is adequate to reduce soil inoculum to economic levels. However, soil type can potentially impact the rate of pathogen decline. Observations suggest that the rate of *Rhizoctonia* decline in sandy soils exceed that of loamy soils with more organic matter. Rotational crops such as sugar beet and dry beans, which are affected primarily by different AG's of *R. solani* tend to raise the general activity of *Rhizoctonia* in the soil thereby increasing the inoculum potential of soil-borne inoculum. Cereal crops such as oat and barley are generally regarded as being good rotations with potato for the reduction of soil-borne inoculum.

Since young developing potato crops are more susceptible to injury by *Rhizoctonia*, practices that encourage rapid emergence and plant development will reduce disease severity. Potatoes planted into soils with excessive amounts of residue from a previous crop, such as corn, tend to have serious *Rhizoctonia* problems. Soils with large amounts of residue remain cold for extended periods and moister, thereby favoring disease development. Therefore, high amounts of crop residue should be dealt with prior to planting.

Cultural practices such as pre-cutting seed, shallow hilling or hill drag-off and delaying planting until soil temperatures are more than 60°F encourage emergence and will also reduce sprout pruning. It is also important to avoid high soil moisture during this period. Early irrigation, prior to emergence, should be kept to a minimum. Potato hills with steep sides and sharp peaks generally warm more rapidly than large, flat beds which can also reduce *Rhizoctonia* disease severity.

Seed selection is also important. Seed tubers free of obvious sclerotia are more desirable than seed with a significant amount of black scurf. Sclerotia coverage as little as 5% can significantly increase disease incidence and severity. In reality, however, much less black scurf sclerotia can significantly increase disease development. One black sclerotium per square inch, which is far less than 1% coverage, can significantly increase disease pressure under environmental conditions that favor disease development. For this reason, disease-free seed is the best option. Harvesting seed tubers within 2–3 weeks of vine desiccation will reduce the amount of sclerotial formation. Tubers left in the soil for more than 3 weeks after vine desiccation will generally have significantly higher levels of black scurf which can lead to deep discounts on the fresh market.

A number of fungicide seed piece treatments have been tested for their effectiveness in controlling the seed-borne *Rhizoctonia* inoculum. Of these, Maxim has consistently provided the best control of tuber-borne *Rhizoctonia*.

Use disease-free seed tubers for planting. Seed treatment with acetic acid +Zn, boric acid or *T. viride* at 3.5 g/kg was found very effective. Combination of soil solarization of black scurf infested fields and seed treatment with *T. viride* further improved the disease control.

Soil application of mustard/groundnut cake at 2.5 MT/ha significantly reduced the number of black scurf diseased tubers and intensity of the disease.

Tuber damage and injury must be avoided during harvest and storage. Tubers should be washed to remove contaminated soil adhering to the tubers, followed by drying under shade reduces the infection substantially. In plains, tubers should be stored in cold stores. In country stores, tubers must be examined periodically and rotting tubers sorted out.

5.1.2.4. Collar Rot, *Sclerotium rolfsii*

(i) Management

Wet tuber treatment with *P. fluorescens* combined with *T. harzianum* at 10 g/kg was most effective.

5.1.2.5. Charcoal Rot, *Macrophomina phaseolina, Rhizoctonia solani*

(i) Management

Grow early maturing varieties. Harvest the crop early before it gets too warm by the middle of April. Keep the harvested tubers in cold stores. If the harvesting is delayed, keep the soil cool by frequent irrigations. Treat the seed with Agallol at 5 g or with Tafasan or Emisan at 2.5 g/litre of water for 10 min.

5.1.2.6. Wart, *Synchytrium endobioticum*

(i) Symptoms: Tumors or warts appear on tubers (Fig 5.9), stems and stolons. The wart pathogen is readily seed and soil-borne. Warted tubers used as seed are the chief means of disease spread. The wart pathogen may also spread through seed of wart immune varieties grown in infested land, contaminated soil carried on the feet of workers, animals or farm implements and manure containing diseased material.

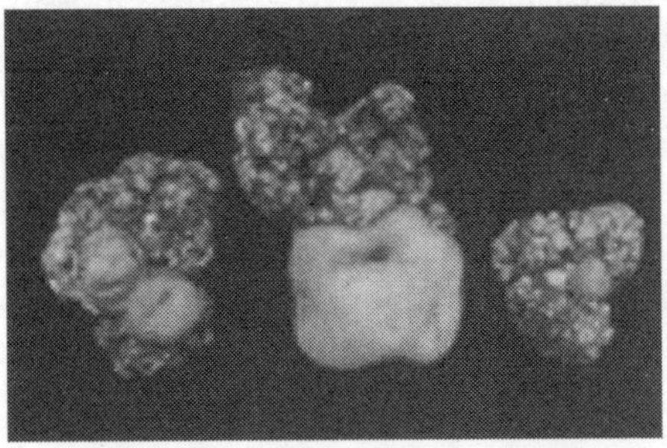

Figure 5.9. Wart on potato tubers

(ii) Management

Cultivate immune potato *cvs.* like Kufri Jyoti, Kufri Sherpa, Kufri Kanchan and Pimpernel. Practice long crop rotation (5 years or more). Rogue out infected plants along with tubers. Rogue out plants of susceptible varieties by burning and destroying warty tubers. Use disease-free seeds for planting.

5.1.2.7. Common Scab, *Streptomyces scabies*

Potato scab is caused by a soil-borne organism that is worldwide in distribution. The same organism also attacks many other root crops as well as the fleshy roots of weeds. It has been reported on beet, turnip, rutabaga, radish, carrot and parsnip. Beets are especially susceptible when soil conditions are favorable for the disease.

(i) Symptoms: Scab spots are more or less circular brown, roughened areas, with irregular margins. Sometimes the ridged portions are in broken concentric rings. The spots may be raised or warty in appearance, level with the surface, or somewhat sunken. Superficial russetting to deep pitting may occur (Fig 5.10). White grubs, wire worms, and millipedes are attracted to this affected tissue so that scab spots may be further enlarged or deepened. Scab begins when tubers start forming. Initially the spots may be so small that they are not noticed. As the tuber continues to grow, the areas of these reddish-brown spots also enlarge. An older tuber has too thick a protective layer on its surface to be invaded readily.

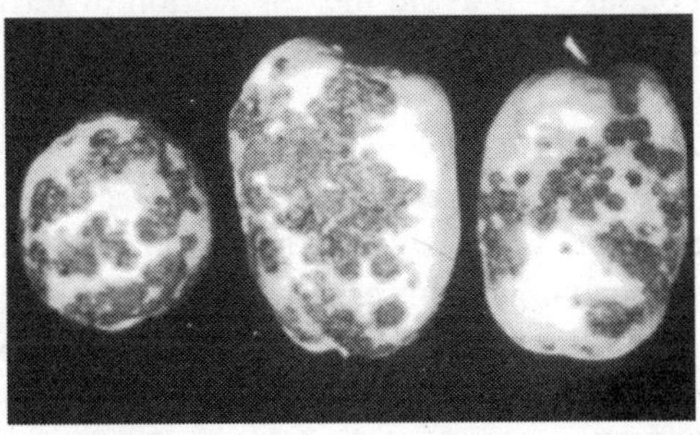

Figure 5.10. Common scab symptoms on tubers

(ii) Disease Cycle: *Streptomyces scabies* can live on decomposing material in the soil and does not require a potato or root crop to remain alive. However, it does become more aggressive on any given crop, such as potatoes, if the same crop is grown year after year without rotation. The organism infects the tuber by means of a tiny thread that pushes directly through the tender skin of the forming potato. It slowly continues to grow until the crop is harvested. The infected areas on the potato respond at once by laying down a corky layer. As each layer is invaded by the scab organism, new cork formation takes place until a mature scab spot is produced. Scab does not develop further after the potatoes are dug, but the organism remains alive all winter in storage.

(iii) Management

Use disease-free seed tubers for planting. Irrigate the crop repeatedly to keep the moisture to the field capacity right from tuber initiation until the tubers measure 1 cm diameter. Follow crop rotation with wheat, peas, oat, barley, lupin, soybean, sorghum, bajra and adopt green manuring to keep the disease in check. Plough the potato fields in April and leave the soil exposed to high temperature during May-June. Grow disease resistant *cvs.* like Patna Red, Kufri Alankar, Kufri Sindhuri, Kufri Dewa and Kufri Lalima.

Management in the vegetable field involves correcting the conditions that are favorable for the scab organism. Some of the important points to be considered are given below:

- Rotations with other than root crops should be employed as the size of the field permits. This way the build up of aggressive scab strains can be avoided.
- Purchase certified potato seed pieces. Seed treatment will help to prevent the introduction of the organism into relatively scab-free soils but is not a replacement for using clean seed. Currently, thiophanate-methyl + mancozeb (Tops MZ) is recommended for seed piece treatment against scab.
- Plant resistant varieties. If scab has been a problem before, test one of the following varieties under local conditions to determine whether it is suitable: Belchip, Centennial Russet, Keuka Gold, Norchip, Norgold Russet, Norland, Onaway, Pike, Russet Burbank, Salem or Superior. Of these, Norchip, Norland, Superior and Russet Burbank are probably the most available as seed tubers.
- In a garden where irrigation is available, scab can be reduced by keeping the soil reasonably wet for several weeks while the young tubers are beginning to form.
- Great care should be taken to avoid the application of fertilizer or other materials that tend to make the soil highly alkaline. At the head of the list are lime, ashes, and farm yard manure.
- One of the best methods for combating scab is the use of acid producing fertilizers, especially those that contain liberal amounts of sulphate of ammonia. If this practice is repeated for three or four successive years, the soil will finally become acid enough (pH 5.2) to keep the scab organism from developing.
- Test the pH of the soil and apply elemental sulphur to lower the soil pH to 5.2. Sulphur is usually slow in action so it may be more effective if it is applied a year ahead of the potato crop.

5.1.2.8. Bacterial Wilt or Brown Rot, *Ralstonia solanacearum*

Bacterial wilt of potato continues to be a major disease since 19th century. In India, the disease was first recorded on potatoes by Cappel (1892). This was the first bacterial disease recorded in India. The disease is more severe in tropical, subtropical and warm climates. In Ranchi district of Jharkhand (India), cultivation of rainfed potato was abandoned due to this disease (Gurfan & Chakravarti, 1960) and similar instances are recorded in Indonesia (Machmud, 1986) and in Peru (French, 1986a). In India, the losses reported are up to 55% in Kumaon hills (Paharia, 1963), 0.33–40% in Maharashtra (Paharia, 1963), 20–25% in Hyderabad (Nath *et al.*, 1958) and over 75% in some localities of Karnataka (Gadewar *et al.*, 1991).

(i) Symptoms: The disease affects both above and under ground parts of the plant. Infected plants show sudden wilting and complete collapse of plants. Vascular browning has been reported to be the characteristic symptom of the disease and the bacterial mass ooze out.

(ii) Epidemiology: Latently infected material (tubers, rhizomes and corms) is the single most important carrier of the bacterium from one place to another. Bacteria spread from plant to plant by root contact or through the water films or flowing water. In general, increase in the temperature enhances wilting. High soil moisture favours survival of the pathogen in the soil, infection and exit of the bacterium from the host and spread in the soils.

(iii) Management

The management practices ought to be directed towards reducing the inoculum in the field and avoiding introduction of fresh inoculum, growing the crop when the environmental conditions are not favourable for the pathogen and developing the resistant varieties or by biological means such as the use of avirulent strains of the pathogens and/or rhizobacteria.

(a) Cultural Methods: Seed production may be undertaken only in disease-free areas (higher hills of Himachal Pradesh and plains of Punjab). Selection of disease-free tubers for planting, soil application of bleaching powder, hot/cold weather cultivation, disinfection of cutting knives with mercurials, adopting crop rotation for 2–3 years with non-host crops (maize, sorghum, finger millet, cereals, garlic, onion, lupin, cabbage, green manuring crops like sunnhemp, etc.) and use of resistant cultivars have been recommended. Restrict post-emergence tillage to minimum. Practice full earthing up immediately after planting. Collect and burn all diseased plant material including tubers after harvest. The pathogen is susceptible to high temperature. Raising the soil temperature by covering the soil with polythene film has been found effective in reducing the wilt. Opening the deep furrows in infested fields at 50–60 cm intervals, spreading the straw and burning (Jhumming cultivation) reduced the wilt incidence by 100%.

In north-eastern and north-western hills, early planting of the summer crop in February and harvesting in June can avoid the wilt. In Shillong and Ooty, planting in last week of July reduced the bacterial wilt and the wilt was negligible to nil in crop planted in first week of September. These practices reduce the cropping period to only 90 days. Therefore short duration varieties can be used to minimize the losses.

Crop rotation with maize, wheat, barley, oat, sunnhemp, finger millet and vegetables like cabbage, onion, garlic reduced the wilt incidence to the extent of 94%. Studies undertaken in north-western hills using 6 cropping sequences revealed that the reduction in wilt incidence ranged between 25 to 81% by following 2 years rotations. Potato-finger millet-finger millet-potato gave maximum reduction in wilt incidence (81%) (Table 5.4). Roots of finger millet were free from *R. solanacearum* infection, whereas paddy, maize, wheat and beans carried the infection. However, rhizosphere population of the pathogen was found to be reduced.

Table 5.4. Effect of crop rotation on bacterial wilt incidence and yield of potato in North Western Hills (Bhowali)

Crop rotation	Wilt incidence	Yield loss	Population of *R. solanacearum* in cfu/root system of different crops
Potato-Potato-Potato-Potato	51.7 (+92)	56.9	9.0×10^6 (Potato)
Potato-Bean-Bean-Potato	19.7 (–25)	18.3	9.2×10^4 (Bean)
Potato-Paddy-Paddy-Potato	12.0 (–53)	29.7	1.0×10^5 (Paddy)
Potato-Wheat-Wheat-Potato	9.2 (–63)	18.8	4.2×10^4 (Wheat)
Potato-Maize-Maize-Potato	8.7 (–66)	13.1	1.0×10^6 (Maize)
Potato-Finger millet-Finger millet-Potato	4.7 (–81)	7.1	0.0 (Finger millet)

Intercropping of potato with maize and cowpea were the most effective in reducing the wilt incidence.

In non-endemic areas (Gujarat, Maharashtra, north-western and north central plains and north-western high hills), use of disease-free seeds alone is adequate to control the disease completely. In mid hills (north-western mid hills up to 2200 MSL, north- eastern hills and Nilgiris hills), use of disease-free seeds, cold weather cultivation and application of bleaching powder at 12 kg/ha will be adequate for the disease control. Application of bleaching powder can be substituted with 2 years crop rotation with wheat, barley, finger millet, cabbage, cauliflower, knol-khol, carrot, onion, garlic. In eastern plains and Deccan plateau, the disease can be reduced to considerable extent by following cultural practices like use of disease-free seeds, adjusting the date of planting, application of bleaching powder, straw burning in infested soil, blind earthing, hot weather cultivation and crop rotation (Table 5.5).

(b) Biological Methods: Promising biocontrol agents include *Bacillus* spp., *B. polymyxa* and *Pseudomonas fluorescens* which have been found to reduce the wilt development and incidence. Root colonization by *P. fluorescens* has been reported. *Bacillus* spp. reduced bacterial wilt incidence up to 79%, while *P. fluorescens* by 43 to 75%. Summer (during May-June in plateau and plains where very hot and dry climate prevails) and winter ploughing (during December-January in hills where cold temperature prevails) alone reduced wilt incidence by 66% and 50%, respectively. Dipping of seed tubers in 0.25% solution of the formulation of *Bacillus cereus* (strain B4) and *B. subtilis* (strain B5) for 20 min. was effective in reducing the wilt by 59–72% and enhancement of tuber yield by 16.7–25.6%, respectively. Tuber treatment with *P. fluorescens/B. subtilis* reduced wilt incidence by 76.31 and 53.15% and increased tuber yield by 220 and 129%, respectively.

Table 5.5. Control of bacterial wilt and brown rot

Treatment	% Reduction		% Yield increase
	Wilt	Brown rot	
Healthy seed	50 – 76	40 – 46	18 – 27
Bleaching powder at 12 kg/ha + Blind earthing	70 – 94	50 – 85	22 – 28
Summer ploughing in plains	50 – 83	50 – 70	5 – 25
Winter ploughing in hills	60 – 75	62 – 78	35 – 40
Early planting in hills	77 – 80	75 – 85	44 – 68

Avirulent and bacteriophage producing strains of *R. solanacearum* induce resistance to bacterial wilt. Tuber bacterization followed by one soil application of avirulent strain of *Ralstonia solanacearum* significantly reduced the disease incidence to an extent of 80 to 82.5% and increased the tuber yield (15.08 to 15.83 t/ha) compared to control (6.04 to 6.94 t/ha).

(c) Host Resistance: Some of the resistant/tolerant *cvs*. under cultivation are Ontario, Sebago, Kathadin and Green Mountain in USA; Prisca in Madagascar; Myrylin and Greta in Mauritius and Caxamarca and Molinera in South America.

5.1.2.9. Soft Rot, *Erwinia carotovora*

(i) Symptoms: Fruits develop sub-epidermal soft-rot, which spreads rapidly reducing the tissue to a pulp in 4 to 6 days.

(ii) Management

Avoid excess irrigation. Provide proper drainage. Adjust planting time to avoid hot weather during plant emergence. Harvest the crop before the temperature raises to 28°C. Harvest the crop only when the tuber skin is fully cured. Avoid injury to tubers and sort out injured tubers. Treat tubers before storage with 3% boric acid for 30 min and dry under shade. Store the produce in well ventilated cool stores or cold stores.

5.1.2.10. Potato Virus X (PVX)

(i) Symptoms: The characteristic symptoms of PVX include interveinal mosaic, mild or barely perceptible mosaic, light and dark green patches, mottling with stunting, crinkling (Fig 5.11) with severe strains and top necrosis in some varieties. An infection of PVX has the potential to reduce the yields up to 10–30%.

(ii) Transmission: Use of infected tubers is the main reason for spread of PVX from one place to another. PVX is contagious and readily spread upon contact of plant foliage and roots in the field as well as upon injury/cutting of the tubers.

PVX is readily transmitted mechanically. It is spread by foliar or sprout contact, seed cutters, pick-type planters, cultivators, spray equipment, and possibly by insects with chewing mouth parts.

Figure 5.11. PVX on potato leaves

(iii) Management

Strict sanitation should be practiced in the field and also in stores. Disinfect all field equipments with 3% trisodium phosphate or 1% calcium hypochlorite solution. Use disease-free seed tubers for planting. Rouging of diseased plants along with their tubers at least twice in growing season helps in reducing the incidence of the disease. Dehaulm the crop before the aphids cross the critical level. Follow thermotherapy by meristem culture for eliminating the virus from seed tubers. Potato *cvs.* having resistance are Kufri Ashoka, Kufri Badshah, Kufri Jawahar, Kufri Jyoti, Kufri Lalima, Kufri Pukhraj and Kufri Sindhuri.

5.1.2.11. Potato Virus Y (PVY)

(i) Symptoms: The characteristic symptoms of PVY include mild to severe mosaic, veinal necrosis, leaf drop streaking, stunting of plants, rugosity, bunching or twisting of leaves, down-turning of leaflet margins and necrosis, dropping and collapse of older leaves (Fig 5.12). Symptoms of PVY may vary by cultivar. Necrosis and veinal necrosis of petioles often cause leaves to drop or cling to the stems. Infected plants have bunchy tops and leafless lower stems or clinging necrotic leaves. An infection of PVY has the potential to reduce the yields up to 60–80%.

(ii) Transmission: PVY is a potyvirus which is transmitted through aphids, *Myzus persicae* and *Aphis gossypii* in non-persistent manner.

PVY is a stylet-borne virus transmitted by at least 30 species of aphids, particularly the green peach aphid and potato aphids. Cereal aphids in high numbers become important vectors of PVY. Aphids landing on infected plants acquire PVY within a few seconds of feeding and can transmit the virus immediately, in a non-persistent manner. While aphid transmission is the most important means of spread in the field, all PVY strains also are mechanically transmissible by leaf contact and may be carried in seed tubers. Current-season infection late in the growing season may not cause foliar symptoms.

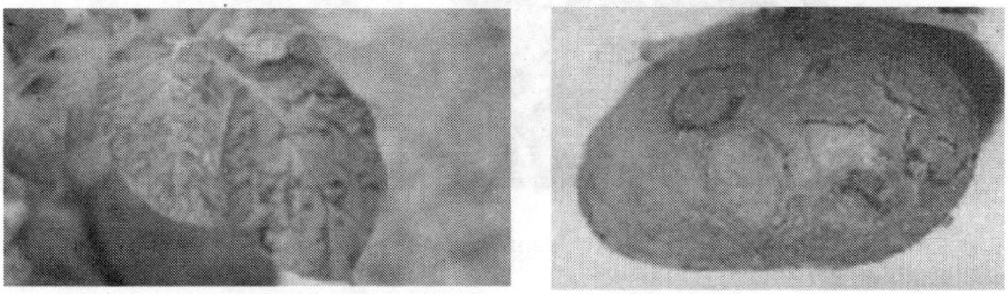

Figure 5.12. PVY on potato leaves and tuber

(iii) Management

A combination of thermotherapy coupled with meristem culture eliminated the virus from seed tubers. Potato *cvs*. Kufri Jyoti, Kufri Kuber, Kufri Lalima, Kufri Sindhuri and Kufri Ashoka have in built resistance/tolerance to PVX, PVY and PLRV.

5.1.2.12. Potato Leaf Roll Virus (PLRV)

Potato leaf roll is caused by the potato leaf roll virus. The severity of symptoms depends on the strain of virus, susceptibility of the cultivar, time of infection, and environment. Yield may be substantially reduced in plants grown from infected tubers. An infection of PLRV has the potential to reduce the yields up to 60–80%. Potato leaf roll severity can be reduced as a result of a comprehensive aphid suppression and monitoring programme.

(i) Symptoms: The characteristic symptoms of PLRV include rolling of upper leaves in primary infections and of the lower leaves in the secondary infection, leaves turning leathery, brittle, pink or brown pigmented, stunting of plants and making rattling noise when shaken.

Lower leaves of plants grown from leaf roll-infected seed tubers become rolled and leathery and may turn light pink or yellow. Infected plants usually are stunted and more upright than healthy plants. Young leaves may become yellow or slightly pinkish and rolled and may stand upright in early current-season infections (Fig 5.13). Current-season infections that develop late in the season may cause no visible symptoms, but tubers usually become infected. Early season infections may reduce yield.

(ii) Transmission: The virus is transmitted by aphids, *Myzus persicae* and *Aphis gossypii* in persistent manner. The green peach aphid, *Myzus persicae* is the primary vector. Aphids acquire the virus after several hours of feeding and require an incubation period before being able to transmit the virus to non-infected plants. Once infected, the aphid usually retains the virus for life.

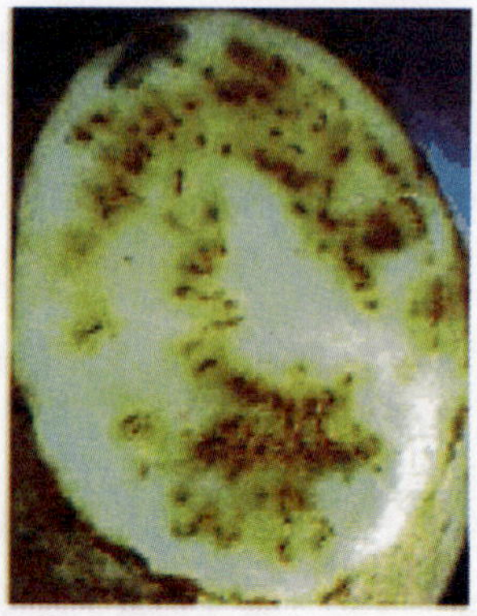

Figure 5.13. Leaf roll virus infected plants and tuber of potato

(iii) Epidemiology: Leaf roll is introduced by planting infected seed or by aphids carrying the virus to the field from an outside source. Leaf roll is transmitted by several species of aphids. The green peach aphid is the primary vector.

(iv) Management

Plant high quality, certified seed. Two hour daily treatment of tubers for 6 weeks at 40°C inactivates PLRV. Treatment of tubers in hot water at 55°C for 15–20 min has also been found effective in eradicating the virus. Control aphid populations in the field, and on other hosts. Rogue infected plants early in the season to control disease sources in the field. Consider early vine-kill when high aphid populations are present.

Potato *cvs.* Kufri Jyoti, Kufri Kuber, Kufri Lalima, Kufri Sindhuri and Kufri Ashoka have in built resistance/tolerance to PVX, PVY and PLRV.

Integrated Pest and Disease Management

- Deep summer ploughing to expose resting stages of pests to sunlight.
- Use disease-free seed material for planting.
- Grow resistant varieties.
- Adjustment of planting and harvesting time for avoidance of pests and diseases. Optimum time for planting is 15th to 30th October in western and central plains and 1st to 15th November in eastern plains.
- Covering the exposed tubers by high ridging.
- Use yellow sticky traps at 10/ha for attracting aphids.
- Use sex pheromone traps at 20/ha for mass trapping of tuber moth.
- Rouging of diseased plants.

5.1.3. Nematodes

Cyst, root-knot, dry rot and lesion nematodes have been recognized as the major nematode pests of potato.

Table 5.6. Management practices for pests and diseases of potato at each crop stage

Crop stage	Action	Reason
Before planting	Clean up crop area and remove weeds and self-sown potatoes.	Weeds and self-sowns harbour viruses, insect pests and diseases which can be transferred to the new crop.
	Avoid continuous cropping with potatoes (minimum 2 years between crops).	Reduces the build-up of pests and diseases in the soil.
Planting	Plant certified seed.	Certified seed is grown under strict hygiene regulations and will reduce the risk of introducing seed-borne pests and diseases to the crop.
	Plant varieties which are less susceptible to powdery scab or common scab.	In paddocks where scabs have been present in the past, less susceptible varieties will reduce disease risk.
	Change planting depth according to relative risk of pests and diseases.	Deep-setting varieties are less susceptible to potato moth attack; shallow planting will help reduce *Rhizoctonia* infection.
	Plant in optimum conditions of soil temperature (16–20°C) and soil moisture.	High temperature and high moisture can lead to seed piece breakdown (especially in Coli ban); low temperature and high moisture favours infection by *Rhizoctonia.*
Hilling	Maintain and improve soil structure and create well-drained hills.	Poorly-drained soils which are easily saturated favour development of diseases like powdery scab, blackleg and home (gangrene).
	Maintain good soil cover throughout season.	Good soil cover acts as a barrier, preventing exposure of tubers to potato moth and greening.
	Monitor pests and diseases throughout season.	Control measures are only required when pest populations reach a critical level.
	Monitor beneficials and release if necessary.	
Irrigation/ rain	Check soil coverage of tubers regularly throughout season.	Rolling and hilling prevents greening of tubers as well as protecting them from pests and diseases attack.
	Manage irrigation to avoid pests and diseases attack.	Paddocks with a history of powdery scab should not have saturated soil during tuber set.

Table. 5.6. (Contd.)

Crop stage	Action	Reason
		Paddocks with a history of common scab should have moist soil during tuber set.
	Avoid water stress.	Maintaining good crop health makes plants less susceptible to pests and diseases attack (especially black dot).
Die-down/ slashing	Maintain soil barrier using irrigation/rolling.	Prevents greening, potato moth attack and late blight.
Harvesting	Ensure tubers are mature before digging.	Mature tubers have a protective skin which reduces the risk of infection.
	Harvest as soon as possible after maturity.	Increased time in the ground increases risk of exposure to pests and diseases (*e.g. Rhizoctonia*, silver scurf, black dot, wireworm).
	Handle tubers carefully during harvest.	Damage to tubers creates an equity point for disease (*e.g. Phoma*, dry rot, soft rot).
	Harvest under cool (12–18°C), moist conditions.	Hot and dry, or cold, conditions increase damage to tubers and the risk of rots.
Post harvest	If tubers were very wet at harvest, place them in a drying tent with circulating warm air until dry.	Free moisture on tubers increases the risk of rots and of seed piece breakdown.
	If tubers were cold at harvest or have just been removed from cool storage, warm them to at least 120°C before grading or cutting for seed.	Handling cold tubers is likely to damage them and increase the risk of disease.
	Handle tubers carefully during grading and inspection.	Damage to tubers increases risk of infection and can make them unsalable.
	If tubers are to be cool stored, cure them beforehand by holding them at 14–16°C and 85–95% relative humidity, with good airflow, for 10–14 days.	Curing heals skin wounds and reduces the risk of storage rots (*e.g. Phoma*, dry rot).
	Don't retain tubers for seed.	Non-certified seed increases chances of pests and diseases (especially viruses) developing in next season's crop.
	Store tubers at optimum temperature and humidity (3–4°C for seed; 4–10°C for ware; 10°C for processing tubers; 85–95% relative humidity for all).	Storing tubers under optimum conditions minimizes disease and prolongs storage life.

5.1.3.1. Root-knot Nematode, *Meloidogyne* spp.

Thirumalachar (1951) noticed root-knot galls on potato tubers from Shimla during 1950 for the first time and since then it has been recorded from all the potato growing regions of the country. The root-knot nematodes, *Meloidogyne* species are by far the most important nematode pests of potatoes in India. Severe infestation of root-knot nematodes leading to crop failure have been noticed in Mahasu district of Himachal Pradesh and Hassan in Karnataka where farmers did not get even as much as planted by them.

An initial level of 2 root-knot larvae per gram of soil resulted in 100% tuber infection with an overall yield reduction of 42.5% (Prasad, 1989).

(i) Distribution: *M. incognita* has been recorded on potato in Himachal Pradesh, Uttar Pradesh, Assam and Karnataka; *M. javanica* in Himachal Pradesh, Uttar Pradesh, Punjab, Haryana, Delhi, Bihar and Maharashtra; and *M. hapla* in Gulmarg (Jammu and Kashmir) and Shimla (Himachal Pradesh).

(ii) Symptoms: Plants affected by root-knot nematodes are generally stunted, slightly yellow and may wilt during hot weather. The diagnostic symptoms of a *Meloidogyne* attack on potatoes are the galls found on both roots and tubers. Tuber surface becomes uneven and warty because of numerous blisters like galls (Fig 5.14). When an infested tuber is transversely cut, the white glistening swollen females of the size of pin head can be easily seen embedded in the potato tissue. Such tubers are invariably small, unfit for marketing, rot quickly and cause considerable yield losses. Galling on tubers may render them unsalable.

Figure 5.14. Potato tubers with blisters incited by *Meloidogyne incognita* (Courtesy: Sasser, 1971)

(iii) Life Cycle: There are up to five generations on the susceptible host under favourable conditions. Both roots and tubers are infected, however, the first generation occurs mainly on the root systems, while the succeeding generations attack tubers.

(iv) Interaction with Other Pathogens: *Meloidogyne* species are associated with the fungi such as *Rhizoctonia solani* and *Verticillium* species and the bacterium, *Ralstonia solanacearum* in forming disease complexes.

Jatala *et al.* (1990) reported the interactions between *R. solanacearum* and nematodes, especially *Meloidogyne* spp. in potato in relation to breeding for combined resistance. Resistance of potatoes to bacterial wilt is broken in the presence of *M. incognita*.

(v) Ecology: Root-knot damage to potatoes is usually associated with light soil. Aggravated attack by *Meloidogyne* species on potatoes under irrigation is usually ascribed to spread of the nematodes in irrigation water. *M. hapla* requires cool climate, whereas *M. incognita* and *M. javanica* require higher temperatures for their development and reproduction.

(vi) Races: Four races in *M. incognita* and 2 races in *M. arenaria* have been identified based on the North Carolina Differential Host Test which employed 5 differential hosts for delineating races (Table 5.7).

Table 5.7. Races of *Meloidogyne incognita* and *M. arenaria*

Meloidogyne		Cotton	Tobacco	Pepper	Watermelon	Peanut	Tomato
M. incognita	1	–	–	+	+	–	+
	2	–	+	+	+	–	+
	3	+	–	+	+	–	+
	4	+	+	+	+	–	+
M. arenaria	1	–	+	–	+	–	+
	2	–	+	+	–	+	+

In India, all 4 races of *M. incognita* have been reported and the race 1 is dominant (Sharma & Gill, 1992).

(vii) Survival and Spread: Since *Meloidogyne* spp. attack a large number of plant species, their population can be maintained on weeds and volunteer crops. They over winter usually in the form of eggs, although the ability of juveniles to go through anhydrobiosis may contribute to the survival of some *Meloidogyne* spp. Infected tubers, plant parts and planting material, as well as movement of infested soil by farm machinery and irrigation water are the main avenues of disseminating *Meloidogyne* spp.

(viii) Management

(a) Physical Methods

Heat Treatment: Incubation of potato tubers at 45°C for 48 hours has been shown to kill about 98.9 per cent *M. incognita* without affecting tuber viability (Nirula & Bassi, 1965). Hot water treatment of potato tubers at 46–47.5°C for 120 min gave effective control of root-knot nematode infection.

Rabbing: In small holdings, burning of trash in the field before taking up tuber planting not only helps in reducing the nematode population but also enriches the soil.

Irradiation: Fassuliotis and Sparrow (1955) demonstrated that irradiation of potato tubers with x-rays inhibited sprouting and also development of *Globodera rostochiensis*. Cysts of this nematode exposed to 20,000*r* contained only brown and dead eggs, while at 40,000*r* the eggs lost their contents completely.

(b) Cultural Methods: Tubers from root-knot infested potato crops should not be used as seed, as even outwardly clean looking seed may be infested. The movement of soil and water from the infested fields should be avoided. The field should be kept free from weeds, since root-knot nematodes have a wide host range and most of the weeds help in the build up of nematodes. Thus, clean cultivation (free from weeds which act as hosts) reduces nematode population to a great extent.

Deep ploughing and drying of soil in the summer months facilitates the drying and death of infective larvae thereby reducing initial inoculum in the soil. Dry fallow with 2 or 3 deep ploughings during the hot summer months gives excellent control of root-knot nematodes.

Intercropping: Potatoes grown with white mustard in a pot of infested soil were less heavily attacked by root-knot nematodes than potatoes growing alone. Potato root diffusate was ineffective in the presence of leachings from the roots of mustard seedlings. Mustard oil increased the yield of potatoes by reducing the severity of nematode attack. The active principle involved in mustard is allyl isothiocyanate which is toxic to the nematodes. The interplanting of marigold with potato has resulted in a 7% increase in tuber yield.

Intercropping of onion and maize reduce galling by *M. incognita* on potato roots. Potatoes intercropped with maize have been found to be extremely vigorous with well developed root systems.

Time of Planting: In long season, warm climate regions, potatoes may be planted during the winter months and harvested before injury occurs in the spring without visible infection. Frequently potatoes are planted in January-February and harvested in April-May before the second generation of nematodes develop and attack the tubers. Planting of potato during third or fourth week of March in Shimla hills would reduce both root and tuber infestation of *M. incognita* without affecting the yield (Ahuja, 1983; Prasad *et al.,* 1983). The yields were maximum which was concomitant with the lowest tuber infestation and lowest larval population in the soil at harvest. Late planting of autumn crop and early planting of spring crop in north-western plains; while in hills, early planting of summer crop in fourth week of March is ideal. At Jalandhar, early planting of spring crop in first week of January and late planting of autumn crop in second and third week of October reduced the root-knot infection in potato.

Crop Rotation: In north-western hills, rotation of potato with French beans, leafy vegetables, maize and wheat helped in minimizing the nematode damage and increasing the productivity of the crop (Raj & Nirula, 1969). Crop rotation with non-host crops like maize, wheat, barley, beans, etc. reduces nematode population. Growing of trap crop like French marigold, *Tagetes patula* in alternate rows with potato reduced the root-knot nematode larval population in soil, root galling and tuber infestation while the yields increased up to

123% over control. Similarly, growing *Cassia* sp. and incorporation of saw dust at 2.6 t/ha also helped in reducing tuber infestations. Potato crop can be followed by peanuts without any damage to peanuts and nematode populations will be reduced.

Botanicals: The oil cakes (neem, groundnut, mustard and castor) were effective in reducing parasitic nematode population (*Hoplolaimus*, *Tylenchorhynchus*, *Meloidogyne* and *Helicotylenchus* species) and in increasing yields of potato. Application of eucalyptus leaf waste from oil distillation plants at 2.5 t/ha helps in reducing the nematode population.

(c) Chemical Methods: Application of carbofuran at 3 kg *a.i.*/ha in heavily infested fields and 2 kg *a.i.*/ha in moderately infested fields in two equal splits, first at planting and second at the time of earthing up, is recommended. Application of carbofuran at 3 kg *a.i.*/ha or aldicarb at 2 kg *a.i.*/ha or ethoprop at 10 kg *a.i.*/ha in two equal splits, first at planting and second at the time of earthing up, increased the yields by 18 to 23% while accounting for 80% reduction in tuber infestation.

(d) Biological Methods: The bacterium, *Pasteuria penetrans* has offered possibilities of biocontrol. Arbuscular mycorrhizal fungi such as *Glomus fasciculatum* and *G. mosseae* have shown promise in reducing the root-knot nematode infestation.

(e) Host Resistance: Potato *cv.* Kufri Dewa was found resistant to root-knot nematodes.

(f) Integrated Methods: Application of neem cake/FYM/compost enriched with *Trichoderma harzianum*/*Paecilomyces lilacinus* gave effective control of root-knot nematodes.

Following 2 years crop rotational sequence of maize-wheat-potato-wheat coupled with summer fallow after 2 or 3 deep ploughings in north-western plains reduced root-knot damage significantly.

5.1.3.2. Cyst Nematodes, *Globodera rostochiensis, G. pallida*

The potato cyst nematode is established as one of the major crop protection problems of the world. The ability of this nematode to build up to damageable levels in a short span of 5–6 years, substantial yield reductions in the crop, lack of inexpensive nematicides for soil treatment capable of providing adequate level of control under field conditions, the relative ease with which the cysts are dispersed with soil adhering to the seed tubers and the long persistence of eggs within the cyst in the absence of the host makes this nematode as probably the most important pest problem on potatoes.

The avoidable yield loss in susceptible potato due to cyst nematodes was 99.5 to 99.8% in summer and autumn crops. Total failure of the crop has been reported under severe infestation conditions. An average loss of about 9% of global potato is accounted to the cyst nematodes amounting to about 45 million tonnes (Prasad, 1989).

(i) Distribution: In India, it was first reported by Jones (1961) from Nilgiri hills, Tamil Nadu. It was also reported from Kodai hills, Tamil Nadu, and Munar hills of Kerala (Ramana & Mohandas, 1998). Out of 9,000 ha under potato, 3,000 ha are infected by this nematode in Nilgiris. In Kodai hills, about 200 ha are infected (Thangaraju, 1983). Tomato and brinjal are also attacked by this nematode.

(ii) Symptoms: The disease usually occurs in patches. The infested plants exhibit typical symptoms of patchy growth of weak and stunted plants. The patches increase in time with continuous potato cropping. Under conditions of severe infestation, the plant growth is stunted and wilting occurs during hot part of the day. Plants show tufting of leaves at the top as the outer leaves turn yellow and die. Root system is smothered, secondary roots are induced at the collar region and the plants can be easily pulled out. Tubers formed are less in number and reduced in size. In extreme cases, tuber formation is arrested. The total photosynthesis per plant is also significantly reduced as a result of reduced leaf area and this is reflected in reduced potato yield (Fig 5.15).

Figure 5.15. Cyst nematode on potato. *Left*-Infected, *Right*-Healthy

(iii) Life Cycle: *G. rostochiensis* is amphimictic, with sessile, globose females and motile, vermiform males preceded by four larval stages. The second stage larva is the resistant, dormant and infective stage remaining inside the egg within the cyst formed by tanning of the female's cuticle. The cysts remain in the soil after death of the host, each containing about 500 eggs. After hatching, the second stage larva enters the root of a host plant, just behind the root tip or at the site of a new lateral root and moves through the root away from the tip before starting to feed on a group of pericycle, cortex or endodermis cells which are modified to form a large syncytial transfer cells on which the nematode feeds throughout the rest of its development to the adult stage. Sex is determined by nutrition, a poor food supply resulting in the formation of a male, a good supply, a female. Thus with heavy infections males predominate, the reverse with light infections.

The adult females swells greatly, rupturing the root cortex and protrudes from the root, exposing the vulval region while the head and neck remain inserted in the root (Fig 5.16). As they develop, males become coiled within the earlier larval cuticles and on maturity they rupture them, then leave the root and are attracted to females by a secretion. Both sexes mate several times. Males do not feed but remain active for up to 10 days, females normally mate soon after rupturing the root cortex and within 50 days of larval invasion.

Figure 5.16. Potato roots infected with *Globodera rostochiensis* (Courtesy: Sasser, 1971)

(iv) Host Range: All known varieties of potato, tomato and brinjal are severely attacked by the cyst nematode. A number of other Solanaceae are recorded as hosts, but the majority of hosts are species and varieties of tuberous *Solanum*. The only susceptible plant outside the family is snap dragon, *Antirrhinum majus*.

(v) Interaction with Other Pathogens: The potato cyst nematode and the black-scurf fungus *Rhizoctonia solani* together caused much greater reduction in yield than did the organisms separately. The cyst nematode also interacts with *Verticillium dahliae*, enhancing the severity of the wilt caused by the fungus. The leaf symptoms appeared earlier and the tuber yields were depressed in the presence of the nematode.

The potato cyst nematodes, *Globodera rostochiensis* and *G. pallida* in association with the wilt bacterium, *R. solanacearum* causes wilt complex in potatoes.

(vi) Ecology: Well-drained, aerated sands, silts and peat soils with a moisture content of 50–75% water capacity are suitable for survival and movement of the free-living stages of the potato cyst nematode. A minimum soil temperature of 10°C is necessary for larval activity.

(vii) Spread and Survival: The cysts can be spread by the movement of soil through seed material for a very long distance. They can also be spread through field tools, implements, seedlings grown in infested soil for transplanting elsewhere, labour's feet, etc. from one field to another. The cysts can remain viable in the soil without a host for 30 years.

(viii) Physiological Races: At present, five pathotypes in *G. rostochiensis* and four pathotypes in *G. pallida* are differentiated using the following differentials (Table 5.8).

Table 5.8. Pathotypes of *Globodera rostochiensis* and *G. pallida*

Differential host	Nematode pathotypes*				
	1	2	3	4	5
Globodera rostochiensis					
Solanum tuberosum	+	+	+	+	+
ex *andigena* CPC 1673	–	+	+	–	+
ex *kurtzianum* 60.21.19	–	–	+	+	+
ex *vernei* 58.1642/4	–	–	–	+	+
ex *vernei* 62.33/3	–	–	–	–	+
ex *vernei* 65.346/19	–	–	–	–	–
Globodera pallida					
Solanum tuberosum	+	+	+	+	
ex *multidissectum* P 55/7	–	+	+	+	
ex *vernei* 62.33/3	–	–	+	+	
ex *andigena* x *multidissectum* D 47/1	–	–	–	+	
ex *andigena* x *multidissectum* D 47/2	–	–	–	–	

* (+) indicates ability to produce females, (–)indicates inability to produce females

In India, potato cyst nematodes are restricted to Nilgiri and Palani hills in south India, where 5 pathotypes – Ro 1, Ro 2 of *G. rostochiensis*, and Pa 1, Pa 2 and Pa 3 of *G. pallida* have been detected (Amalraj & Rao, 1992).

(ix) Management

(a) Regulatory Methods: In India, there is a Quarantine Act against the cyst nematode of potato (*G. rostochiensis*) in Nilgiris. Infected potatoes are not allowed to be transported from Nilgiris to other parts of India for seed and table purposes in order to prevent the spread of potato cyst nematode from Tamil Nadu to other states and Union Territories.

Potato seed pieces free of the cyst nematode can be produced commercially by seed certification. Use of certified planting material is fundamental to the prevention of cyst nematodes spreading into uninfested regions.

(b) Physical Methods: Fassuliotis and Sparrow (1955) demonstrated that irradiation of potato tubers with x-rays inhibited sprouting and also development of *G. rostochiensis*. Cysts of this nematode exposed to 20,000*r* contained only brown and dead eggs, while at 40,000*r* the eggs lost their contents completely.

Careful washing of soil adhering to potato tubers can considerably reduce the risk of spreading cyst nematodes. Commercial potato tuber washing apparatus has been developed.

(c) Cultural Methods: The potato cyst nematode (*G. rostochiensis*) can be eliminated by selecting nematode-free planting material.

Early Maturing Varieties: Early maturing varieties of potato like Irish Cobbler usually suffer less than do late maturing varieties. These varieties start growth in spring at soil temperature too low for much nematode activity and may be harvested before many of the nematodes can reproduce.

Crop Rotation: Continuous cropping with potatoes increases the cyst population of *G. rostochiensis*. More than three year rotations with wheat, strawberry, cabbage, cauliflower, peas, maize and beans reduces the nematode population to a safe level. Growing non-host crops like radish, garlic, beet root, French bean, cruciferous vegetables, turnip or green manuring crops bring down the cyst nematode population by more than 50%. A crop rotation pattern (using potato-French bean-peas-potato), wherein potato is grown once after 3–4 other crops, decreased the nematode population by 98 to 99% and increased yields of potato by 90%. In Nilgiris, crop rotation with cabbage and carrot gave effective control of cyst nematodes. Growing of vegetables and resistant potato in a long term rotation trial brought down the cyst propagule population up to 92% and was effectively used for nematode management at Nilgiris.

Botanicals: Incorporation of eucalyptus litter at 10 t/ha at the time of preparation of land reduces the population of the potato cyst nematodes. Application of eucalyptus distillation waste at 2.5 t/ha helps to reduce the cyst nematode population in soil.

(d) Chemical Methods: Application of carbofuran at 3 kg *a.i.*/ha or aldicarb at 2 kg *a.i.*/ha or ethoprophos at 10 kg *a.i.*/ha was found to reduce the infestation of root-knot and cyst nematodes in potato crop. The yield increase ranged from 18 to 27%. The efficacy of these treatments further increased when applied in equal split doses.

Application of aldicarb at 2 kg *a.i.*/ha as spot treatment was the most economical dose which ensured reasonable level of nematode control while increasing yields (14 to 219% in summer and 15 to 115% in autumn), followed by carbofuran (18 to 206% in summer and 10 to 115% in autumn) and phenamiphos (15 to 105% in summer and 10 to 90% in autumn). The least build up of cyst nematode population was noticed in aldicarb and carbofuran (Rf = 0.5 to 1.2) followed by phenamiphos (Rf = 1.5 to 1.9) and control (Rf = 4.7 to 9.2).

(e) Biological Methods: Application of *Pseudomonas fluorescens* at 10 g/m^2 gives good control (Mani *et al.*, 1998). Application of neem cake at 1 t/ha or FYM at 20 t/ha or compost at 10 t/ha enriched with *Trichoderma harzianum*/*Paecilomyces lilacinus* gave effective control of cyst nematodes.

(f) Host Resistance: In the United States, both USDA and Cornell University scientists have evolved two commercial potato varieties – Peconic and Wauseon – which are resistant to the cyst nematode.

In India, Potato variety Kufri Swarna developed by the Central Potato Research Institute Substation at Ooty is resistant to both the species of cyst nematode and the late blight disease. The variety is also draught resistant and has shorter dormancy which is advantageous to Nilgiris region where potato is taken up mainly as a rain fed crop almost throughout the year and thus covers about 40% of potato area (Anon,1985).

(g) Integrated Methods: The combination of disease escape (by planting early maturing varieties) and the use of nematicides gave good control of the potato cyst nematode. Hygiene in the form of seed certification, combined with crop rotation in seed growing areas is effective. Use of a resistant variety followed by a nematicide would kill 99% of the nematode population. The potato cyst nematode problem in Nilgiris is being managed by chemical treatments, crop rotations and utilizing the available sources of resistance (Kufri Swarna). The most effective management combines crop rotation, use of nematicides and resistant varieties to keep the nematode at an economically acceptable level (Jones, 1969) (Table 5.9).

Table 5.9. Integrated management of the potato cyst nematode

Sl. No.	Management method(s)	Resulting kill population (% initial popn)	Kill (%)	Popn. after growing & harvesting a susc. *cv.*, calculated at 2 assumed nematode multiplication rates (% initial population)	
				30x	70x
1.	4 years without potato	3	97	90	< 100
2.	1 year with resistant Potato	20	80	> 100	> 100
3.	Nematicide treatment (Fumigant)	25	75	> 100	> 100
4.	1 and 2	0.6	99.4	18	42
5.	1 and 3	0.75	99.25	22.5	52.5
6.	2 and 3	5	95	> 100	> 100
7.	All 3 methods	0.15	99.85	4.5	10.5

Better control of *G. rostochiensis* by *P. penetrans* was possible with soil solarization combined with application of FYM or karanj cake which resulted in greater kill of nematodes (Sitaramaiah & Naidu, 2003).

A well known integration of methods can control *G. rostochiensis* in potato in a 4-year rotation combining soil fumigation (causes 30% reduction in nematode population), cultivation of non-host (50% reduction), cultivation of resistant potato variety (30% reduction), cultivation of non-host (50% reduction) and then cultivation of susceptible variety of potato (Oostenbrink, 1972).

5.1.3.3. Rot Nematode, *Ditylenchus destructor*

D. destructor is the most important pest of potato tubers and is responsible for dry rot of tubers. High yield losses occur in the areas where climatological conditions favour establishment of the potato rot nematodes. The effect of nematodes will manifest itself at harvest or storage when infected tubers will rot.

(i) Symptoms: *D. destructor* enters potato tubers through lenticels and initially causes small white mealy spots just below the surface that are only visible if the skin is removed. Infested areas enlarge and coalesce and light brown lesions, consisting of dry granular tissue, may be visible beneath the skin. As the infestation progresses, the tissues dry and shrink and the skin becomes cracked and papery (Fig 5.17). Internal tissues gradually darken and there are often secondary invasions of fungi, bacteria, mites, etc. If stored in moist conditions, a general rot may ensue and spread to neighbouring tubers.

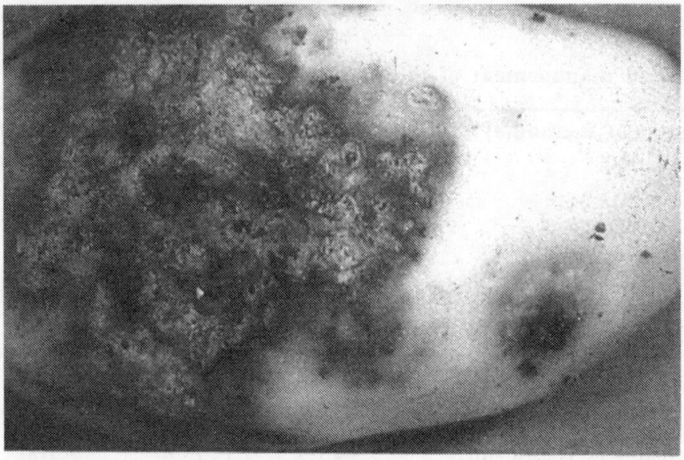

Figure 5.17. Dry rot of potato tuber incited by *Ditylenchus destructor* (Courtesy: Sasser, 1971)

(ii) Life Cycle: The life cycle is very similar to *D. dipsaci*. *D. destructor* does not seem to have a resistant stage, as it does not form 'nematode wool' or readily withstand desiccation but apparently survives in soil on alternative weed and fungal hosts. Thorne (1961) suggests that it may over winter as eggs.

(iii) Survival and Spread: *D. destructor* has a wide host range, can survive on weeds and on a wide range of soil-inhabiting fungi. It can also survive on infected tubers left in the field. Spread occurs by introduction of infected tubers and in soil adhering to seed pieces. Irrigation water and cultivation by infested farm tools and machinery are other sources of inoculum dissemination.

(iv) Management

(a) Cultural Methods: The use of healthy seeds, destruction of infected plant parts in the field, control of weed hosts, late planting when soil moisture is inadequate for optimum invasion and early harvest reduces losses. Two to three years rotation of potato with beans, corn and fallow kept free of weeds gave a very marked reduction in nematode populations. Satisfactory control of *D. destructor* can also be obtained by crop rotation with small grains, vetch and lupine provided potatoes are grown once in 3–4 years. Rotation of potatoes with sugar beet and other non-host crops can reduce nematode populations.

The application of ammonium nitrate and ammonium sulphate (at 144 and 118 kg/ha, respectively) while planting potato reduced infection from 7% to 0.2–0.5%, and raised the yield from 10.6 t/ha to 16.3 and 25.2 t/ha, respectively.

(b) Chemical Methods: Darling (1959) found that heavily infested land treated with ethylene dibromide at 35 litres/ha in June followed by 17.5 litres/ha in August, gave clean potato crops in the following two years.

(c) Integrated Methods: The control of potato rot nematode was achieved by the combination of disease escape, hygiene, in the form of seed certification and crop rotation as follows (Winslow & Willis, 1972):

- Healthy seed, planted late, harvested early and stored as cool and dry as possible.
- Proper rotation of potatoes with non-host crops and growing of potatoes not more frequently than once in 3 or 4 years.
- Field hygiene in the form of removal of old infested tubers and weed control.

5.1.3.4. Lesion Nematode, *Pratylenchus penetrans*

(i) Symptoms: High populations of lesion nematodes cause areas of poor growth; plants are less vigorous, turn yellow and cease to grow. Damage is often caused by direct feeding and, usually, only cortical tissues are affected. Large nematode populations cause extensive lesion formation and cortex destruction of unsuberized feeder roots (Mai *et al.*, 1981). Tubers are often attacked and small lesions are formed on the surface.

(ii) Survival and Spread: Infected tubers are sources of nematode inoculum and aid in survival of the nematodes.

(iii) Management

(a) Physical Methods: Hot water treatment of infected tubers at 50°C for 45 to 60 min aid in reducing nematode spread.

(b) Cultural Methods: In areas infested by *P. penetrans*, beet should precede the susceptible potato in the crop rotation system. The cultivation of legumes should be avoided since nematodes breed well in them.

(c) Host Resistance: Potato *cv*. Russet Burbank is tolerant, while Peconic appears to be less susceptible to the lesion nematode.

5.1.3.5. Other Nematodes, *Quinisulcius capitatus* and *Helicotylenchus dihystera*

(i) Symptoms: The pathogenicity of the stunt nematode, *Quinisulcius capitatus* frequently occurring in hill tracts was established on potato *cv*. 'Kufri Jyoti' at Shimla (Prasad & Sharma, 1985). The nematode build up ranged from 5 to 8 times at initial inoculum levels of 10 to 1000 at 45 days. Besides, the plant growth parameters were also reduced which affected the tuber production. The tuber reduction ranged from 14 to 29% and was negatively correlated with nematode build up.

The spiral nematode, *Helicotylenchus dihystera* was also pathogenic to potato accounting for 9–27% reduction in yield in 90 days with 2–4 times nematode build up (Prasad, 1990).

5.1.3.6. Root-knot Nematode, *Meloidogyne* spp. and Wilt, *Verticillium dahliae* Disease Complex

The root-knot nematodes often interact with *V. dahliae* in development of disease complex in potato.

(i) Management

(a) Cultural Methods: In fields infested with root-knot nematodes and *V. dahliae*, Scholte (1990) found that a five-year rotation of maize-sugar beet-barley-barley-potato had significantly higher potato yields than continuous potato.

5.1.3.7. Lesion Nematode, *Pratylenchus penetrans* and Wilt, *Verticillium dahliae* Disease Complex

(i) Management

(a) Chemical Methods: Methylisothiocyanate has been successfully used for soil fumigation to reduce the populations of soil-borne fungus, *V. dahliae* and alleviate the effects of disease complex involving wilt-inducing fungus (*V. dahliae*) and the lesion nematode (*P. penetrans*) on potato (Rowe *et al*., 1987).

5.1.3.8. Cyst Nematode, *Globodera rostochiensis* and Wilt, *Verticillium dahliae* Disease Complex

The association of fungi like *Rhizoctonia solani* and *V. dahliae* with *G. rostochiensis* has been reported on potato indicating that the cyst nematode is involved in the disease complex.

(i) Management

(a) Chemical Methods: Soil fumigation with methyl bromide resulted in yield increases of 68% for potato fields infested with *Globodera* spp. and *V. dahliae* (Hide & Corbett, 1974).

In *V. dahliae-G. rostochiensis* disease complex on potato, a mixture of methyl bromide and chloropicrin was found to have comparatively more fungicidal and nematicidal efficacy (Hide & Corbett, 1974).

5.2. YAM

Yam (*Dioscorea* spp.) belongs to the family Dioscoreaceae, which is widespread throughout the tropics and subtropics. Normally yams are consumed as boiled, baked or fried vegetables. They contain 18-20% starch with mucilaginous substance. Yam also contains alkaloids, tannins and steroids which have pharmaceutical value. *Dioscorea alata* (greater yam), *D. rotundata* (white or white Guinea yam), *D. cayensis* (yellow Guinea yam) and *D. esculenta* (lesser yam) produce edible tubers. *D. alata* was first cultivated in Southeast Asia and is widely grown in Asia and the Pacific, is less favoured in Africa, but much grown in the Caribbean. *D. rotundata* is the most important species in W. Africa, where it is originated. *D. cayenensis* is a native of W. Africa, where it is extensively cultivated.

D. esculenta originated in Indo-China and is extensively grown in Southeast Asia and elsewhere in the tropics.

In India, greater yam and lesser yam are under cultivation. Madhya Pradesh, north-eastern states, West Bengal, Bihar, Orissa, Uttar Pradesh, Kerala, Tamil Nadu, Gujarat and Maharashtra are major yam-growing states.

5.2.1. Pests

5.2.1.1. Scale, *Aspidiella hartii*

(i) Distribution: It is widespread in distribution and present in Old World Tropics and Pacific.

(ii) Damage: Adult and nymphal scale insects (whitish yellow in colour) are present on the tubers and sometimes on the aerial portions of.the plant which do not thrive. Stored tubers may be thickly covered by scales and shrivel. The problem is perpetuated by the use of infested tubers as planting material.

(iii) Management

(a) Cultural Methods: Scale insects on tubers may be controlled by using clean planting material. Yams should not be planted after ginger, as this crop also supports the pest.

(b) Chemical Methods: Dipping the tubers in 0.3% aldrin or in a mixture of 5.6 ml diazinon 20% EC and 425 ml white oil in 13.5 litres of water before planting has been recommended. Fumigation of tubers with carbon disulphide has also been suggested.

5.2.1.2. Spiraling Whitefly, *Aleeurodicus dispersus*

(i) Management

The field release of endoparasitoid, *Encarsia* sp. gave 69.53% parasitization during August and 76.22% during September. The predators like *Cybocephalus* sp., *Cryptolaemus montrouzieri* and *Mallada astur* can also be effectively used for the biocontrol of spiraling whitefly.

5.2.2. Diseases

5.2.2.1. Anthracnose, *Glomerella cingulata*

(i) Distribution: The disease is distributed in Caribbean, East Africa, Far East, India and West Africa.

(ii) Symptoms: In *Dioscorea alata*, the first signs of infection are small brown spots (2–5 mm in diameter), sometimes with a yellow edge, which appear mainly on the lower leaves (Fig 5.18) and stems. The spots enlarge and spread giving the stem a blackened appearance and causing leaves to wither. Brown to black microscopic acervuli (fruiting bodies) may be observed on the lesions of the leaves and stem, particularly during the rains, giving the appearance of dark concentric rings on the lesions. Sometimes secondary infection occurs and the leaf is destroyed.

Figure 5.18. Anthracnose on yam leaves

Stems are also attacked, but less frequently. If the infection reaches stele, death of the plant can result. Attacks on the stem generally occur as a result of direct penetration near the terminal bud which leads to rapid death of the plant. Direct infection can also occur at ground level.

The reduction in yield caused by anthracnose can be particularly serious in the more humid regions. *D. alata* is more susceptible but *D. esculenta* is more resistant. *D. cayenensis* and *D. rotundata* are only slightly susceptible, even in humid forest regions, their leaves only develop a few brown spots and stems are not attacked.

(iii) Development and Spread: The pathogen overseasons on crop debris as acervuli. Infection is favoured by wet and humid conditions. Many acervuli of the *Gloeosporium* type which become erumpent later are produced on lesions. Acervuli remain devoid of setae when humidity is over 95% but at lower humidity sterile setae mixed with conidiophores convert the acervuli to the *Colletotrichum* type. In very dry conditions, only setae are formed. Spores are mainly dispersed by rain splash and the disease is therefore most severe when there are heavy rains.

(iv) Management

Crop debris should be removed as it is a major source of inoculum. Spraying maneb at 3.32 g *a.i.*/litre at 10 days interval is effective. Benomyl alone or applied alternately with mancozeb on a weekly cycle has also been shown to give good control and increase yields. Benomyl + propineb are also effective. Zineb, ferbam and ziram sprays were shown to give 50%, 41% and 30% yield increases, respectively.

5.2.2.2. Leaf Spots, *Cercospora brasiliensis, C. carbonacea, C. pachyderma, Phaeoramularia dioscoreae, Pseudocercospora contraria*

(i) Distribution: The disease is worldwide in distribution on all species of yams.

(ii) Symptoms: Several *Cercospora* spp. and some species which were formerly classified as *Cercospora* are recorded as causal agents of leaf spots on yams. Generally the symptoms are brownish spots which darken with age and coalesce as the disease progresses. *Phaeoramularia dioscoreae* causes small brown spots which sometimes coalesce. *C. carbonacea* is the most commonly recorded causal agent and causes large blackish-brown to brown spots which may spread over the whole leaf.

(iii) Development and Spread: The fungus overseasons on crop debris. Conidia are produced on both surfaces of infected leaves and are spread by rain. The disease is more common during warm and wet weather.

(iv) Management

Spraying of maneb or captan at 2 g/litre of water every 10 days was found effective in the management of the disease.

5.2.2.3. Tuber Rot, *Fusarium oxysporum, F. solani*

(i) Distribution: The disease is distributed in China, India, Ivory Coast, Japan, Nigeria and Puerto Rico.

(ii) Symptoms: Nematode infestation facilitates the entry of *Fusarium* spp. into the tubers. Infected tissues develop a brown coloured dry rot. Often the infection is not noticed until after harvest or until the tubers are cut opened. If the tubers are cut transversely, zonations of infection can be seen. Brown dry rot caused by *Fusarium* occurs centrally and there may be a blackish zone round this where there is considerable damage and secondary bacterial infection.

(iii) Management

Cut tubers used as planting material should be allowed to dry in order to prevent the entry of the pathogen. The incidence of infection can be greatly reduced by treating the cut surfaces with Bordeaux mixture or coating with lime, chalk or wood ashes. Crop rotation helps to reduce the disease incidence. Also avoid planting yams in nematode infested soil.

Dipping the tubers in systemic fungicides such as benomyl and thiabendazole has also been found effective. A 5 min dip in 3000 ppm benomyl followed by dusting the cut ends with 0.3% benomyl in talcum powder effectively controlled tuber rot. A 30 min dip in 1000 ppm benomyl or thiabendazole gave 80 to 90% control of pre-harvest rotting in the field.

The set treatment (30 g/litre for 15 min) and soil application (1 kg/ha) of *Trichoderma viride* recorded the lowest tuber rot incidence (7.33%) and highest harvest index (66.9%).

5.2.3. Nematodes

The most important nematode problem is the dry rot nematode. Lesion and root-knot nematodes are also known to cause root rotting in yam crop.

5.2.3.1. Yam Nematode, *Scutellonema bradys*

The most important nematode problem on yams is the dry rot of tubers caused by *S. bradys*. Nadakal and Thomas (1967) reported the occurrence of dry rot of *Dioscorea alata* and the association of *S. bradys* from Kerala. *Dioscorea alata, D. cayenensis, D. esculenta* and *D. rotundata* are good hosts of the yam nematode. Sesame and cowpea support high root populations and melon can increase soil populations.

(i) Economic Importance: Dry rot of yams alone causes a marked reduction in the quality, marketable value and edible portions of tubers and these reductions are more severe in stored yams. When dry rot is followed by wet rot in stored yams, losses of whole tubers can be as high as 80–100%. Weight differences between healthy and diseased tubers harvested from the field have been estimated to be 20–30%.

(ii) Symptoms: *S. bradys* causes a disease of tubers commonly known as 'dry rot'. In early stages of infection, small yellow lesions can be observed just below the skin; these become dark brown and eventually form a continuous dark, dry rot layer just beneath the surface in heavily infested tubers. Outwardly the infected tubers have cracks in the skin, which are often deep, can be malformed, or the skin flakes off exposing dark brown, dry rot tissue underneath. *S. bradys* reproduces and builds up large populations in stored yam tubers and causes severe damage during storage.

Nematode damage to the tubers results in considerable reduction in the edible portion and marketable value of the tubers. Moisture loss is also significantly greater from diseased tubers during storage.

(iii) Life Cycle: *S. bradys* is an endoparasite of plant roots and yam tubers. The nematodes attack the primary roots growing from the rhizomatous head of the tuber and can enter the developing tubers through their growing points. The life-cycle is simple. Eggs are laid in soil or plant tissues (roots and tubers) where they hatch and the juveniles develop into adults by subsequent molting. All stages seem to be infective.

(iv) Histopathology: The nematodes were found in the intercellular spaces of the tissues lying within the periderm where extensive cell destruction was noticed.

(v) Survival and Spread: Sizeable populations of the nematode are maintained in the absence of yams probably on other host plants. Yams are propagated from whole tubers or pieces of tuber which are the principal means of spread of *S. bradys*. Infested seed tubers rather than soil are probably the main source of nematode inoculum in yam fields.

(vi) Interaction with Other Pathogens: The more extensive internal decay of tubers known as "wet rot", 'soft rot" or "watery rot" is associated with fungal (*Botryodiplodia theobromae* and *Fusarium* spp.) and bacterial (*Erwinia* sp.) pathogens. This general decay of tubers, which is a serious problem in stored yams, is increased when tubers are wounded or damaged. The damage caused by nematodes can predispose the tubers to invasion by decay organisms resulting in complete rotting of the tubers.

(vii) Management

(a) Physical Methods: Treatment of yam tubers at 50–55°C for 40 min has given over a 98% kill of *S. bradys* without damage to *D. rotundata* tubers.

(b) Cultural Methods: To check the spread of *S. bradys* in yams, Decker *et al.* (1967) suggested use of healthy seed tubers. The bottom or distal portions of the tubers have the least nematodes and can be selected for planting. Mixing cow dung in yam mounds before planting at 1.5 kg/mound (1886.3 kg/ha) can increase yields of tubers and significantly decrease nematode numbers.

Nematodes will decline in the absence of host plants and a fallow period of several months has been suggested to control *S. bradys* in *D. alata* tubers.

Soil populations of *S. bradys* will be reduced if a non-host or poor-host crop, such as, peanut, chilli pepper, tobacco, Indian spinach, cotton, maize or sorghum is grown prior to yams.

NPK fertilizers can reduce *S. bradys* populations in tubers of *D. alata* to a very low level. N alone can increase both populations of *S. bradys* and the percentage of infested tubers of *D. rotundata*, whereas P alone can reduce percentage of infested tubers. These results support observations by farmers in certain yam growing areas of Nigeria that yams fertilized with N alone do not store well, but yams fertilized with mixtures that contain P store longer.

(c) Chemical Methods: Bare root-dip treatment of diseased yams in fensulfothion (Dasanit), thionazin (Nemafos) and DBCP (Nemagon) at 1250 ppm for 15 min and 625 ppm for 30–60 min was found effective in eliminating the nematode infection (Ayala & Acosta, 1971). Significant increases in yield have been obtained by soaking tuber pieces of *D. alata* infected with *S. bradys* for 30 min in 1000 ppm aqueous solutions of carbofuran and oxamyl.

Application of aldicarb, oxamyl, carbofuran and miral or isazophos each at 2 kg *a.i.*/ha two weeks after planting reduced soil populations of *S. bradys* to very low levels with remarkable yield increases recorded.

(d) Host Resistance: Yam *cv.* Florido of *Dioscorea alata* did not seem to be susceptible to nematode attack (Ayala & Acosta, 1971).

5.2.3.2. Lesion Nematode, *Pratylenchus coffeae*

P. coffeae also attack yams and cause dry rot symptoms.

(i) Symptoms: Symptoms of damage to yam tubers by *P. coffeae* are similar to those caused by *S. bradys*. The nematodes are restricted to the outer layers of tuber tissue, cause severe necrosis or dry rot, cracking of the skin, and are associated with complete deterioration of tubers.

(ii) Life Cycle: *P. coffeae* is a migratory endoparasite of yam roots and tubers. The life cycle is completed in 3–4 weeks on yams.

Host Range: *P. coffeae* is a parasite of *D. alata, D. cayenensis, D. rotundata, D. bulbifera* and *D. trifida*. In addition to yams, *P. coffeae* has an enormous host range covering almost all plant families.

(iii) Management

(a) Physical Methods: Immersion of *D. rotundata* tubers in hot water maintained at 46–52°C for 15–30 min. has given good control of *P. coffeae*.

(b) Cultural Methods: Using plant material which is free of nematodes is an effective means of controlling or reducing damage by *P. coffeae*. Collection of central or distal tuber pieces, which generally contain least *P. coffeae*, are recommended for propagative material.

(c) Chemical Methods: Immersion of *D. rotundata* tubers in DBCP and fensulfothion at 1250 ppm for 15–30 min effectively reduced *P. coffeae* present to a depth of 6 mm in the tuber tissues.

Aldicarb as a single application at planting at 5.4 kg *a.i.*/ha can give 72% control of *P. coffeae* and significantly increase high quality tuber yields of *D. rotundata*. Significant increases in yield of *D. rotundata* have also been obtained by a combination of foliar and seed tuber treatments with oxamyl.

(d) Host Resistance: *D. alata cv.* Florido is not susceptible to attack by *P. coffeae* (or *S. bradys*). *D. esculenta* is possibly less susceptible to *P. coffeae* because of its different growth habit.

5.2.3.3. Root-Knot Nematodes, *Meloidogyne* spp.

The root-knot nematode is another important nematode occurring on yams and is present world wide. Yams have been found infested by *M. incognita, M. javanica, M. arenaria* and *M. hapla. M. javanica* populations of 30,000 nematodes/plant can reduce yields of *D. opposita* by over 50%. It is estimated that there is a reduction of 39–52% in the price of galled tubers compared to healthy ones.

(i) Symptoms: Yam seedlings infested with root-knot nematodes can be severely stunted and chlorotic; very young seedlings may be killed by severe infestation. The roots swell in size at the site of infection producing the characteristic and typical knots on the roots. Tuber formation was severely affected, often infested tubers are smaller and most of the tuber forming roots does not develop. Rotting of tubers has been reported.

(ii) Life Cycle: The life cycle of *M. incognita* in *D. rotundata* or *D. alata* tubers is 35 days. In *D. alata*, most nematodes are concentrated to a depth of 2 mm with none beyond the 8 mm depth. In *D. rotundata* they are concentrated at depths between 4 to 6 mm with few at 14 mm.

(iii) Host Range: Susceptible yam hosts of *M. incognita* are *D. alata, D. cayenensis, D. esculenta, D. bulbifera, D. composite, D. floribunda, D. trifida, D. praehensilis, D. spiculiflora* and *D. rotundata*. Hosts of *M. javanica* are *D. alata* and *D. rotundata*.

(iv) Survival and Spread: Where *Meloidogyne* juveniles and/or eggs survive in stored tubers, they will be spread in propagative material. However, *Meloidogyne* spp. have extremely wide host ranges and damaging populations will come from field soil having survived on other weed hosts, or be introduced into yam fields on infested seedlings of other crops.

(v) Management

(a) Physical Methods: Hot water treatment of corms at 51°C for 30 min eliminates root-knot nematode infection.

(b) Cultural Methods: The root-knot nematode can be controlled by growing sweet potato *cv*. Shree Bhadra as a trap crop.

(c) Chemical Methods: Application of carbofuran at 3 kg *a.i.*/ha gave effective control of root-knot nematodes. Granular oxamyl at 3 or 6 kg *a.i.*/ha applied at planting and at 3, 4-week intervals can control *M. javanica* on *D. rotundata*. In the presence of both *M. javanica* and *Pratylenchus brachyurus*, tuber yields can be increased by over 40% when granular oxamyl at 3 kg *a.i.*/ha applied at planting is combined with subsequent applications of calcium nitrate or ammonium sulphate incorporated at 3, 4-week intervals, each at 60 kg N. These treatments also reduce the incidence of rot in stored yams associated with nematodes.

(d) Host Resistance: The only yam species consistently found to be resistant to attack by *M. incognita* is the cluster yam, *D. dumentorum*. *D. alata cv*. Obunenyi is reported to be resistant to *M. incognita* and *D. cayenensis* can be resistant to *M. incognita* and *M. javanica*. Yam *cvs*. Sree Latha and Sree Kirthi were found resistant to root-knot nematodes.

5.3. SWEET POTATO

Sweet potato (*Ipomea batatas*) is an economically important member of the family Convolvulaceae and probably originated in Central America, where it has long been cultivated. It is now grown extensively throughout the tropics and subtropics. It is one of the life sustaining crops rich in carbohydrates that stand between man and starvation. Sweet potato ranks fourth and sixth on the list of dry matter production per hectare and edible energy production per hectare, respectively. It is also used for making starch, syrup and alcohol. It occupies a prime place in terms of calories production/unit area and time. It is grown as a starchy food crop throughout the tropical, sub-tropical and frost-free temperate climatic zone in the world. It is being cultivated in 0.129 million hectares producing 1.148 million tonnes of sweet potatoes with an average yield of 8.87 tonnes per hectare (Table 5.10). India ranks sixth in area but the productivity is very low (8.9 t/ha), lower than the world productivity level. Orissa, Uttar Pradesh, West Bengal, Maharashtra, Karnataka, Chhattisgarh, Tamil Nadu, Assam, Madhya Pradesh, and Meghalaya are the leading states in sweet potato cultivation.

Table 5.10. Area, production and productivity of leading sweet potato growing states in India (2005–06)

State	Area (in '000 ha)	Production (in '000 MT)	Productivity (in MT/ha)
Orissa	47.1	394.3	8.4
Uttar Pradesh	20.0	240.8	12.1
West Bengal	25.8	208.0	8.1
Maharashtra	5.0	70.0	14.0
Karnataka	3.6	42.1	11.6
Chhattisgarh	3.0	38.2	12.6
Tamil Nadu	1.4	29.6	21.1
Assam	8.5	29.5	3.5
Madhya Pradesh	4.7	28.2	6.0
Meghalaya	5.4	18.9	3.5
Others	4.9	49.1	10.0
Total	**129.4**	**1148.6**	**8.87**

(Source: Indian Horticulture Database 2006)

5.3.1. Pests

The most common pest of sweet potato is weevil, *Cylas formicarius* which reduces tuber yield and quality, although there are simple cultural techniques that reduce the losses due to this pest. Young plants are susceptible to foliage feeders, such as hornworm and it may be necessary to treat the unestablished plants with insecticide. *Cylas* spp. and *Alicodes* spp. continue to damage the tubers in storage.

5.3.1.1. Leaf Folder, *Brachmia* spp.

(i) Distribution: *Brachmia convolvuli* is distributed in Canary Island, Mauritius, Comoro Island, Southeast Asia and Jawa. *Brachmia macroscopa* has been recorded in India.

(ii) Symptoms: Cream and black larvae spin webs close to the main veins of the leaves, on the underside, while older lager larvae spin webs across the upper surface of the leaf and draw the two sides together. The leaves eventually dry up, die and drop off.

(iii) Biology: *Brachmia convolvuli* eggs are laid on the lower surface of leaves and hatch after 3 days. There are 5 larval instars. The mature larva is 11–12 mm long and cream coloured with dorsal black bands on segments 8–10 of the abdomen; the head, prothoracic shield and all of abdominal segment 5 are black. The larvae pupate for 7 days in the folded leaf. Adults of *B. convolvuli* are 4 mm long with grayish brown forewings, marked with white hind wings fringed with grey. The entire life cycle, egg to adult takes 24 days.

(iv) Management

Sprays of 0.2% carbaryl or 0.025% fenitrothion are recommended in India for the control of *B. convolvuli*.

5.3.1.2. Hornworm, *Agrius convolvuli*

(i) Distribution: Hornworm is distributed in most of Africa, Bangladesh, Burma, South China, Southern Europe, India, Indonesia, Iran, Malaysia, New Zealand, Pacific Islands and Papua New Guinea.

(ii) Symptoms: Leaves are eaten by hornworms.

(iii) Biology: Spherical eggs (1 mm in diameter) are laid singly on the host plant, usually on the upper leaf surface. The larvae, which concentrate on the young shoots are variable in colour from green to brown and distinctively patterned. There are 5 larval instars and larvae mature after 3–4 weeks and reach a length of about 95 mm. Pupation takes place in the soil and may take 3 weeks or several months depending upon the climate. The adult hawk moths (wing span 80–120 mm) are grey with black lines on the wings; the abdomen is marked laterally with pink bands. They feed upon nectar in flowers.

(iv) Management

(a) Cultural Methods: Ploughing the land between crops to expose the pupae reduces infestation. Hand picking of the larvae may be quite effective in small areas.

(b) Chemical Methods: A 0.2% solution of carbaryl in a high volume spray, 0.05% endosulfan or trichlorphon (7 days waiting period before harvest) at 100 g *a.i.* in 100 litres of water should be applied when the larvae are first seen. Early instars can be controlled more easily than fifth instars. Particular care should be taken to spray the young leaves.

5.3.1.3. Bihar Hairy Caterpillar, *Diacrisia obliqua*

(i) Distribution: Bihar hairy caterpillar is distributed in South and East Asia.

(ii) Symptoms: Numerous larvae covered in yellow hairs skeletonise the leaves. The entire vines may be defoliated.

(iii) Biology: Eggs are laid on the leaves in rows or clumps and after 4–9 days the larvae emerge. Mature larvae are 30–40 mm long and 7 orange bands across the body on which grow tufts of yellow hairs. At first, the larvae feed in groups but when they are about a week old, they disperse to new feeding sites. The larval period lasts 28–34 days. Pupation takes place in the soil in reddish brown cocoons and may take 51 days.

(iv) Management

Endosulfan and fenitrothion both at 0.05% are recommended for the management of the pest. Diazinon or dichlorvos or carbaryl at 0.2% are also suggested for the control of larvae.

5.3.1.4. Leaf Miners, *Bedellia* spp.

(i) Distribution: *Bedellia ipomoeae* from Fiji, *B. somnulentella* from Papua New Guinea and Queensland, and a number of *Bedellia* spp. from Africa have been reported attacking sweet potato.

(ii) Symptoms: Narrow mines widening to blotches appear on the leaves. The plants have a withered appearance and growth is reduced.

(iii) Biology: Eggs are laid on the leaf. The emerging larva is distinctly segmented with a rather pointed head and abdomen. Larvae may be red or yellowish grey in colour. After making several mines, they pupate on a web of silken threads among the leaves. When moving outside the mines, the larvae travel by a looping movement. The life cycle can be completed within 4 weeks. Adult *Bedellia* spp. are small (3 mm long with a wing span of 8 mm) tan to grey moths.

(iv) Management

Generally *Bedellia* spp. are controlled by predators, but these may be killed by foliar sprays of insecticide. A serious breakdown can cut down the effective leaf surface and lead to reduced yields of tubers. Carbaryl at 1 kg *a.i.*/ha, chlorfenvinphos at 0.6 to 0.9 kg/ha, diazinon at 55 ml/ha, dimethoate at 14 ml/ha, endosulfan at 70 ml/ha, malathion at 14 g/ha or trichlorphon at 100 g/ha have been recommended for the management of the pest.

5.3.1.5. Tortoise Beetles, *Aspidomorpha* spp., *Conchyloctaenia punctata*

(i) Distribution: *Conchyloctaenia punctata* from South Africa; *Aspidomorpha adheares, A. miliaris, A. quadriradiata* and *A. socia* from Papua New Guinea; *A. miliaris* from India and *A. dissentanea* from Angola, Cameroon and Guinea have been reported attacking sweet potato.

(ii) Symptoms: Adults and larvae feed on the leaves making large round holes and eventually skeletonising the leaves.

(iii) Biology: The eggs are laid on the underside of the leaves in batches cemented to the leaf. The larvae are flattened, spiny and usually carry excreta and the skins of earlier instars on anal projections. Adult tortoise beetles (Sub-family Cassidinae) are broadly oval in shape and hence the common name. They are bright, sometimes patterned or metallic in colour.

(iv) Management

(a) Cultural Methods: Weeds of the family Covolvulaceae should be controlled as tortoise beetles can breed on them.

(b) Chemical Methods: Carbaryl (0.2%) at 1000 litres/ha is recommended in India and Papua New Guinea, while endosulfan at 500–750 g *a.i.*/ha elsewhere.

5.3.1.6. Weevil, *Cylas* spp.

(i) Distribution: *Cylas formicarius* from Pan Tropical, *C. punctcollis* from Africa (countries south of Sahara) and *C. brunneus* from West Africa have been recorded on sweet potato.

(ii) Symptoms; Tubers and vines are tunneled by small white, legless larvae. The tunnels may be partially filled with frass. Ant-like weevils, either having glassy blue black elytra, reddish brown legs and thorax and a black head (*C. formicarius*) or entirely shiny black (*C. punctcollis* or *C. brunneus*) may be found on the leaves or in the tunnels. Yield, storage life and plant vigour are reduced.

(iii) Biology: The total length of development is 4–5 weeks. Eggs are laid in holes in the vines or tubers and hatching takes place within a week if the temperature is between 20 and 30°C. The larvae tunnel in the vines and tubers for about 14 days. When fully developed (fourth instar) the larvae are 3 mm in diameter. Pupation takes place inside the plant and lasts about one week.

(iv) Management

(a) Cultural Methods: Sweet potatoes should be planted only in fields that have had no *Cylas* infestation within the last 12 months and preferably more than 1 km away from any infested land, and only *Cylas*-free planting material should be used. Earthing-up of the plants, particularly those varieties with a tendency to push out of the ground reduces attack. After harvest, the field should be cleared of trash, which should be burnt or fed to livestock.

Crop rotation (Paddy – Sweet potato – Cowpea) can minimize weevil damage. Boemeryl acetate, a kairamone present in the periderm of sweet potato (Coleus) can attract both male and female weevils which offers immense scope for the management of weevils.

(b) Chemical Methods: Some of the chemicals recommended for the management of weevils are presented in Table 5.11.

Table 5.11 Recommendations for chemical control of weevils

Insecticide	Dose/Conc.	Application method
Carbaryl	0.1%	3 times at 4 weekly intervals.
Carbofuran	1.4 kg/ha	Soil application around vine cutting at planting.
Carbophenothion	0.035%	——
Chlorfenvinphos dust	600–900 g/ha	30 days waiting period before harvest.
Chlorfenvinphos EC	240 g/ha	——
Diazinon	0.3%	Dip for planting material or foliar spray.
Dieldrin	1 kg/ha	4 times at weekly intervals beginning 40 days after planting; five weeks waiting period.
Disulfoton	1.8 kg/ha	4 weeks after planting.
Endosulfan EC	2.25 kg/ha	Application at planting time.
Endosulfan granules	2.25 kg/ha	Soil application at planting.
Fenitrothion	0.05%	——
Fenthion	0.1%	3 times at 4 weeks intervals
Malathion EC	28.5 ml	Dip for cuttings or foliar spray.
Mephosfolan	1.4 kg/ha	Soil application around cutting at planting.

Table. 5.11. (Contd.)

Insecticide	Dose/Conc.	Application method
Methomyl	1 kg/ha	Four times at 2 week intervals beginning 40 days after planting.
Methyl-demeton	0.05%	Spray at beginning of 2nd, 3rd and 4th month of growth.
Monocrotophos	0.05%	4 times at monthly intervals.
Parathion	—	
Phorate	1.4–2.0 kg/ha	Soil application around cutting at planting and 4 weeks after planting.
Phosphamidon	0.05%	Spray at beginning of 2nd, 3rd and 4th month of growth.

(c) Host Resistance: Deep rooting varieties with long necks between the tubers and the stems being less susceptible because of adult weevil cannot burrow downwards more than 1 cm. Early maturing varieties escape serious damage.

5.3.1.7. Striped Weevil, *Alcidodes* spp.

(i) Distribution: *Alcidodes dentipes* from tropical Africa, *A. erroneous* from Kenya and *A. fabricii* from India have been recorded attacking sweet potato.

(ii) Symptoms: *Alcidodes dentipes* and *A. erroneous* attack both stems and tubers. Adults girdle the stems causing wilting, while larvae bore into stems and tubers making tunnels slightly larger than those of *Cylas* larvae and causing galls to form on the stems. *A. fabricii* has been recorded breeding inside the seed capsules of sweet potato.

(iii) Biology: The life cycle of the weevils is similar to that of *Cylas*. *Alcidodes dentipes* is about 14 mm long and brown with 4 white streaks along its elytra. *A. erroneous* is similar in size but brown-black in colour with an irregular white patch on each elytrum. *A. fabricii* is reddish brown with four longitudinal white stripes on the elytra. The eggs are laid at the base of the fruit; the larvae bore into the fruit and feed on the seeds. Pupation takes place within the fruit.

(iv) Management

The methods outlined for *Cylas* should control *Alcidodes* spp.

5.3.1.8. Stem or Vine Borer, *Omphisa anastomosalis* (Lepidoptera: Pyraustidae)

(i) Distribution: Generally widespread in Southeast Asia; China, India, Japan and Sri Lanka.

(ii) Symptoms: Stems wilt and die and black-headed yellowish-white larvae, marked with brown tubercles can be found boring in the stems. Frass can be found close to the attacked stems. These larvae may also bore into the tubers.

(iii) Biology: The flattened yellow, oval eggs are laid in groups of 2 or 3, usually on the stem and occasionally on the leaves. The leaves emerge after an incubation of 5–8 days and eventually reach 30 mm in length. They tunnel in the stem for 28–50 days and pupate in a thin cocoon within the tunnel, close to the tubers. Adults emerge after about 14 days. The moth is 15 mm long with a wing span of 33 mm, with a reddish brown body and reddish brown markings on the wings, which are white.

(iv) Management

Crop hygiene and crop rotation reduce infestation.

5.3.2. Diseases

Sweet potatoes are affected by several different diseases in the field, some of which also affect the storage life of the tubers. Virus diseases are probably responsible for the greatest field losses and some fungus diseases can also cause economic losses. The use of clean planting material is the most important disease control measure and crop rotation aids control of many fungal diseases.

5.3.2.1. *Cercospora* Leaf Spot, *Cercospora ipomoeae, Psudocercospora timorensis, Phaeoisariopsis bataticola*

(i) Distribution: *Cercospora ipomoeae* is very common throughout the tropics. *Psudocercospora timorensis* is distributed in Brunei, Fiji, Formosa, Ghana, Guinea, Hong Kong, India, Malaysia, Mauritius, Nepal, New Guinea, New Hebrides, Sabah, St. Lucia, Sterra Leone, Solomon Islands, Sudan, Tanzania and Uganda. *Phaeoisariopsis bataticola* is present in India, USA and Venezuela.

(ii) Symptoms: Leaf spots are small (about 6 mm in diameter) and angular with brownish colouration. Leaves may drop from the lower parts of the stem in severe attacks. Damage to leaves is most severe during wet weather conditions.

(iii) Development and Spread: Conidia are produced abundantly on the leaf spots and are spread by wind and rain.

(iv) Management

Three sprays of 0.25% Dithane M-45 at fortnightly intervals was found effective in reducing the disease intensity (19.89% as compared to 50.78% in control) and in increasing the yield of tubers (21.76 MT/ha as compared to 12.91 MT/ha in control).

5.3.2.2. Wilt/Stem Rot, *Fusarium oxysporum* f. sp. *batatas*

(i) Distribution: This disease appears to occur in temperate regions and found in USA, Japan and Northern India. Stem rot is widespread in USA and 10–50% losses in susceptible varieties may occur. Losses up to 95% have been recorded. In India, the disease is of common occurrence in north and north-east India.

(ii) Symptoms: The fungus causes vascular wilt. The first symptom of the disease is yellowing of the youngest leaves. The vascular system becomes blackened causing stems to darken internally. This darkening may extend several feet among stems. Infected plants wilt, collapse and die. Infected tubers have a blackened ring about 6 mm under the skin and develop a surface rot with shallow, sunken, circular lesions in storage. Plants produced from these tubers are likely to be diseased.

(iii) Development and Spread: The main sources of inoculum are infected tubers and plants. The fungus can live for several years on plant remains in the soil and can spread locally by water, wind, animals, machinery etc. Diseased plants develop as a result of either diseased planting material or healthy plants becoming infected from the soil. Conidia, which are readily dispersed, are produced on the dead leaves and stems.

The incidence of the disease varies with the soil type, which may influence disease severity more than the inoculum density.

(iv) Management

(a) Cultural Methods: The most important means of controlling stem rot is to use healthy planting material and crop rotation. Crop rotation will not eradicate the fungus but it will help to reduce the losses. Tobacco should be avoided in the rotation as some varieties are attacked by the stem rot fungus. Sweet potato should not be grown in the same field more often than once in 3–4 years.

(b) Chemical Methods: The systemic fungicide thiobendazole has shown some promise as a dip treatment of tubers.

(c) Host Resistance: Sweet potato *cvs*. Pusa Safed and Kalmegh have been reported resistant to wilt/stem rot disease.

5.3.2.3. Black Rot, *Ceratocystis fimbriata*

(i) Distribution: The fungus occurs in North and Central America, parts of South America, West, Central and Southern Africa, India, Southeast Asia, Japan and the Pacific.

(ii) Symptoms: Black rot is an important field and storage disease of sweet potato. The fungus infects underground parts of the plant. Infection may extend a short way up the stems causing blackening above the surface of the soil giving rise to the common name 'black shank'. The first symptoms are small, slightly sunken black spots on the underground parts of the plant. The spots enlarge and may cover almost the whole tuber. Fungal spores are often produced in the centre of these areas of infection. If the infected tubers are stored, the rot spreads rapidly causing heavy losses. Infected sweet potatoes have an unpleasant taste when cooked. Plants produced from infected tubers often have the disease.

(iii) Development and Spread: The fungus persists in the soil between sweet potato crops. The sources of inoculum are soil and infected tubers. Plants may become infected in the seedbed or in the field. The disease spreads particularly quickly in stored sweet potatoes.

(iv) Management

(a) Cultural Methods: The best method of control is to use disease-free planting stock. Healthy seed sweet potatoes can be grown from cuttings, which will be free from the disease as only underground parts are affected. Cuttings should be planted in the soil free from black rot pathogen to ensure the production of disease-free roots for field planting. As the fungus persists in soil, sweet potatoes should only be grown once in every 2–3 years.

Sweet potatoes should not be manured with dung from animals that have eaten black rot infected tubers as it may contain spores of the fungus.

(b) Chemical Methods: Tubers to be used for planting can be surface sterilized with fungicides. Thiabendazole has shown promise in trials. When inoculated sweet potatoes were dipped in a suspension of 3.32 g *a.i.*/litre, development of the disease was prevented. The control given by thiabendazole and benomyl dips was greatly improved by raising the temperature of the dips to 44–54°C. Good control of black rot was also obtained with ferbam dips at 44–54°C.

5.3.2.4. Soil Rot, *Streptomyces ipomoeae*

(i) Symptoms: A heavy infection results in dwarfed plants which only produce one or two short vines with small, pale green and thin leaves. These symptoms result from damage to the underground parts of the plant. Lateral roots are attacked and many are destroyed, others are often distorted. The most obvious symptom is the development of pits on the tubers, the first signs of which are water-soaked dark spots.

As the spots on the tubers develop, the overlying skin breaks and the necrotic tissue underneath falls out leaving the characteristic pits with jagged margins. The pits range in diameter from 5–30 mm. The tubers may become girdled by lesions but continue to grow on either side of the constricted infected area.

(ii) Development and Spread: The organism which causes soil rot can survive in the soil for several years and the disease is most severe in soils that are alkaline or only slightly acidic. The causal organism is spread through farm implements, by people and animals and by transport of diseased tubers. Dry seasons and poor soils are conducive to severe losses.

(iii) Management

The most effective method to control this disease is to prevent the transfer of infected materials (soil and tubers) to uninfected fields. Planting material should be healthy. Fields known to be infested with the disease should be used for other crops for longest period possible. Addition of sulphur to the soil as an acidifying agent will reduce the disease incidence.

5.3.3. Nematodes

The root-knot and the reniform nematodes are the two most widely occurring and important parasites of sweet potato throughout the world. *Radopholus similis* caused significant reduction in root weight, tuber weight and tuber length even at a low inoculum level of 100 nematodes/plant (Koshy & Jasy, 1990).

5.3.3.1. Root-Knot Nematodes, *Meloidogyne* spp.

M. incognita is the major nematode species damaging sweet potato, but the crop is also attacked by *M. javanica, M. hapla* and *M. arenaria*. In India, *M. incognita* and *M. javanica* are reported on sweet potato (Ray *et al*., 1990). Gapasin and Valdez (1979) reported tuber reduction to the extent of 47.7 and 50.6% with an initial inoculum level of 20,000 eggs of *M. incognita* and *M. javanica*, respectively.

(i) Symptoms: Nematode infection in primary roots cause swelling (knotting) of the entire root, and heavy infection can inhibit apical growth. The most obvious symptoms of damage on enlarged roots are longitudinal cracking and general rough appearance of the skin. In many cases lenticels are noticeably enlarged. Both the roots and tops of infected plants are reduced. In fibrous roots, the nematode infestation produces small swellings or knots which are visible in most cases to the naked eye. Infected feeder roots are usually shorter and have fewer secondary roots and root hairs. Symptoms such as reduction in vine growth, yellowing or abnormally abundant production of flowers could also be due to infestation by the nematode.

(ii) Life Cycle: The nematode may complete several generations during the cultivation of this crop. Feeder and storage roots are attacked at the same rate. All races of *M. incognita* can attack sweet potatoes at varying degrees.

(iii) Survival and Spread: *Meloidogyne* juveniles and/or eggs survive in storage roots and can be disseminated in root, but not stem, propagative material. Irrigation water and unclean farm tools and machinery can aid dissemination of the nematodes. Nematodes can survive on many alternate weed hosts.

(iv) Ecology: *Meloidogyne* species seem to do well in light, friable, sandy loam soil which happens to predominate and constitute the major portion of the world's sweet potato growing areas. *M. incognita* requires warm temperature for completion of its life cycle. During a normal growing season it can undergo 4–5 generations. Therefore, it is capable of increasing its population to a level of economic threshold in a short period.

(v) Interaction with Other Pathogens: *M. incognita* interacts with *Fusarium* spp. and *Ralstonia solanacearum* causing severe wilting and premature death. Although there are several *Fusarium* resistant cultivars, their resistance is broken in the presence of *M. incognita*.

(vi) Management

(a) Physical Methods: Hot water treatment at 45°C for 30 hr, 46.7°C for 65 min or 50°C for 3–5 min eliminates root-knot nematode infection without seriously impairing the viability of roots. Hot air treatment of roots, 5–8 cm in diameter, at 50°C for 4–8 hr eliminates most root-knot nematodes without significantly affecting their keeping qualities.

(b) Cultural Methods: The clean cuttings of seed roots from the apex are less prone to root-knot damage when planted out than cuttings from the basal end.

Groundnut or maize in rotation with sweet potato decreases numbers of *M. incognita* to low levels.

(c) Chemical Methods: Application of carbofuran granules at 6.7 kg *a.i.*/ha, 1 to 4 days before planting; fensulfothion/ethoprophos granules each at 3.4 kg *a.i.*/ha in a 30 cm band 2 weeks before planting have shown to control root-knot nematodes and increase the quality and yield of sweet potato.

Total yield and number of marketable roots can also be increased by treating nursery beds with ethoprophos and aldicarb granules at 1 g/30 cm^2.

Bare root-dip treatment of the propagating material in a solution of oxamyl or side dressing with nematicides at the time of planting will allow the establishment of the crop by providing early protection against nematodes.

(d) Host Resistance: Sweet potato *cvs.* Heartogold, Nemagold, Redmar, Jersey types and Kyushu No. 52 were reported to be resistant to root-knot nematodes (Giamalava *et al.*, 1960). The *cv.* Jasper is moderately resistant, while *cv.* Painter shows some tolerance. Mohandas and Palaniswami (1990a) reported the availability of high degree of resistance in high yielding released varieties of sweet potato, *viz.*, Sree Vardhini and Sree Nandni. Mohandas and Ramakrishnan (2002) reported that sweet potato *cvs.* Sree Vardhini, Sree Nandni, Sree Bhadra and Kanjangad were resistant to *M. incognita.*

(e) Integrated Methods: Control of root-knot damage to sweet potatoes involves integration of at least 3 methods:

- The selection of seed roots that are either free of nematodes or freed of nematodes by hot water treatment.
- Planting of the seed roots into beds of sand or coarse textured soil that is either nematode-free or, if infested, is pre-plant fumigated.
- Transplanting of the clean 'slips' into nematode-free soil or pre-plant fumigated soil.

5.3.3.2. Reniform Nematode, *Rotylenchulus reniformis*

R. reniformis is another important nematode species reported from sweet potato from a large number of countries. Gapasin and Valdez (1979) reported 60.6% yield reduction with an initial inoculum level of 5,000 larvae of *R. reniformis*.

(i) Symptoms: The reniform nematode is associated with leaf yellowing and reduction in yield and size of swollen tubers. The most serious injury occurs when nematode population levels are high at planting time and there is a seasonal increase of the nematodes. Infestation by *R. reniformis* cause cracking of storage roots.

(ii) Interaction with Other Pathogens: *R. reniformis* interact with *Fusarium* spp. in development of disease complex.

(iii) Management

(a) Physical Methods: Hot water treatment of sweet potatoes at 50°C for 3 to 5 min eliminated the reniform nematode (Martin, 1970).

(b) Chemical Methods: Good control of *R. reniformis* and increase in sweet potato grade have been achieved by application of granules in 30 cm band of fensulfothion/aldicarb each at 5.6 kg *a.i.*/ha or ethoprophos at 3.4 to 6.7 kg *a.i.*/ha.

(c) Host Resistance: The sweet potato variety Goldrush is less susceptible to *R. reniformis* than other varieties, but is highly susceptible to root-knot nematodes. Selection P-104 is reported resistant to cracking.

5.4. CASSAVA (TAPIOCA)

Cassava (*Manihot esculenta*), popularly known as tapioca, is a member of the family Euphorbiaceae, which has long been cultivated in Central America and Brazil and was later carried to Africa (where it is now an important subsistence crop), Asia and the Pacific Islands. An economic crop can be obtained even under marginal conditions or on exhausted soils. The crop requires very little attention after planting. It has draught tolerance (mainly due to in-built mechanism to shed or drop the leaves under adverse moisture conditions) and produces more calories/unit area. The tubers contain 25–30% starch, 1.5% protein and 2.1% minerals. It is being cultivated in 0.252 million hectares producing 7.925 million tonnes of cassava with an average yield of 31.36 tonnes per hectare (Table 5.12). The Salem belt of Tamil Nadu and Samalkot belt of Andhra Pradesh are known for cassava as an industrial crop. The major cassava producing states include Tamil Nadu, Kerala, Andhra Pradesh, Meghalaya, Pondicherry, Nagaland, Assam and Karnataka.

Table 5.12. Area, production and productivity of leading cassava growing states in India (2005–06)

State	Area (in '000 ha)	Production (in '000 MT)	Productivity (in MT/ha)
Tamil Nadu	127.1	4857.4	38.2
Kerala	95.0	2585.1	27.2
Andhra Pradesh	19.0	388.7	20.5
Meghalaya	4.2	23.1	5.5
Pondicherry	0.7	19.5	27.9
Nagaland	1.3	15.9	12.2
Assam	2.9	14.0	4.8
Karnataka	1.1	12.0	10.7
Others	1.4	9.2	6.6
Total	**252.7**	**7924.9**	**31.26**

(Source: Indian Horticulture Database 2006)

5.4.1. Pests

5.4.1.1. Whitefly, *Aleeurodicus dispersus*

(i) Damage: Whiteflies suck the sap and devitalize the plant (Fig 5.19).

Figure 5.19. Whiteflies on leaves of cassava

(ii) Management

The field release of endoparasitoid, *Encarsia* sp. gave 69.53% parasitization during August and 76.22% during September. The predators like *Cybocephalus* sp., *Cryptolaemus montrouzieri* and *Mallada astur* can also be effectively used for the biocontrol of spiraling whitefly.

5.4.1.2. Stem Muscle Scale, *Aonidomytilus albus*

(i) Distribution: The pest is distributed in West and East Africa, Argentina, Brazil, India, Madagascar, Mexico, Taiwan, West Indies and USA.

(ii) Damage: Grey, elongate adult scales 2.5–3.0 mm long and pale nymphs, infests the lower stems of the plant (Fig 5.20). Infestation may become severe in cassava cuttings kept for planting; the consequent drying up reduces germination success.

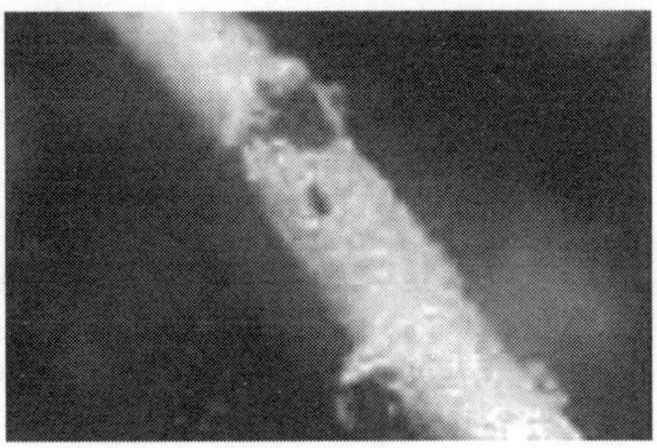

Figure 5.20. Scale infestation on stem of cassava

(iii) Biology: About an hour after hatching the nymphs begin to feed, during the next two weeks they secrete a cottony wax cover beneath which molting and then secretion of a new scale takes place. After another two weeks a second molt occurs. Males subsequently enter the pupal phase and females mature. Male pupation takes less than one week to complete. The entire life cycle takes 5–6 weeks.

(iv) Management

(a) Cultural Methods: Clean healthy cuttings should be used for planting.

(b) Chemical Methods: Spraying or dipping the cuttings in 0.1% demeton-s-methyl, 0.5% parathion or 0.1% malathion is recommended.

5.4.1.3. White Peach Scale, *Pseudaulacaspis pentagona*

(i) Distribution: It is cosmopolitan in distribution.

(ii) Damage: Circular papery female scales (2 mm in diameter) and long (0.75 mm) male scales can be seen on the stem and leaf bases. Weak or old plants are particularly susceptible.

(iii) Biology: The nymphs emerge from eggs beneath the female scale and begin to feed. They secrete cottony threads but these disappear when a thin skin is secreted over the nymph. At the first molt, the insects become disc-like. Cast skins at each molt are added to the scale. Males molt 3 times and then pupate. Females mate after the second molt and begin egg laying.

(iv) Management

Spraying of oil emulsion or 0.05% parathion will control this scale.

5.4.1.4. Black Scale, *Parasaissetia nigra*

(i) Distribution: It is distributed in most warm countries.

(ii) Damage: Colonies of dark brown to black domed soft scales 2–4 mm long can be seen on the stems. *Saissetia coffeae, S. miranda* and *Ceroplastes cirripediformis* have also been reported on cassava stems and are similar in appearance to black scale, *i.e.*, hemispherical and waxy in appearance.

(iii) Biology: The scale is the female insect, beneath the scale she lays eggs which hatch into crawlers that migrate over the plant before setting down to feed. Although females do not move around they retain their legs. Female crawlers molt twice before becoming adult and males, which have two wings, molt 3 times.

(iv) Management

Cassava cuttings must not be allowed to develop heavy infestations of scale insects since they will produce poor plants when planted or do not germinate. Healthy plants will not suffer heavy yield loss due to scale insects. Kerosene emulsion and parathion have been recommended for soft scale control.

5.4.1.5. Mealybugs, *Phenacoccus gossypii, P. manihoti, Ferrisis virgata*

(i) Damage: Leaves and shoots become distorted and yellow due to mealy bugs feeding on them. Adult female mealy bugs are segmented soft-bodied insects without wings, about 2.5 mm long and covered in white, waxy meal.

(ii) Biology: Generally the female lays groups of 200–400 eggs, which are protected by waxy filaments. In a few hours, nymphs very much like females in appearance but much smaller, emerge. The nymphs pass through several immature stages, if males are present, then they pupate, emerge and mate with the females. The life cycle, egg to adult, may be completed within 6 weeks.

(iii) Management

(a) Cultural Methods: Clean cuttings should always be used for planting.

(b) Chemical Methods: Infected cuttings should be dipped or dusted with aldrin, BHC or DDT. Mephosfolan and parathion sprays or dusts may be applied to the growing point if it is seriously affected by mealy bugs and should control the pest.

5.4.1.6. Red Spider Mite, *Tetranychus cinnabarinus*

(i) Distribution: The pest is distributed throughout the semitropics and in protected cultivation in temperate zones.

(ii) Damage: Lower leaves develop yellow dots along the veins, these dots become reddish and if the dry season is prolonged, the infestation spreads to higher leaves (Fig 5.21). Leaves dry up and fall and some plants die. The mites responsible are brownish-red with dark lateral internal markings, they produce webs.

Figure 5.21. Red spider mite infestation on leaves of cassava

(iii) Biology: Reproduction takes place during dry weather; the optimum temperature for development is 32°C. Single eggs are laid on the under surface of the leaves or on the webbing. Development of egg to adult takes about 17 days and there may be 19 broods per year.

(iv) Management

Spraying of 0.04% binapacryl, 0.05% chlorobenzilate, 0.03% dimethoate or parathion, 0.01% lime sulphur, 0.16% mancozeb, 0.001% naled or 0.07% phoslane is recommended for the management of spider mites.

5.4.2. Diseases

Even though cassava is a hardy crop, it has been shown that diseases can cause serious losses. As cassava is propagated vegetatively, the use of disease-free planting material is the most important single factor in disease control. Under traditional farming systems, yields have been more than doubled by using disease-free planting material and improving cultural practices. Cassava bacterial blight is the most important disease of cassava and virus diseases are also of considerable threat to the crop.

5.4.2.1. Tuber Rot, *Phytophthora drechsleri*

(i) Symptoms: *Phytophthora* root rot is prevalent in water-logged soils. The disease causes yield losses up to 80%. The pathogen attacks both mature and young plants causing sudden wilting and severe soft rotting of the swollen roots. Diseased roots develop a brown discolouration and disintegrate rapidly. Roots exude a fowl smelling liquid when partially rotted. In severe cases, plants may die.

(ii) Management

The disease can be controlled by cultural practices such as providing good drainage, use of lighter soils, early harvest, crop rotation with cereals and fallowing. Wide spacing tends to discourage the spread of root pathogen. Crop residues should be removed and burnt and healthy planting material should be used for planting.

The set treatment (30 g/litre for 15 min) and soil application (1 kg/ha) of *Trichoderma viride* recorded the lowest tuber rot incidence (7.33%) and highest harvest index (66.9%).

5.4.2.2. Brown Leaf Spot, *Cercosporidium henningsii*

Brown leaf spot (Fig 5.22) is a major fungal disease in heavy rain fall areas. It causes considerable premature defoliation and loss in tuber yield up to 30% in susceptible variety like Malayan-4.

Figure 5.22. Brown leaf spot on tapioca leaves

(i) Distribution: Brown leaf spot disease is widespread fungal leaf disease in all cassava growing areas.

(ii) Symptoms: Older leaves tend to be more susceptible than younger ones and the disease tend to be more widespread and severe in crops over 5 months old. The disease cause characteristic brown spots (3–12 mm in diameter) on both sides of the leaves. Spots on the upper surface are angular and uniformly brown with a distinct dark border. Spots on the lower surface have less distinct margins and appear grayish in the centre because of the presence of fruiting bodies of the fungus. The lesions expand and become irregular and angular in shape as they are limited by leaf margins and veins. Commonly an indefinite halo appears round the lesions. As the disease develops the infected leaves turn yellow, become dry and fall off. Susceptible varieties can be severely defoliated during the rainy season.

(iii) Development and Spread: Warm, wet conditions favour the development and spread of *C. henningsii*. Conidia of the fungus on older or fallen leaves are spread by rain splash and wind to younger leaves and to new plantings. The conidia need moisture to germinate on the leaf surface. The fungus penetrates through stomata and grows trough the intercellular spaces. The fungus survives during the dry season in old lesions, often on fallen leaves.

(iv) Management

(a) Cultural Methods: Measures designed to reduce the humidity within the stand of cassava, such as wider spacing and the varieties with a low leaf area index are recommended. The severity of the disease can be reduced if the planting is timed so that the most susceptible growth stage (over 5 months old) does not coincide with the wet season.

(b) Chemical Methods: Spraying of copper fungicides, thiophanate, benomyl and carbendazim at monthly intervals starting from third month was effective in controlling the disease.

(c) Host Resistance: Cassava *cvs.* Sri Prakash and Sri Visakam released from Central Tuber Crops Research Institute, Thiruvanathapuram, Kerala, India, are resistant to brown leaf spot disease.

5.4.2.3. Cercospora Leaf Blight, *Cercospora viscosae*

(i) Distribution; The disease occurs as commonly as brown leaf spot in warm cassava growing areas of Brazil and Colombia and also Southeast Asia.

(ii) Symptoms: The disease is characterized by large brown spots without definite borders. Individual spots may cover a fifth or more of the leaf lobe. The lesions are uniformly brown on the upper surface but on the lower surface the centres of the spots are grayish due to the presence of fruiting bodies of the fungus. The lesions usually have concentric rings on the lower surface. *Cercospora* leaf blight usually occurs during the rainy season and can cause heavy defoliation in susceptible cultivars with resulting yield loss of over 30%. Plants over 6 months are more severely affected.

(iii) Management

The same management methods used for brown leaf spot also control *Cercospora* leaf blight.

5.4.2.4. Concentric-Ring Leaf Spot, *Phoma* sp., *Phyllosticta manihobae, P. manihoticola*

(i) Distribution: The disease occurs in Africa, Brazil, Colombia, India and Philippines.

(ii) Symptoms: The disease occurs in cooler cassava growing areas during the rainy season when temperatures are below 22°C. Susceptible varieties may be severely defoliated and occasionally dieback or death may occur. The disease is characterized by large brown spots with indefinite margins. Lesions are usually found at the tips or edges of the leaf lobes or along the main veins or midrib. At first the upper surface of the lesions consists of concentric rings formed by the brown pycnidia (fruiting bodies) of the fungus. These rings are often absent on the older lesions as the epidermis disintegrates allowing mature pycnidia to be washed off by rain so that the lesions resemble those caused by *Cercospora viscosae*. Few pycnidia are produced on the lower surface of the lesions, which appear uniformly brown.

When humidity is high grayish-brown hyphae can be observed on the surface of the lesions. On the underside of leaves, veins and vein lets become necrotic in the region of the lesions. The lesions expand causing leaf blight. Eventually the necrosis spreads to the whole leaf and petiole. Leaves then wilt and drop, sometimes resulting in extensive defoliation. Dieback of young shoots can also be caused and in extreme cases the plant may die. Diseased stems become brown and are often covered with pycnidia.

(iii) Development and Spread: The young upper leaves are more susceptible to the disease. The disease is more prevalent at high altitudes or in low land areas during the rainy season, because conditions are best for spore germination (20–25°C).

(iv) Management

Field resistant cultivars have been observed and these may be used in a breeding programme for resistance.

5.4.2.5. Powdery Mildew, *Oidium manihotis*

(i) Distribution: The disease is distributed worldwide and endemic in almost all cassava plantations.

(ii) Symptoms: The first sign of the disease is a white fungal mycelium growing over the surface of the leaf. Yellowish lesions are formed where the fungus penetrates the leaf. The lesions are indefinite and contain pale brown angular spots of necrotic tissue.

The disease occurs most commonly during the dry season in warmer cassava growing areas. Mature leaves appear to be the most susceptible.

(iii) Management

Resistance has been observed in some varieties.

5.4.2.6. Anthracnose, *Glomerella manihotis*

(i) Distribution: The disease is widespread in distribution and appears to be serious in Southeast Asia, Africa and Brazil.

(ii) Symptoms: Anthracnose tends to appear after long periods of rain. Young shoots seem to be the most susceptible. Leaf border blight and spots appear on young leaves, which become distorted. Infected tissue becomes necrotic and whole leaves may die. The disease also attacks stems, causing a dieback of young stems and the production of cankers on older stems. In the centre of the cankers, pinkish areas of the fungal fruiting bodies can be observed. Anthracnose tends to disappear at the onset of the dry season.

(iii) Management

Planting material taken from infected stems is of poor quality, does not germinate well and results in weak plants and reduced yields. The best preventive measure is to use healthy planting material and to avoid planting at the height of the rains.

5.4.2.7. Bacterial Blight, *Xanthomonas manihotis*

(i) Distribution: The bacterial blight is the most important disease of cassava and is widely distributed in Argentina, Brazil, Cameroon, Colombia, Ghana, India, Indonesia, Madagascar, Malawi, Malaysia, Mauritius, Mexico, Nicaragua, Nigeria, Panama, Philippines, Taiwan, Thailand, Togo, Uganda, Venezuela and Zaire.

(ii) Symptoms: It is particularly devastating during wet seasons. The characteristic symptoms are leaf spotting and blight (Fig 5.23), wilting, dieback, gum exudation and stem and root vascular necrosis. Primary symptoms, shown by infected planting material, are wilting of the young germinated shoots followed by dieback.

Figure 5.23. Bacterial blight on tapioca leaf

Symptoms of secondary infection are angular leaf spotting, followed by blight, defoliation, wilting and dieback. At first the leaf spots are small, angular and water-soaked but they enlarge, covering all or part of the leaf, and turn brown or dark brown. The initial spots are clearly visible on the undersides of leaves. The necrotic leaves roll and dry up and after remaining attached to the stem for a short while they fall off. Gum is often exuded from young infected stems, petioles and leaf spots, and collects as droplets. On leaves these droplets occur mainly on undersides along veins or veinlets. On petioles and stems the gum is exuded from cracks. The gum forms a yellowish glistening scab when dried.

Vascular tissue of infected petioles and stems becomes necrotic and appears as brown strings. Leaves supplied by these vascular tissues wilt and young stem tissues rot. The young green stems rot faster than the older greenish-brown ones which maintain a healthy appearance. The rotting of the young stems causes characteristic dieback symptoms.

Root rot may appear when susceptible cultivars with a good water balance are infected. Roots have dry vascular necrosis.

On young immature plants, complete destruction of the aerial portions may be caused. New shoots are usually produced but these rapidly become infected, especially during the rainy season.

All symptoms appear mainly during the rainy season and vary according to the susceptibility of the variety and the period of infection.

(iii) Development and Spread

Primary infection is due to the use of infected planting material. Secondary infection results from spread of the bacteria in the gummy exudates by rain splash. Movement of man, animals and insects through the crop and use of contaminated tools are also likely to spread the disease. Rain splash is the most important means of spread over short distances.

Development of the disease is retarded during dry periods and little gum is exuded thus halting the spread of the disease. However, the bacteria remain viable in the plant and are reactivated in later wet periods.

The bacteria enter the plant through stomata or wounds and penetrate to the vascular system from where they can spread to the rest of the plant. Symptoms generally develop within 7 to 13 days after infection.

(iv) Management

Only healthy planting material should be used for planting. Clean seed material can be obtained by pruning most of the above ground portion of infected plants. This method is most effective with moderately resistant cultivars which are only lightly infected. Tools used for cutting planting material should be regularly disinfected to avoid mechanical dissemination of the pathogen.

The bacteria can only survive in the soil for a month or two enabling land to be cleared of infection by a 6-month rotation or fallow period before planting disease-free material.

5.4.2.8. Cassava Mosaic Virus

(i) Distribution: The disease occurs in East, West and Central Africa and India. Estimate of losses caused by mosaic ranges from 30–80%.

(ii) Symptoms: The virus disease causes characteristic mosaic symptoms. Young leaves have chlorotic areas and are often deformed. Leaves may also be reduced in size. When the infection is severe, leaves are small, misshapen, distorted and have bright yellow areas with normal green tissue in between (Fig 5.24). In milder cases of infection, leaves may show only a mild chlorotic mosaic. The disease is transmitted by whitefly. Insect transmission alone does not cause much spread of the disease-man appears to be the main factor.

Figure 5.24. Cassava mosaic virus infection on cassava leaves

(iii) Management

The use of disease-free planting material and rogueing of infected plants can give excellent control of the disease.

5.4.3. Nematodes

Important nematodes parasitizing cassava are root-knot and root lesion nematodes. Ray *et a*l. (1990) reported *Helicotylenchus multicinctus* and *Tylenchorhynchus mashhoodi* from Orissa.

5.4.3.1. Root-Knot Nematodes, *Meloidogyne* spp.

Meloidogyne incognita, M. javanica and *M. arenaria* have been reported from cassava from Brazil, Colombia, Dominican Republic, Hawaii, India, Ivory Coast, Nigeria, Rhodesia, Tanzania, Togo, Trinidad and USA. Nirula (1963) reported *M. javanica* for the first time from India. Attack by root-knot nematodes was considered serious when cassava was intercropped with other susceptible hosts like brinjal and roselle. Jatala (1988) reported considerable yield reduction in cassava due to *M. incognita* and *M. javanica* infestation. *M. incognita* race 2 reduced fresh storage root weight by 49% and dry top weight by 22%.

(i) Symptoms: Root-knot nematodes produce the typical swelling or knotting of cassava roots. *M. incognita* race 2 and *M. javanica* significantly reduced stalk length, stalk weight and storage root weight of cassava after 15.5 months growing period (Caveness, 1981).

(ii) Management

(a) Cultural Methods: The summer fallowing which is already being practiced by farmers in many parts of cassava growing areas in Kerala due to non-availability of water is an excellent means of controlling nematode population. It is also essential to avoid planting susceptible crops either in the previous season or as intercrops.

Cassareep, a by-product of the cassava industry, was apparently effective in controlling *M. incognita* and *M. javanica* on cassava.

(b) Host Resistance: Cassava *cvs*. Sree Sahya and Narayaniyakappa were found resistant to root-knot nematodes (Sreeja *et al.*, 2001). Freitas and Moura (1986) reported that cassava *cv*. Mandiocol to be resistant to *M. incognita* and *M. javanica*.

5.4.3.2. Lesion Nematode, *Pratylenchus brachyurus, P. sefaensis*

Pratylenchus brachyurus and *P. sefaensis* were found attacking cassava. *P. brachyurus* is regarded as a serious pest of cassava causing damage to cortical tissues of roots and also predisposing plants to infection by secondary pathogens. *P. sefaensis* reduced fresh storage root weight by 45% and dry top weight by 21%.

(i) Symptoms: The lesion nematode produces root lesions and subsequent rotting of roots. Affected plants are stunted with yellowing of leaves, die-back of twigs and produce small or no tubers. Leaves also show nutrient deficiency symptoms (Mohandas *et al.*, 1990).

(ii) Management

(a) Chemical Methods: A 7.2% increase in tuber yield was recorded after application of DBCP at 50 l/ha.

(b) Host Resistance: A number of cassava varieties (Atitogen, Ba Pou II, Agba Tiega, Agba Boquia, Sodjievi) show some resistance to the lesion nematode.

5.5. COLOCASIA

Colocasia (*Colocasia esculenta*), also known as cocoyam, taro, dasheen and eddoe, is a member of the family Araceae and occurs wild in Southeast Asia. It is cultivated throughout the humid tropics and is of great importance in the Pacific Islands. It is mostly a staple food or subsistence crop grown commercially in some countries.

5.5.1. Pests

5.5.1.1. Aphid, *Aphis gossypii*

(i) Distribution: The pest is cosmopolitan in distribution, but absent from the cooler pats of Asia and Canada.

(ii) Damage: Nymphs, winged and wingless adult aphids are feeding on leaves, causing specking, leaf curl and a generally sickly appearance. A sooty mold may develop on the honeydew secreted by the aphids in addition ants are attracted to the honeydew.

(iii) Biology: Adults may give birth to two nymphs/day. Winged adults (all female) have a black head with two green bands. The abdomen is green with black cornicles. Wingless adults (also all female) are yellowish green to black.

(iv) Management

Avoid planting colocasia close to alternate hosts such as melon, cucumber or other cucurbits or cotton. The following chemicals have been recommended for the management of aphids (Table 5.13).

Table 5.13. Recommended chemicals for the control of aphids

Insecticide	Dose	Application method
Carbaryl WP	1 kg	In 450 litres (high volume) or 112–168 litres (low volume) of water.
Diazinon EC	11 ml	In 13.5 litres of water, 14 days minimum waiting period.
Dimethoate EC	4.4 ml	In 13.5 litres of water, 14 days minimum waiting period.
Malathion EC	14.2 ml	In 13.6 litres of water, 7 days minimum waiting period.

5.5.1.2. Whitefly, *Bemisia tabaci*

(i) Distribution: The pest is recorded in most parts of the tropics and subtropics.

(ii) Damage: Adults, nymphs and pupae are found on the lower leaf surface. When infestation is heavy, the leaves turn yellow, wilt and die. The growth of the colocasia may be stunted.

(iii) Biology: Yellow elliptical eggs, attached by filaments to the leaf, are laid on the undersurface of the leaves. The first instar nymphs emerge and feed. After the first molt, they are fixed to one place on the leaf as yellow elliptical scale-like insects. The third molt gives rise to the puparium which is less flat than the previous nymphal stages and the red eyes of the adult can be seen through the integument. Egg to adult development takes 2–3 weeks. The adult has a yellow body, white waxy broad tipped wings and is 1 mm long with a wing span of about 2 mm.

(iv) Management

The following chemicals have been recommended for the control of whiteflies (Table 5.14.).

Table 5.14. Chemical control of whitefly on colocasia

Insecticide	Dose (*a.i.*/13.6 litres of water)	Application
Carbaryl WP	34 g	7 days minimum waiting period before harvest.
Malathion WP	14 g	7 days minimum waiting period before harvest.
Trichlorphon WP	23 g	7 days minimum waiting period before harvest.

5.5.2. Diseases

5.5.2.1. Leaf Blight, *Phytophthora colocasiae*

The leaf blight of colocasia is a major and widespread disease causing 50% loss in tuber yield.

(i) Distribution: This disease occurs widely in India, Indonesia, Malaysia, Sarawak and the Pacific. It has also been recorded from parts of Africa and the Caribbean.

(ii) Symptoms: Leaf blight is the most destructive leaf disease of colocasia. Lesions are initially small, dark and round but rapidly enlarge to 2.5–5.0 cm in diameter and become purplish to brownish in colour (Fig 5.25). Drops of a clear liquid exude from the spots, and turn yellow, orange or purple when dry. There is usually chlorotic halo round the spots. As the disease progresses, the spots coalesce and have characteristic rings of yellow or brown colour. Eventually the whole leaf may be affected and may die.

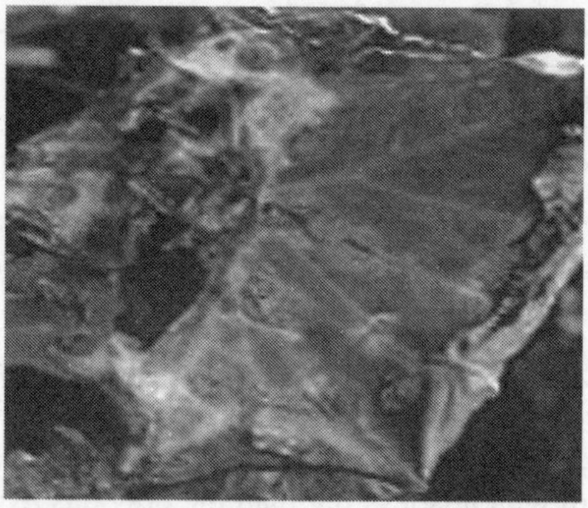

Figure 5.25. Leaf blight damage on leaves of colocasia

(iii) Development and Spread: Leaf blight is common in wetland and is favoured by humid, cloudy conditions and poor soil fertility. Losses up to 50% have been reported. Spores are produced on the leaf spots and are readily spread by rain.

(iv) Management

(a) Cultural Methods: Wide spacing minimizes spread of the disease due to rain splashing. Wide spacing together with interplanting and open sides to the colocasia field help to improve the air flow and thus keep the humidity down. Farmers should avoid walking through the field when it is wet as this encourages the spread of spores. Where possible the colocasia should be cultivated on higher land. Crop remains should be burnt after harvest.

In case of mild infection of young plants (up to 2–3 months), leaf rogueing can be effective in controlling or delaying the build-up of the disease. Rogued leaves should be burnt.

Strict hygiene should be observed at planting and infected leaves should be removed twice a week for 80–90 days after planting, or until a dense canopy is formed.

(b) Host Resistance: Colocasia *cvs*. Jhankri and Muktakeshi are tolerant to leaf blight disease.

(c) Chemical Methods: Tribasic copper sulphate and copper oxychloride have been used successfully at intervals of 7–15 days. Zineb and mancozeb have been shown to give effective control. The best control was obtained with sprays of mancozeb at 2.24–4.48 kg/ha at intervals of 5–7 days. It has been suggested that copper oxychloride sprays and phytosanitary measures could be used together in a practical control programme.

Four sprays of 0.25% zineb at 15 days interval was found effective in reducing the disease intensity (23.24% as compared to 56.15% in control) and increasing the tuber yield (20.91 t/ha as compared to 15.49 t/ha in control). Two sprays of ridomil at 0.25% was also found effective in management of the disease.

5.5.2.2. Brown Leaf Spot, *Cladosporium colocasiae*

(i) Distribution: The disease is distributed in Borneo, Ethiopia, Fiji, Formosa, Guinea, Hawaii, Hong Kong, India, Nepal, New Caledonia, New Guinea, Nigeria, Pakistan, Sarawak, Salmon Islands and Tahiti.

(ii) Symptoms: Superficial dark reddish-brown spots of 5–10 mm in diameter develop on both sides of the leaves. The spots are diffuse in appearance. The disease occurs in wetland and upland colocasia and is particularly severe on old leaves.

(iii) Management

Fungicidal sprays applied for leaf blight control will also control brown leaf spot.

5.5.2.3. Corm Rot, *Pythium myriotylum, P. splendens, Phytophthora cinnamoni, P. nicotianae* var. *parasitica.*

(i) Distribution: The disease occurs in India and Pacific area.

(ii) Symptoms: If young plants are attacked, they may be killed without producing leaves or they may become stunted. In plants over 3 months old, the first symptoms observed are wilting, chlorosis and stunting which may be followed by collapse of the whole plant. The rotting of the corms usually starts at the base and progresses upwards so that eventually the whole corm may be affected. The rotted tissue is firm and may be whitish-yellow to grey and blue or dark purple. Healthy and diseased tissues are clearly divided. In later stages, the diseased tissue crumbles like cheese and has an offensive smell.

Roots decay in association with the corms. In some cases there is little damage to the corm but many roots are killed leaving the plant loose in the ground. Rotting may also spread from the main corm to lateral suckers.

(iii) Development and Spread: The fungi are encouraged by plentiful moisture, warm temperature and poor soil conditions. Any physical damage to colocasia plant predisposes them to infection. If colocasia is left in the field after maturity, it causes a build-up of the disease. These fungi survive between crops in crop residues. Damage may be confined to a few corms or may be widespread and cause 100% loss depending upon the conditions. Damage is particularly severe in wetland areas and poorly drained soil.

(iv) Management

(a) Cultural Methods: Colocasia should be harvested as soon as possible after it is mature and crop debris should be removed from the fields. Diseased plants should be rogued from the growing crop taking care to remove as much of the corm and decayed roots as possible. If a field is badly infected with the disease, colocasia should not be cultivated for 2 or 3 seasons and the soil should be allowed to dry out by deep cultivation. In irrigated wetland colocasia, water should be kept flowing and not be allowed to stagnate. Dryland colocasia should be planted in well drained soil. Healthy planting material should always be used for planting.

(b) Chemical Methods: Dipping of planting material in copper oxychloride or captan has been recommended. Soil drenches with these fungicides have also been found to be effective.

5.5.2.4. *Sclerotium* Rot, *Corticium rolfsii*

(i) Distribution: The disease is widespread in distribution.

(ii) Symptoms: The pathogen causes a rot of dying petioles and a corm rot. The rotted tissue is brown and soft but not watery and tends to be stringy. Characteristic white mycelial growth and numerous dark brown sclerotia which resemble cabbage seed are observed. The disease may be more severe in colocasia that is over mature.

(iii) Development and Spread: *Corticium rolfsii* is a common pathogen and attacks many crops; sources of inoculum are therefore readily available. The disease is more common in upland colocasia.

(iv) Management

Prompt harvesting of colocasia and crop rotation help to avoid the disease. Infected plants should be removed and burnt. Soil drenches with quintozone or dicloran have been recommended for control.

5.5.2.5. Dasheen Mosaic Virus

(i) Distribution: The disease occurs in Egypt, Hawaii, India, Japan, Papua New Guinea, Puerto Rico, Trinidad, USA and Venezuela.

(ii) Symptoms: The virus has long flexuous rod-shaped particles measuring about 75 nm in length. The infection results in a feathery mosaic pattern adjacent to the veins. Yellow-green streaking of the petioles may also be observed in *Xanthosoma* spp. In severely affected plants, petioles are short and leaves are dwarfed and fail to unfurl properly.

The virus is transmitted non-persistently by the aphids, *Myzus persicae, Aphis gossypii* and *Pentalonia nigronervosa*. This virus is also sap transmissible.

(iii) Management

Only certified healthy planting material should be used for planting. The healthy planting material may be obtained by meristem tip culture.

5.5.3. Nemotodes

The root-knot nematode is an important limiting factor in successful production of colocasia crop.

5.5.3.1. Root-knot Nematode, *Meloidogyne incognita, M. javanica*

M. incognita and *M. javanica* are widely reported on colocasia. Nirula (1959) was the first to report root-knot nematode on colocasia from India. *M. javanica* causes severe losses to colocasia in India. *M. incognita* was responsible for 24% loss in tuber yield of colocasia (Anon, 1990b).

(i) Symptoms: The foliage of nematode infested plants at first becomes yellow and then turns brown and ultimately dies back. No corms were found in plants attacked in the earlier stages and late stage attack resulted in deformed and galled corms (with blister-like swellings varying in size from 2–15 mm) of little market value. Rotting of corms associated with the nematode, occur during storage.

(ii) Survival and Spread: *Meloidogyne* spp. can be carried over from one *Colocasia* crop to next in the wide range of other host crops and weeds. As the nematodes feed and reproduce in corm tissues, they can be spread in corms and cormels if infested material is used for propagation.

(iii) Management

(a) Physical Methods: Root-knot can be controlled in corms by dipping in hot water at 50°C for 40 min.

(b) Cultural Methods: *Meloidogyne* populations could be suppressed when colocasia is grown in very wet or flooded conditions.

Use of nematode-free planting material will prevent spread into the field. Seed corms or cormels should be free of root-knot damage.

(c) Chemical Methods: Srivastava *et al.* (1971) found that application of DBCP at 3.6 l/ha as soil drench was more effective than aldicarb (2.5 kg *a.i.*/ha), fensulfothion (3.0 kg *a.i.*/ha) and dichlofenthion (2.0 litres *a.i.*/ha) for the control of *M. javanica* in India.

(d) Host Resistance: Mohandas and Palaniswami (1990b) reported resistance in colocasia *cv.* Sree Reshmi and very high resistance in C-9 (a popular variety among farmers). In *cv.* C-9, no nematode could be detected either from root or from tubers. *Cv.* Dodare was found to be completely resistant to both M. *incognita* and *M. javanica*, while *cv.* Samra is described as moderately susceptible to these two species.

5.6. ELEPHANT FOOT YAM/*AMORPHOPHALLUS*

Elephant-foot yam (*Amorphophallus paeoniifolius*) is a popular tuber crop of humid tropics. Among root and tuber crops, *Amorphophallus* is one of the most important tuber crops that occupy special significance as high value cash crop due to its high production potential and popularity as a vegetable. The genus *Amorphophallus* is indigenous to tropical Asia and Africa and *A. paeoniifolius* is found widely distributed in thickets and secondary growth forests at low and medium altitude in the Philippines, Malaysia, Indonesia, Sri Lanka and the Southeast Asia subcontinent. In India, it is widely grown in Andhra Pradesh, Bihar, West Bengal, Assam, Jharkhand, Gujarat, Tamil Nadu, Kerala, Uttar Pradesh, Maharashtra, Chhattisgarh, Meghalaya and Andaman and Nicobar Island. Chemical composition of Elephant-foot yam corms: moisture 60-80%, starch 25–28%, sugars 0.5–0.6%, crude protein 1.0–2.8%, fat 0.1–0.3%, ash 0.5–2.6%, Vitamin A-0.8 mg, Vitamin B1-0.08–0.09 mg, Vitamin B2-0.02–0.03 mg, Vitamin C-5–12 mg.

5.6.1. Diseases

5.6.1.1. Collar Rot, *Sclerotium rolfsii*

(i) Management

Trichoderma harzianum was highly effective in suppressing *S. rolfsii* in elephant foot yam under field conditions. The conidial mixture of 2 bioagents *T. pseudokoningii* (effective in reducing the mycelial growth) and *T. virens* (good colonizer of sclerotia) was effective in reducing the incidence of storage rot. Application of *T. harzianum* as corm treatment + soil treatment after 30 DAP significantly managed the disease. Mulching with paddy straw/ polythene sheet reduced the collar rot disease.

5.6.1.2. Bacterial Leaf Blight/Leaf Rot, *Xanthomonas amorphophalli*

(i) Symptoms: The leaf blight/leaf rot is a major and widespread disease of elephant foot yam causing considerable loss in yield (Fig 5.26).

Figure 5.26. Leaf blight damage on leaves of elephant foot yam

(ii) Management

Three sprays of 100 ppm Agrimycin at 15 days interval after the first appearance of the disease significantly reduced the disease incidence (24.65% as compared to 50.88% in control) and increased the yield (36.25 t/ha as compared to 17.81 t/ha in control).

Seed tuber dip treatment with 0.05% Agrimycin for 12 hrs decreased the disease incidence (19.73% as compared to 65.91% in control) and increased the tuber yield (23.95 MT/ha as compared to 7.81 MT/ha in control).

Integrated Disease Management: Elephant-foot yam is affected by three major diseases namely, Collar rot caused by *Sclerotium rolfsii*, leaf blight caused by *Phytophthora colocasiae* and mosaic caused by virus.

Use of mosaic virus free healthy planting material (corm) + corm treatment with *Trichoderma viridae* at 5 g/litre or cow dung slurry mixed with mancozeb (0.2%) along with soil application of neem cake at 250 g/pit followed by one foliar spray with neem gold at 0.5% at 60 days after planting effectively controlled almost all the field diseases. Corm treatment with cow dung slurry mixed with mancozeb (0.2%) along with dimethoate (0.1%) has been found most effective against post harvest diseases and also enhancing sprouting. But care should be taken to dry the treated corms in shade before planting. Corm treatment was found most effective when cut pieces are used as planting material than whole corms. If diseases appeared in the field, then 2–3 foliar spray of mancozeb (0.2%) at fortnightly interval has been recommended.

Use of mosaic virus free tubers + mulching after planting and two foliar sprays with mancozeb (0.2%) and monocrotophos (0.05%) at 60 and 90 days after planting has proved very effective in reducing the incidence of major diseases and pests of Elephant-foot yam.

5.6.2. Nematodes

The root-knot and the lesion nematodes are important pests of amorphophallus (*Amorphophallus* spp.).

5.6.2.1. Root-knot Nematode, *Meloidogyne incognita*

Amorphophallus spp. are highly susceptible to *M. incognita*. The nematode infestation is very high in Kerala, Andhra Pradesh and Tamil Nadu. The root-knot nematodes have been recorded in high numbers from Tamil Nadu leading to crop failure in many farms.

(i) Symptoms: The roots produce typical root-knot symptoms. In corms, the galls appear as irregular projections which harbour adult females and eggs. The area of infestation in tuber was discoloured when the infestation was severe, the infested area dried up resembling dry rot. Infested tubers were deformed and smaller. The degree of infestation was very high in cormels compared with that in corms.

5.6.2.2. Lesion Nematodes, *Pratylenchus* spp.

The lesion nematodes have been recorded in high numbers from Tamil Nadu leading to crop failure in many farms.

(i) Symptoms: The surface of the infested tubers were black in colour with cracks. The blackening extended further deep inside the tubers reducing the edible portion. The infested tubers were smaller in size.

5.6.2.3. Stunt Nematode, *Tylenchorynchus indicus*

Tylenchorynchus indicus was found to infect the roots and corms and cause severe post harvest loss in combination with fungi like *Fusarium* sp. and *Sclerotium* sp.

5.7. LESSER YAM

The lesser yam, *Dioscorea esculenta* originated in Indo-China and is extensively grown in Southeast Asia and elsewhere in the tropics, although it is still rare in Africa. It produces large numbers of small ovoid tubers resembling sweet potatoes in appearance. It is high yielding and is well adapted to mechanical harvesting. Tubers are rich in diosgenin (from which steroidal drugs can be made). It is also used for rheumatic and ophthalmic diseases.

5.7.1. Pests

5.7.1.1. Yam Beetle, *Heteroligus meles*

(i) Damage: The yam beetle is a pest that starts as a curled grub that changes into a beetle. Yam beetles may spoil the tubers by eating and making holes into them. Then the tuber rots. Dusting of holes or soil surface with aldrin, endosulfan or grammalin gives effective control of beetles.

(ii) Management

- If the yam beetle is infesting the tubers, wash the planting pieces in lead arsenate and Bordeaux mixture. This will stop the grubs from eating into the tubers and spoiling them. Follow crop rotation with non-host crop.

5.7.1.2. Chrysomelid Beetle, *Lilioceris livida*

(i) Management

Control by removing larva by hand or by spraying with 0.2% carbaryl.

5.7.1.3. Mealybugs, *Planococcus citri*

(i) Management

Control by using a clean planting stock.

5.7.2. Diseases

5.7.2.1. Dieback, *Colletrotrichum* sp.

(i) Symptoms: This disease starts when water from heavy rains splashes the small spores of the pathogen from the soil up onto the young leaves. The fungus does not grow so well on the older leaves. The first signs of this disease are small round brown dead spots on the leaves. There is a yellow colour on the outside of the spots. Later the dead spots grow in size until the whole leaf has died. Then the whole branch with its stem and leaves may dieback. The tubers will not grow properly.

(ii) Management

- Do not use the tubers from diseased plants as planting material.
- Do not plant yams in the infested soil, but use crop rotation with non-host crop.
- Spray the plants when they are very young with 0.1% benomyl.
- Plant the yams at a time when there are no rains. The danger of attack is less once the yams have grown higher, because the older leaves do not get the disease so easily.

5.7.2.2. Collar Rot, *Rhizoctonia* sp.

(i) Symptoms: This fungus gets into the roots of tubers of the yams and rots them. There may be signs of the disease low down on the stem near the soil. It makes brown marks low down on the stems. Dark coloured spots on this part of the stem are called "collar rot". This fungus is carried by the tuber. If there are any marks of this disease on the leaves, they are on one side of the leaf only.

(ii) Management

The most important way of stopping this disease is not to use tubers from diseased plants as planting material. The big tubers are probably good to use for planting.

5.7.2.3. Brown Leaf Spot, *Cercospora* sp.

(i) Symptoms: The lesser yam is affected by many foliar diseases of which the brown leaf spot is widespread and recurring in nature. The damage due to the leaf spot is generally quite serious.

(ii) Management

Two to three sprays of 0.05% carbendazim at 15 days interval starting after the first appearance of the disease was found effective in decreasing the disease incidence (24.61% as compared to 62.64% in control) and increasing the tuber yield (19.75 MT/ha as compared to 7.72 MT/ha).

5.7.2.4. Tuber Rots, *Botryodiplodia theobromae, Rhizopus nodosus, Fusarium oxysporum, F. solani*

(i) Management

Control through careful selection of tubers for planting. Cut seed should be dipped in Bordeaux mixture or with a systemic fungicide. Also controlled with crop rotation.

5.8 VERTEBRATE PESTS OF TUBER CROPS

(i) Damage: Wherever cassava, yams, taro or sweet potatoes are grown, they shall be attacked and eaten by rodents. The damage may be seen immediately when the roots are eaten at ground level or may not be apparent until the crop is harvested if rats have burrowed and eaten away the roots from beneath. Losses of the order of 5–15% in West and Central Africa and up to 50–60% in the Cayman Islands in the West Indies were reported. Rat attacks on sweet potatoes begin when the crop is about 100 days old. Rodent attack may continue after harvest in farm and storage. The species responsible for damage in storage include *Bandicota bengalensis* in Indian sub-continent, *Mastomys natalensis* and *Rattus exulans* in Far East and South Pacific region.

(ii) Management

There are three main methods to control rodents in the field. Trapping has been employed since the earliest times, and much skill has been developed in its use. Poisoning and environmental alteration have the biggest contribution to make to increasing crop yields.

6 ALLIED TOPICS IN PEST MANAGEMENT

6.1. USE OF POWER SPRAYERS

Power sprayers require less spray fluid/unit area as compared to high volume sprayers like gator, back pack, etc. While high volume spray require about 800–1000 litres of spray fluid per hectare, low volume spray with power sprayers require only 200–300 litres. Hence, while using power sprayers, spray solutions should be at least three times more concentrated than high volume sprays and spray solutions should also be reduced three times. For example, if 450 ml of monocrotophos is to be mixed with 300 litres of water for using Gator sprayer, then mix 450 ml of monocrotophos in 100 litres of water for power sprayer.

6.2. USE OF STICKERS/SPREADERS

The addition of stickers/spreaders to the spray solution improves the retention, even spreading of the spray solution over sprayed surface. Commonly used stickers and spreaders are as follows:

Sandovit	:	0.5 ml/litre
Teepol	:	0.5 ml/litre
Triton	:	0.5 ml/litre
Citowet	:	0.5 ml/litre

6.3. PREPARATION OF 1% BORDEAUX MIXTURE

Dissolve one kg of copper sulphate (98% purity) in 50 litres of water in wooden barrel or plastic drum. Dissolve one kg of fresh hydrated lime in 50 litres of water. Both the solutions should be sieved. Pour copper sulphate solution slowly and lime water simultaneously by continuous stirring into a tank.

It should be ensured that there is no free copper in the mixture. To check this, a clean iron knife/blade should be held/dipped in the mixture for a few minutes, removed and observed for any reddish brown deposit of copper on it. If the copper deposit is noticed, add a little lime solution to the mixture to neutralize excess copper. The best way is to use litmus paper and check pH. The Bordeaux mixture should be neutral or slightly alkaline, otherwise it will injure the leaves/plants. Bordeaux mixture should always be prepared fresh and should not be stored for more than one day. However, mixture can be stabilized by adding 200 gm of jaggery per 100 litres of mixture.

6.4. PREPARATION OF 4% NEEM SEED KERNEL EXTRACT

4 kg kernel of neem seed is required to prepare 100 litres of spray fluid. Grind 4 kg of kernel with 20 litres of water to get a fine paste. For grinding, a mixer or wet grinder may be used. This fine paste is kept for 12 hours and repeatedly filtered, washed and squeezed through a rough cloth and made up to 100 litres, which is used for spraying. About 800–1000 litres of solution is required to spray one hectare of cabbage or other vegetable crops.

6.5. SEED TREATMENT FOR STORAGE

Pests and diseases affect seeds in storage. Following seed treatments may be adopted for protecting seeds during storage (Table 6.1):

Table 6.1. Seed treatment for the management of pests/diseases in storage

Causal organism	Pesticide	Dose
Insects	Carbaryl	4 g/kg of seed
Diseases	Subeej/derosal	2 g/kg of seed
	Chlorothalonil/captan/iprodione	2 g/kg of seed
	Thiride/vitavax	2 g/kg of seed

6.6. CONTROL OF RODENTS

There are three main methods to control rodents in the field. Trapping has been employed since the earliest times, and much skill has been developed in its use. Poisoning and environmental alteration have the biggest contribution to make for increasing crop yields.

6.6.1. Control in the Field

6.6.1.1. Poisons

There are three main kinds of poisons for rodent control, multi-dose, single-dose and gassing poisons. Warfarin, coumatetralyl, chlorophacinone, coumachlor, difenacoum and calciferol are multi-dose poisons. All except the last are anticoagulants, which have an effect on the blood system of the animal. These compounds will normally give up to 100% kill. The single dose poison is zinc phosphide. Generally both multi-dose and single-dose poisons have to be mixed with bait and have a lethal effect when eaten, but gassing poisons release poisonous gases which are breathed by the rodents. In most situations where there is a rodent problem in growing crops, the use of bait containing a multi-dose poison is the best method of control.

Detecting live burrows (where rodents are active): Locate all burrows in the field and close them with mud in the evenings. Next morning, check-up and find out the burrows which have opened. They are live burrows.

Baits: Rats and mice shall usually eat cereals (milled or whole grain) and often the cheapest locally available will be accepted as baits by them. Rodents tend to like sugar and a small amount of sugar is added to the bait to make it more attractive. Sometimes a mixture of cereals will give good bait.

Pre Baiting: To familiarize the rats with the food used for poisoning, pre-baiting is required. For this, prepare plain bait of a mixture of wheat, bajra, jowar or maize flour and groundnut chops mixed with vegetable oil for 2–3 days. When rats start accepting the bait, they can be poisoned.

Bait Mixing: It is very important that the correct amounts of poison may be mixed throughout the cereal so that it is all poisoned. The concentration of rodenticides has been calculated to produce the maximum kill of rats. The instructions on the label should be followed. Many of the anticoagulants such as warfarin are sold as a master-mix or concentrate which require to be mixed in the proportion of one part to 19 parts of bait. Mixing of zinc phosphide should be done just before use.

Suitable Mixtures: If a cereal flour or meal is chosen as bait, a satisfactory mix can be obtained by mixing the poison and the meal together for about 1 min.

Quantities for Warfarin Bait

Fine cereal meal or flour	—	19 parts (1900 g)
Warfarin concentrate (0.5%)	—	1 part (100 g)

Quantities for Zinc Phosphide Bait

Fine cereal meal or flour	—	39 parts (975 g)
Zinc phosphide	—	1 part (25 g)

If whole grain bait is chosen, it is necessary to add a 'sticker' to ensure that the poison is evenly distributed. Corn oil or other vegetable oils are the best choice although a mineral oil is also acceptable.

Quantities for Warfarin Bait

Wheat or other grain	—	18 parts (1800 g)
Corn oil	—	1 part (100 ml)
Warfarin concentrate (0.5%)	—	1 part (100 g)

Quantities for Zinc Phosphide Bait

Wheat or other grain	—	36 parts (1800 g)
Corn oil	—	2 parts (50 ml)
Zinc phosphide	—	1 part (25 g)
or		
Zinc phosphide	—	3 parts
Flour	—	80 parts
Groundnut chops	—	16 parts
Edible oil	—	Sufficient to make a lump of the bait

Making Wax Blocks: A very useful way of making long lasting baits is by mixing the poison and the bait with melted paraffin wax. The wax sticks the bait and poison together and small blocks can be prepared by the following method.

Fine cereal meal or whole grain	—	10 parts (1000 g)
Sugar	—	1 part (100 g)
Warfarin concentrate (0.5%)	—	1 part (100 g)
Semi-refined paraffin wax (m.p. 60–65°C)	—	8 parts

To prepare the bait blocks, first heat the paraffin wax just enough to melt it. Next, stir the cereal, sugar and poison until all are thoroughly mixed. Then pour the mixture into flat tray and leave it to set. When set hard enough, cut into blocks each weighing about 100 g. The advantage of wax blocks over loose baits lies in their greater resistance to the weather. They remain palatable to rats even when exposed to rain and sun.

Prepare the bait as described above and keep some quantity in each live burrow and close it with mud. Check next day. If burrow remains closed, the rat is dead. If burrow is opened, rat is not dead. Hence, repeat the bating.

Gassing Poisons: Gassing powders act by releasing poisonous gases which are breathed in by the rodents. There are two main ones which have been used against rats; sodium cyanide and aluminium phosphide. They produce poisonous gases such as hydrogen cyanide and aluminium phosphine respectively, when they come in contact with the moisture.

Insert 6 grams (2 tablets) of cellophos (Aluminium phosphide) into each live burrow and close it with mud. Check next day. If burrow remains closed, the rat is dead. If burrow is opened, rat is not dead. Hence, repeat the fumigation.

6.6.1.2. Trapping

Even when most skillfully used, traps will rarely control rats as effectively as poisons. There use is probably confined to larger rodents being caught for food or to the few survivors of a poison treatment. Many different local traps will catch rats and those such as the break back trap which will kill the animal, solve the problem of having to kill rats that are caught in the traps. Any locally available food that the rats eat can be used for bait which should be placed in the centre of the treadle.

6.6.1.3. Weed and Bush Clearance

There is some evidence that weed control both within the crop and around the field edges will limit the numbers of rodents by reducing the sources of food and living spaces. Many species of rodents do not actually live in the crops they attack, but in the undergrowth, banks and grassland surrounding it. Bush clearance at intervals may well be less costly than poisoning. This may be a very useful additional benefit of weed control done primarily for improving crop yields directly.

6.6.2. Control in Storage

Making the store rodent-proof is probably the most satisfactory approach, but it is not always practicable and regular poisoning may be necessary to keep rodents in check. A very important and inexpensive method of minimizing a rodent problem is scrupulous attention to hygiene in and around the store. All potential rodent living areas–piles of unwanted building materials, wooden packing cases, waste paper and sweepings – should be removed completely and their further accumulation prevented. In addition, all undergrowth should be cut down and kept at a height which will enable any new rat holes and runs to be readily detected.

6.6.2.1. Poisons

The use of multi- or single-dose poisons outlined in the field can be used. Aluminium phosphide tablets will kill rodents in stores, as will other fumigating gases.

6.6.2.2. Trapping

Traps may be useful to get rid of the survivors of a poison treatment or to catch a few rodents getting into a store.

6.7. CONTROL OF TERMITES:

- Locate termite colony, destroy mound, locate the queen and kill it.
- Most effective method of killing the mound is poisoning the mounds by fumigating with cellophos at 2 tablets/metre diameter mound and plastering all the openings with wet mud.
- Pour 60 ml Coroban 20 EC/Dursban 20 EC through a hole made in the centre of closed mounds or through open chimney in case of open mounds.

6.8. CONTROL OF ROOT GRUBS

Incorporate in the soil 2 kg *a.i.* phorate (20 kg Thimet 10 G) or 1 kg *a.i.* carbofuran (33 kg Furadan 3G) or 30 kg Solvirex 5 G or 25 kg Ekalux 5 G per hectare or 100 ml Dursban/ Coroban in 10 litres of water (500 litres to cover 1 ha.) after first summer rains (April-May) for maximum kill of root grubs.

6.9. READY RECKONER FOR MAKING PESTICIDE SPRAY SOLUTIONS

Table 6.2. Ready reckoner for making pesticide spray solutions

% *a.i.* of pesticides in commercial products	Quantity of pesticide formulation to be added per litre of water in gram or ml to get the desired strength of solution (%)											
	0.01	0.02	0.03	0.04	0.05	0.06	0.07	0.08	0.09	0.1	0.2	0.3
2.5	4.00	8.00	12.00	16.00	20.00	24.00	28.00	32.00	36.00	40.00	80.00	120.0
5	2.00	4.00	6.00	8.00	10.00	12.00	14.00	16.00	18.00	20.00	40.00	60.0
10	1.00	2.00	3.00	4.00	5.00	6.00	7.00	8.00	9.00	10.00	20.00	30.0
15	0.67	1.30	2.00	2.60	3.30	4.00	4.66	5.20	6.00	6.66	13.30	20.00
20	0.50	1.00	1.50	2.00	2.50	3.00	3.50	4.00	4.50	5.00	10.00	15.00

Table. 6.2 (Contd.)

% *a.i.* of pesticides in commercial products	**Quantity of pesticide formulation to be added per litre of water in gram or ml to get the desired strength of solution (%)**											
	0.01	**0.02**	**0.03**	**0.04**	**0.05**	**0.06**	**0.07**	**0.08**	**0.09**	**0.1**	**0.2**	**0.3**
25	0.40	0.80	1.20	1.60	2.00	2.40	2.80	3.20	3.60	4.00	8.00	12.00
30	0.33	0.66	1.00	1.30	1.65	1.98	2.31	2.64	2.97	3.33	6.60	9.90
35	0.29	0.58	0.87	1.16	1.45	1.74	2.03	2.32	2.61	2.85	5.70	8.55
40	0.25	0.50	0.75	1.00	1.25	1.50	1.75	2.00	2.25	2.50	5.00	7.50
50	0.20	0.40	0.60	0.80	1.00	1.20	1.40	1.60	1.80	2.00	4.00	6.00
75	0.13	0.27	0.40	0.53	0.66	0.79	0.92	1.07	1.20	1.33	2.66	4.00
80	0.12	0.25	0.37	0.50	0.63	0.75	0.88	1.00	1.13	1.25	2.50	3.75
85	0.12	0.25	0.36	0.48	0.59	0.70	0.82	0.96	1.06	1.18	2.35	3.54
100	0.10	0.20	0.30	0.40	0.50	0.60	0.70	0.80	0.90	1.00	2.00	3.00

Note: The pesticides should be measured in g or ml. For preparation of malathion 0.05% spray solution from 50% malathion, the value at the point of intersection of 10th row and 5th column should be taken *i.e.*, 1.0 ml should be dissolved in 1 litre of water.

Preparation of Spray Solution: The spray solution must be prepared to the exact concentration as recommended. The quantity of the commercial pesticide formulation required for preparation of the spray solution is calculated by using the following formula:

$$\text{Quantity of the pesticide formulation} = \frac{\text{Quantity of spray solution required} \times \text{Conc. of solution desired}}{\text{\% of } a.i. \text{ of pesticide in the formulation}}$$

Prepare spray solution first in small quantity and then increase the volume to desired level by adding water. In case of wettable powder, take required quantity of pesticide, add a little quantity of water, mix it thoroughly to prepare the paste and then add remaining quantity of water to this paste with constant stirring.

6.10. COMPATIBILITY CHART

See the Table 6.3 as given on next page.

6.11. PEST AND DISEASE MANAGEMENT IN NURSERIES

Among numerous diseases which cause severe damage, damping-off is the most important disease. Besides this, seed rotting, bacterial blight, downy mildew, black rot, termite, leaf miner, flea beetle, whitefly and nematodes are other diseases and pests in nursery. By the time tender radical and plumule are put forth in the nursery bed, they are killed by several seed and soil-borne fungi and bacteria. The pre-emergence damping-off and seed rotting are the usual causes for seedling stand and cause hidden losses. The post-emergence damping-off is a more serious problem. Seedlings just after emergence get infected near the soil surface (collar region) and give wire stem symptoms. The tissue near collar region become soft, water-soaked, constricted and weak causing the seedlings to collapse and

Table 6.3. Compatibility Chart: Precautions to be taken while using insecticides, fungicides and other agrochemicals

No.	Pesticides/Agrochemicals	1	2	3	4	5	6	7	8	9	10	11	12	13	14	15	16	17	18	19	20	21	22
1	Plant origin insecticides@	a	a	a	a	a	a	a	a	a	a	b	a	a	b	a	a	a	a	b	b	c	a
2	Carbaryl	a	a	a	b	a	c	a	d	a	a	a	d	a	d	a	d	a	a	a	d	a	a
3	DDT group	a	a	a	a	a	a	a	a	a	a	a	d	a	a	a	b	a	a	a	b	a	a
4	Monocrotophos	a	b	a	a	a	a	a	a	a	a	a	b	a	a	a	a	a	a	a	b	a	a
5	Dimethoate	a	a	a	a	a	a	a	a	a	c	a	c	a	a	a	b	a	a	a	b	a	a
6	Endosulfan	c	a	a	a	a	a	a	d	a	a	a	b	a	c	a	b	a	a	a	c	a	c
7	Fenitrothion	a	a	a	a	a	a	a	a	a	a	a	b	d	a	a	b	a	a	a	b	a	a
8	Kerosene	a	d	a	a	a	d	a	a	a	a	a	b	b	a	b	b	b	b	b	b	a	c
9	Malathion, parathion, demeton	a	a	a	a	a	a	a	a	a	a	a	b	d	a	a	a	d	a	a	a	a	a
10	Oxy-demeton methyl	a	a	a	a	c	a	a	a	a	a	a	a	c	a	a	a	c	c	a	a	c	c
11	Phoslane	d	a	a	a	a	a	a	a	a	a	a	a	d	a	a	a	a	a	b	b	b	b
12	Bordeaux mixture	b	b	d	b	b	b	b	b	b	b	d	a	b	d	a	b	b	a	a	a	b	a
13	Captan, difolatan	a	a	a	a	a	a	d	b	d	a	a	b	a	d	a	b	a	d	a	b	a	a
14	Copper oxychloride	a	b	a	a	a	c	a	a	a	c	a	a	d	d	d	b	b	a	b	a	a	a
15	Dithiocarbamate group#	a	a	a	a	a	a	b	a	a	a	a	a	d	a	b	d	a	a	a	a	a	—
16	Lime sulphur	b	b	b	b	b	b	b	b	b	b	d	b	b	b	b	a	b	a	a	a	a	b
17	Karathane	a	a	a	a	a	a	a	b	a	a	a	b	e	b	a	b	a	b	c	c	b	c
18	Wettable sulphur	a	a	a	a	a	a	a	b	a	a	a	a	d	a	a	a	a	a	a	a	a	a
19	2,4-D	b	—	a	a	a	c	a	b	a	c	c	a	a	b	a	a	c	a	a	a	c	a
20	Lime	b	b	b	b	b	c	b	b	b	c	a	b	a	a	a	a	a	a	a	a	a	a
21	Streptomycin	c	a	a	a	a	a	a	c	a	c	c	a	a	a	a	a	a	a	c	c	a	c
22	Urea	b	a	a	a	a	c	a	a	a	c	c	a	a	a	b	a	a	a	a	b	c	a
23	Carbendazim	a	a	a	a	-	a	-	—	a	—	—	a	a	—	a	—	—	a	a	—	—	a

@ = Plant origin insecticides: Rotenene, Pyrethrum. # = Dithiocarbamates: Maneb, Zineb, Nabam, Thiram, Ziram, Mancozeb.
a = Can be mixed. b = Not to be mixed. c = Information not available. d = Use immediately.
e = Usually not mixed.

later on die. Damping-off is caused by several species of soil fungi such as *Rhizoctonia, Pythium, Fusarium, Phytophthora, Phomopsis, Sclerotium, Colletotrichum, Phoma* and *Alternaria*. Heavy soil, excessive rain, high soil moisture, dense planting, low light intensity and high dose of nitrogen favour damping-off. Following practices have been recommended for the management of pests and diseases in crops (Pandey, 2000):

6.11.1. Soil Solarization

Soil solarization by the use of white transparent polythene film during hot summer is one of the most effective approaches for the management of soil-borne diseases particularly in nursery beds. The main objective is to eliminate pathogens, weeds, toxic metabolites, insects, etc. The dominant soil pathogens like *Macrophomina phaseolina, Fusarium oxysporum, Meloidogyne* spp., *Pseudomonas* spp., *Streptomyces scabies, Verticillium dahliae, Phytophthora* spp. and *Pythium* spp. which are responsible for nursery diseases were effectively reduced by soil solarization.

6.11.2. Cultural Methods

Nursery beds should be properly drained and pulverized. The soil should be light and made porous by adding equal quantity of sand and compost. Sowing of seeds should be thin enough to provide proper aeration and light to check damping-off infection. Germinating seeds and juvenile seedlings should be protected from heavy rain, scorching sunlight through agronet, muslin cloth, shading net or polythene portable tunnel arch.

6.11.3. Organic Amendments

Decomposition of organic matter in amended soil has a major impact on disease suppression. Reduction in disease is largely induced by high soil temperatures during decomposition of organic amendments. Many types of micro-organisms (*Trichoderma* spp., *Streptomyces* spp., *Pseudomonas* spp., *Bacillus* spp., *Enterobacter* spp., *Flaviobacterium* spp.) present in compost function as biocontrol agents and their population is substantially increased after peak heating. Organic amendments are known to reduce the population of nematodes like *Meloidogyne* spp., *Helicotylenchus* spp. and *Pratylenchus* spp (Hoitinik & Peter, 1986).

6.11.4. Biological Control

Biological control includes practices that create an environment favourable to antagonists and mass introduction of antagonists, non-pathogenic strains of other beneficial organisms. Three antagonistic mechanisms *viz.*, competition, antibiosis and mycoparasitism are involved in biological control of soil-borne pathogens. Competition mechanism works in the control of *Fusarium oxysporum* by *Trichoderma harzianum*. Antibiosis occurs when the biocontrol agent produce antibiotics which have direct effect on the pathogen. The antagonists like *T. virens* produce gliotoxin, viridin by *T. viride* and trichodermin by *T. viride* and *T. polysporum* which are effective against *Pythium ultimum* and *Macrophamina phaseolina*.

Soil-borne pathogens causing root rot, damping-off, collar rot and wilt (*Rhizoctonia solani, Pythium* spp.*, Fusarium* spp.*, Phytophthora* spp.*, Macrophamina phaseolina, Sclerotinia sclerotiorum, Phoma* spp. and *Aspergillus* spp., *Botrytis* spp., *Armillaria* spp., *Verticillium* spp. and *Erysiphe* spp.) are the major targets of the biocontrol agents. Many foliar diseases like blight (*Alternaria* spp.) and die-back/anthracnose (*Colletotrichum* spp.) are also suppressed by the *Aspergillus niger* V isolate (Pandey & Pandey, 2001).

6.11.5. Chemical Control

The easiest control measure for damping-off is to treat soil and seed. Five litres of 0.3% captan should be uniformly applied/m^2 area of nursery bed. Seed treatment with 0.25% carbendazim should be preferred over captan before sowing, considering the internal seed-borne inoculum of the pathogen. In general, seeds should be treated with 0.25% imidachloprid for protection against sucking pests in nursery. After complete germination of all the seeds, spray 0.3% copper oxychloride followed by 150 ppm streptocycline in such a way that seedlings and their roots are covered by the chemicals.

6.12. SOIL SOLARIZATION

Soil solarization is a method of killing soil-borne pathogens (fungi, bactereia, nematodes), soil insects and weed seeds by heat. This technique consists of covering moist soil with a plastic film during periods of intense sunshine and heat, thereby capturing radiant heat energy from the sun, causing physical, chemical and biological changes in the soil. The soil temperature is increased (by 5–15°C) to levels lethal to many soil-borne fungi, bacteria, nematodes and weeds. The short wave incident solar radiation penetrates the polyethylene sheet but the long wave radiation is prevented from soil, thus trapping the heat resulting in thermal inactivation and production of heat shock proteins and irreversible heat injury. Solarization also improves plant nutrition by increasing the availability of N and other essential elements. This practice can be profitably used on a small scale and is especially useful for nursery beds.

Soil solarization using thin transparent polyethylene mulches for six weeks caused as much as 72.6 and 88.5% decrease in the population densities of *Tylenchorhynchus vulgaris* and *Hoplolaimus indicus*, respectively in nursery beds. Application of mahua cake before solarization usually caused slightly greater reduction in nematode population (Kamra & Gaur, 1998).

6.13. MASS PRODUCTION OF BIOAGENTS

6.13.1. *Pseudomonas fluorescens*

A loopful of *P. fluorescens* should be transferred to 100 ml of King's B broth in a 250 ml conical flask under aseptic conditions. The flask can be incubated at room temperature for 48 hours. The population of *P. fluorescens* in the broth can be assessed (a minimum of 9×10^9 cfu/ml is necessary).

The pH of the substrate (talc powder) can be adjusted to 7.0 by addition of 150 g calcium carbonate/kg of talc powder and sterilized at 1.1 kg/cm^2 for 30 min for 2 successive days. One kg of the sterilized substrate (talc powder) and 5 g of carboxy methyl cellulose should be placed in a polythene bag or any sterile container under aseptic conditions, 400 ml of *P. fluorescens* suspension to be added and mixed thoroughly. The formulation can be stored in a polythene bags for one month. The product should contain a minimum of 2.5×10^8 cfu/g.

The product should have the following quality parameters:

- Fresh product should contain 2.5×10^8 cfu/g.
- Three months after storage at room temperature, the population should be 8-9 $\times 10^7$ cfu/g.
- Storage period is 3–4 months.
- Minimum population load for use is 1×10^8 cfu/g.
- The product should be packed in white polythene bags.
- Moisture content of the final product should not be more than 20%.
- Population in broth should be $9 \pm 2 \times 10^8$ cfu/ml.

The composition of King's B medium is as follows:

Peptone	—	20.00 g
Dipotassium hydrogen phosphate	—	1.50 g
Magnesium sulphate	—	1.50 g
Glycerol	—	10.00 ml
Agar	—	15.00 g
Distilled water	—	1000 ml
Cycloheximide	—	100 ppm
Ampicillin	—	50 ppm
Chloramphenicol	—	12.50 ppm

(King's B broth is prepared without the agar and antibiotics)

Liquid Formulation Technology to extend the Shelf Life of *P. fluorescens*

Polyvinyl Pyrrolidone (PVP) (2%) was tested as an effective preservation agent in protecting the cells of *P. fluorescens* for more than a year without affecting the natural properties of bacteria. The antagonistic efficacy of preserved cells was also confirmed *in vitro* and *in vivo*.

6.13.2. *Trichoderma viride, T. harzianum*

Among the various fungal biological control agents which are effective against nematode and fungal diseases of horticultural crops, *Trichoderma viride* and *T. harzianum* have a good market. They can be easily mass produced on several organic substrates.

6.13.2.1. Small Scale

The mycelial discs grown in potato dextrose agar medium was transferred to 250 ml conical flasks containing sterilized molasses yeast medium (Yeast extract –5 g, Molasses

–30 g, Distilled water –1 litre sterilized at 1.42 kg/cm^2 for 20 min) at 100 ml /flask and incubated for 10 days. The mycelial mat along with the broth was homogenized and mixed with talc powder (500 mesh) at 1: 2 ratio. Carboxy methyl cellulose (CMC) was added at 5 g/kg of the product as sticker. The product was shade dried until it attained 12% moisture level. The clumps were broken, homogenized and packed in polythene bags of 100g capacity. The required population level at the time of packing should be 28 × 10^7 cfu/g of the product. The shelf life of the product was 4 months. At the end of expiry, the cfu should not be less than 20 × 10^6 cfu/g of the product.

6.13.2.2. Large Scale: Add 70 ml of molasses yeast medium in each 250 ml conical flasks and plug with non-absorbent cotton wool. Autoclave at 1.42 kg/cm^2 for 1 hr. Transfer 8 ml mycelial disc of *T. viride/T. harzianum* to each flask and incubate for 10 days. This mother culture is used for inoculating the medium in fermentor.

Add 50 litres of molasses yeast medium in 50 litre fermentor, sterilize at 1.42 kg/cm^2 for 20 min. and cool to room temperature. Add 2.5 litres of inoculum (from mother culture) into the sterilized medium and incubate for 10 days. Provide aeration in the fermentor for 4–8 hr/day by agitation. Presence of greenish fungal growth indicates *T. viride/T. harzianum*. This broth culture contains fungal mat, spores and antibiotics. Draw samples 10 days after incubation and assess the spore load in the broth which should be at least 10^8 spores/ml. Mix one litre of broth culture with 2 kg talc powder, add CMC at 5 g/kg of the product, air dry (12% moisture content) and sieve. The population should not be less than 20 × 10^6 cfu/g of the product. It is packed in opaque polythene bags and stored in a cool dry place before use. The product can be safely stored for 4 months without losing its viability.

6.13.3. Arbuscular Mycorrhizal Fungi

6.13.3.1. In Glasshouse

Cement pots or tanks are filled with sterilized sand: soil (1: 1 by volume) or soilrite: peat (1: 1 by volume) mixture. A standard starter culture of AMF obtained from any authentic source is placed as a band 3 cm below the soil surface. The seeds of Rhoades grass/finger millet/maize are sown on them and covered with soil. Pots are irrigated normally. When the plants are about 10–12 weeks old, the top shoots are cut off and the root system colonized by the fungus along with the soil containing chlamydospores is used as the inoculum.

6.13.3.2. On-farm Production

A suitable land (even waste land) is selected near the field where the fungus inoculum has to be used. The land is brought to the fine tilth by ploughing, harrowing and cultivator operations. A raised bed (20 cm height) of soil: sand (1: 1 by volume) mixture is prepared on an even land surface. The raised bed is sterilized by soil solarization for 2 to 3 weeks using polythene sheet. The starter culture of AMF is applied at a depth of 3–4 cm from the top by removing the top soil. The seeds of Rhoades grass/finger millet/maize are sown on them and covered with soil. Irrigation is done normally. When the crop attains a height of 10–15 cm, 12 weeks after sowing, the shoot portion is cut off at ground level. The soil is dug up to a depth of 20 cm along with root system is used as the inoculum.

6.13.3.3. Method of Inoculation

Crops can be inoculated with AMF in the nursery beds. The AMF inoculum consisting of chlamydospores, soil and infected root bits is either spread uniformly in the nursery beds or placed in rows over which seeds of crops are sown. The seedlings get colonized by the fungus in about 4–5 weeks after which they are transplanted in the main field. Inoculated plants show better growth and improved nutrient content.

6.13.4. NPV of *Spodoptera litura*

6.13.4.1. Production and Utilisation

- Collect 7–9 days old larvae (4th instar) and starve them for 8 hours.
- Prepare *Sl* NPV suspension (10^8 POBs-Polyhedral Occlusion Bodies/ml) in 250 ml of water with teepol in a bottle.
- Dip clean castor leaves in *Sl* NPV suspension, shade dry for 15–20 min and feed the larvae for 2 days.
- Provide fresh non-infected leaves for the larvae thereafter.
- Alternatively, expose larvae to semi-synthetic diet treated with *Sl* NPV suspension.
- Larvae die after 7 days and their under surface turns pink and later white.
- Collect dead larvae and allow it to putrify in water for a week.
- Grind dead larvae and filter through muslin cloth.
- Remove the supernatant and suspend the polyhedra in water.
- Centrifuge at 500 rpm for 5–10 min and discard the supernatant retaining the POBs.
- Centrifuge further at 2500 rpm for 5–10 min and suspend the white POBs in water.
- Count using Neubauer haemocytometer and express on larval basis or per unit of larval weight.
- Dry the POBs over calcium chloride or precipitate by acetone.
- Formulate by adding permitted spreaders/wetting agents.

6.13.4.2. Nursery and Field Sprays

- Spray *Sl* NPV at 250 larval equivalent (LE) mixed in 125 litres of water added with crude sugar at 1 ml/l of water and teepol at 1 ml/10 litres of water in the evening hours for one ha of nursery.
- Spray thrice at fortnightly intervals from the day neonate larvae are sighted.
- Continue spraying up to three weeks after planting.
- Spray when eggs start hatching in castor at weekly intervals.
- The dosage varies from 250 LE to 450 LE/ha.
- Larvae feeding on sprayed plants turn pinkish white, move to top of the plant, hang upside down and die (Fig 6.1).

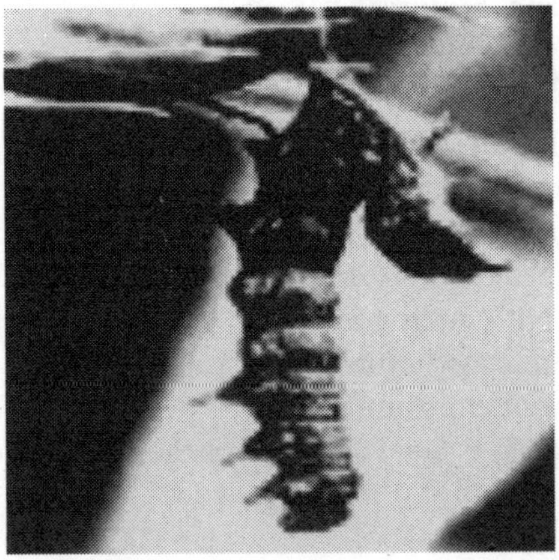

Figure 6.1. Infected larva hanging upside down (Tree top disease)

6.13.4.3. *Spodoptera litura* Culture

- Allow a pair of newly emerged moth in a well ventilated plastic container (20 × 15 cm) lined (inner surface) with paper to obtain eggs (Fig 6.2).
- Cover the bottom sponge lining with blotting paper.
- Provide 50% honey solution and water in small plastic cups.
- Cut out the egg batches on the paper.
- Dip eggs in 10% formalin for 30 min, wash in running water for 30 min and dry using blotting paper.
- Hatch in sterilized glass vials (20 g capacity).
- Transfer 3 days old larvae in bouquets of sterilized castor leaves for feeding.
- Collect the pupae, sterilize with sodium hypochlorite, wash in water and dry with blotting paper.
- Identify the sex with the help of ovipositor and last abdominal segment and keep over a wet sponge in an adult emergence cage (22 × 15 cm).
- Adults emerge after 10 days.

Figure 6.2. *Spodoptera* rearing unit

6.13.5. NPV of *Helicoverpa armigera*

6.13.5.1. Production and Utilisation

- Prepare the viral solution (virus at 18×10^6 POBs/ml of distilled water).
- Feed 5–7 day old larvae with virus infected diet.
- Incubate at 26°C for 7 days.
- Collect the dead infected larvae and macerate in distilled water.
- Standardize the *Ha* NPV with number of POBs/ml in terms of $LC_{50.}$
- Store POBs in plastic cans/bottles with proper instructions on the containers even up to 6 months without exposure to sun.
- 1 larval equivalent (LE) = 6×10^9 POBs or their equivalent in activity.

6.13.5.2. Field Spray

- *Ha* NPV is sprayed to control *H. armigera* on chickpea, pigeonpea, field bean, and tomato.
- Spray 3–4 times of *Ha* NPV 250 LE/ha with crude sugar at 5 ml/l of water, groundnut oil cake at 1 ml/100 ml of water and chickpea flour at 1 ml/100 ml of water.
- After 3–5 days, dead larvae will be visible hanging upside down.

6.13.5.3. *Helicoverpa armigera* Culture

- Collect the adult insects in light traps or from endemic crop areas.
- Place the adults in oviposition cages (cylindrical iron frame-50 cm height × 30 cm diameter and black or white cloth cover) at 100 pairs/cage.
- Hang 2 cotton swabs dipped in 50% honey and water over a circular plastic mesh, 5 cm from frame base.
- Place the frame on aluminium tray (40 × 40 × 5 cm) with a 3 cm thick sponge at the bottom soaked in water.
- Remove eggs daily from the inner side of the cloth cover.
- Sterilize the eggs with 10% formalin, wash in distilled water for 10 min and dry in laminar air flow chamber.
- Place the eggs in a moist cotton inside a flask and wrap with aluminium foil.
- Larvae emerges after 3–4 days and place them in multi cellular trays.
- Larvae are fed with chickpea based nutritious semi-synthetic diet.
- Harvest pupae after 18–20 days, sterilize, wash and dry.
- Separate the male and female pupae and place in adult emergence cages.
- Adults emerge after 7–8 days.

6.13.6. Trichogrammatid Egg Parasitoids

6.13.6.1. Database

- Trichogrammatid egg parasitoids are multiplied on eggs of *Corcyra cephalonica,* a factitious host.

- One cc of *Corcyra* eggs contains 20,000 eggs. This is supplied in a card.
- Eight releases of Trichogrammatids and each release with 1,40,000 are recommended to contain first and second generation eggs/ha.
- Releases are spaced at weekly intervals.
- 56 cc of *Corcyra* eggs are to be produced/ha (7 cards/release × 8 releases).
- 56,000 cc of *Corcyra* are required/1000 ha. *i.e.*, 56,000 cards.
- 150 g of sorghum grains are required to produce 1 cc of *Corcyra* eggs.
- 8,400 kg of sorghum grains are required as feed to produce 56,000 cc.
- Approximately 10 t of sorghum grains are required as feed to produce natural enemies to cover 1000 ha.
- The parasitised *Trichogramma* egg card should be handed over to the growers within three days after despatch.
- It cannot be stored in dealers/retailers shop like antagonastic fungi/ biofertilizers/ pesticides/fertilizers.

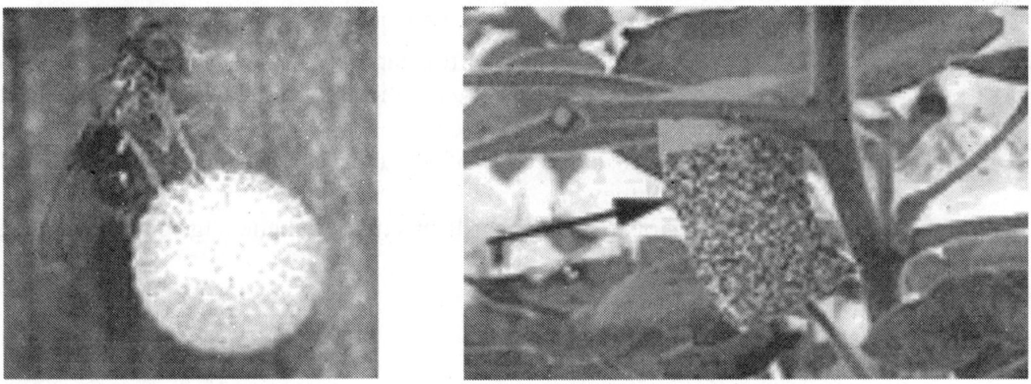

Figure 6.3. *Tricho* adult oviposit on *Corcyra* egg, *Tricho* card

6.13.6.2. Production: *Corcera cephalonica*

Ideal Conditions for Rearing

- Temperature: 27 ± 1°C.
- RH: 60 ± 5%.
- Light: 200–250 lux and avoid direct sun light.

Rearing

- In each box, put 2.5 kg broken cereal or millets, 10 tablets of dry yeast and 0.5 g streptomycin sulphate.
- Keep 0.5 cc *Corcyra* eggs and secure the lid of the charge box for 40 days.
- On 40th day, moths start to emerge and continue to emerge for 2 months with a peak emergence from 45th to 75th day.

- Emergence rate is 5–15 moths/box/day in the beginning and it reaches 75–100 moths/box/day during the peak.
- Each box produces 0.5 cc eggs/day during the peak emergence.
- Collect the moths daily using mechanical moth collection devices and transfer them to special oviposition cages.
- Discontinue the use of box and burn the contents properly after 80 days of infestation when less than 10 moths/box emerges.
- Pass the collected eggs through 15, 30 and 40 mesh sieves and run over a slope of paper or by automatic egg cleaner to eliminate the dust particles.
- Treat the eggs with UV rays (30 W UV tube for 45 min at a distance of 60 cm) to prevent eggs from hatching, if they are meant for *Trichogramma* production.

Note: Eggs meant for multiplication of *Corcyra* should not be treated with UV rays.

6.13.6.3. Production and Utilisation

- Collect cleaned and UV treated *Corcyra* eggs.
- Glue the eggs to "tricho" cards measuring 15 × 10 cm, which are prepunched into 16 pieces of 4 × 3 cm leaving some space for stapling.
- Expose the eggs to adult female *Trichogramma* in the ratio of 30: 1 for 24 hr and allow it to parasitise.
- Expose females to the eggs in the ratio of 40: 1 on cards inside polythene bags, where female oviposits till they die.
- Transfer 3–5 days old parasitised egg cards in safe polythene bags and place dried honey for emerging parasitoids.

6.13.6.4. Field Release

- Cut "tricho cards" into small bits.
- Release cards 8 days after parasitisation for emergence of adults.
- Staple the bits to lower side of the leaves at 100 spots/ha in the evening hours at weekly intervals.
- Release is done based on the appearance of eggs of host.

6.13.7. Chrysopids

6.13.7.1. Production and Utilisation

Ideal Conditions for Rearing

- Temperature: 27 ± 1°C.
- RH: 70%.
- Light: Normal illumination by a fluorescent tube.
- Room area: 6 × 6 m.

Rearing

- Rear the adults with artificial honey based diet and larvae on *Corcyra* eggs separately.

- Keep 200 adults in the oviposition cage measuring 75 × 30 × 30 cm with a sliding top cover.
- Cover the sides of the cage with smooth nylon lining mesh of '0' size which is unsuitable for egg laying.
- Fit a black cloth on the top cover for laying of eggs.
- Replace the top cover with eggs on alternate days starting from 5th day.
- One batch is kept for 30–35 days and dead adults are removed every alternate day.
- Wait for 24 hr till the eggs harden and gently rub with a soft sponge.
- Expose *Corcyra* eggs to UV rays 30 W tube light for 45 min at 60 cm distance and inactivate them.
- Mix 3 day old 120 chrysopid eggs with 1 cc *Corcyra* eggs in a plastic container (27 × 18 × 6 cm).
- On hatching, *Chrysoperla* larvae feeds on *Corcyra* eggs.
- Remove the larvae on the 3rd day and transfer them into individual 2.5 cm cubical cells of plastic louvers.
- Each louver can hold 192 larvae.
- Sprinkle *Corcyra* eggs in all cells.
- Feed the larvae in 2 doses, first with 1.5 cc and second with 2.5 cc for 100 larvae with a gap of 3–4 days.
- Total quantity of *Corcyra* eggs required for rearing 100 *Chrysopid* larvae is 5 cc.
- Secure each louver on one side by brown paper and after transfer of larvae cover it with acrylic sheet. Brown paper would facilitate clear visibility of *Corcyra* eggs and also formation of cocoon by Chrysopid larvae.
- Collect cocoons after 24 hours of formation by removing paper from one side.
- Place them in adult oviposition cages for emergence.
- One set of louver remains in use for 13–15 days, then they are completely cleaned before reuse.

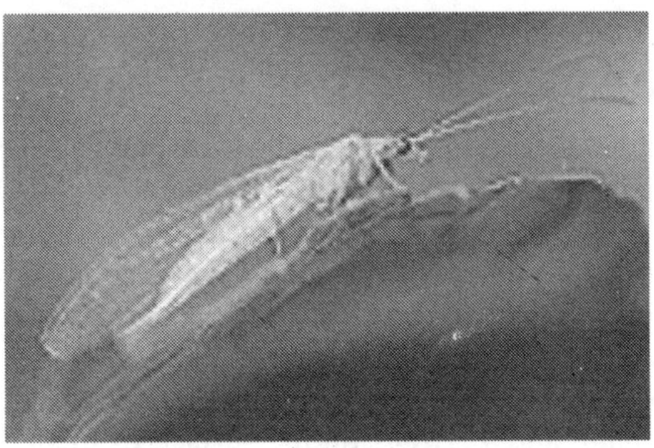

Figure 6.4. Chrysopid adult

6.13.7.2. Field Release

- Mix viable *Chrysoperla* eggs with *Corcyra* eggs. Add saw dust and paper strips. This is done before shipment. Once larvae emerge, they are released in the field.
- Release in the early morning by broadcasting larvae on the plants along with saw dust and paper strips.
- 10,000–20,000 larvae/ha is required in 2 releases which varies with crop, pest predator load.

6.14. ENRICHMENT OF FYM WITH BIOAGENTS (*TRICHODERMA HARZIANUM, PAECILOMYCES LILACINUS)*

Commercial formulation of the bioagent (*T. harzianum/P. lilacinus*) (1 kg/ton) was mixed with moist FYM along with 50 kg of neem/pongamia cake. Mixture was spread 15 cm high heap under shade, covered with polyethylene sheet and incubated for 3 weeks at 25 to 30$^{\text{Ë\%}}$C and sprayed with water regularly to maintain moisture.

Colonized FYM can be used at 6 g/kg soil in pot culture experiments (12 g/2 kg soil in 15 cm diameter pots). It can also be used for main field treatment at 5 tonnes/ha before sowing of seeds or transplanting of seedlings. It can be applied at 2 kg/plant of banana, acid lime, papaya, pomegranate, grapes and roses along with 400 g of neem/pongamia cake once in every 6 months after planting for the management of nematodes and other soil-borne pathogens.

T. harzianum colonized compost significantly enhanced germination and growth of crops as compared to non-colonized FYM. Population of *T. harzianum* in colonized compost was several folds higher (48.6×10^{12}) than non-colonized FYM (3×10^{4}).

Colonization of FYM by *T . harzianum* helps to improve the quality of compost. It significantly enhanced total and water soluble humic matter. A number of macro and micro-nutrients like P, K, S, Zn, Cu and Fe (both total and water soluble content) were significantly higher in *T. harzianum* colonized compost. Six times increase in water soluble humic content due to colonization of *T. harzianum* seems quite important as it is reported to enhance the plant growth.

6.15. SEED BIOPRIMING/SOLID MATRIX PRIMING

Seed biopriming (treatment of seeds with a slurry of biocontrol agents mixed with an organic carrier and incubated for 4 days under warm and moist conditions with relative humidity of 80–100% until just prior to radical emergence) has potential advantages over simple coating of seeds as it results in rapid and uniform seedling emergence. *Trichoderma* required less moisture for growth. *Trichoderma* conidia germinate on the seed surface and form a layer around bioprimed seeds. Such seeds tolerate adverse soil conditions better. Biopriming could also reduce the amount of biocontrol agents that is applied to the seed. Population of *T. harzianum* and *T. virens* on the surface of treated seeds increases from 10^{4} cfu per seed up to 10^{6}–10^{7} cfu per seed after biopriming.

6.16. FORMULATION OF ALGINATED PELLETS OF BIOCONTROL AGENTS

Alginate is a water soluble polysaccharide gum that has excellent gel-forming properties. Living organisms that are capable of biologically controlling a variety of pathogens can easily be entrapped in a calcium or sodium alginate matrix in aqueous form at room temperature. A pellet formulation of biocontrol agents was prepared by mixing sodium or calcium alginate, clay and streptomycin sulfate. The mixture was homogenized in a blender and then mixed with bioagent propagules. The gel beads, granules or seed coatings that are produced by the process are biodegradable. The encapsulation in Ca or Na alginate maintained the viability of the bioagents for 12 months at 7°C and 5 months at room temperature. Antagonistic fungi (species of *Paecilomyces, Trichoderma, Gliocladium, Talaromyces* and *Pencillium*) and bacteria (species of *Pseudomonas* and *Bacillus*) that can control soil-borne pathogens can be formulated in Ca or Na alginate (Connick, 1988). Alginated pellets can be used for soil application at 400 to 800 pellets/sq.m.

6.17. PRODUCTION OF BIOAGENT COLONIZED SEEDLINGS

The horticultural crops are invariably attacked by soil borne pathogens (nematodes, fungi, bacteria) at the seedling stage itself. Securing healthy seedlings of horticultural crops is essential to ensure freedom from soil-borne pathogens, optimum plant stand, good growth and higher yields. The following methods may be adopted to produce bioagent colonized seedlings of horticultural crops:

6.17.1. Seed Treatment

Seeds have to be treated with talc based formulations of *Pseudomonas fluorescens, Trichoderma harzianum* and *T. viride* at 10 g/kg seed before sowing.

6.17.2. Nursery Bed Treatment

The nursery beds may be treated with *P. fluorescens/ T. harzianum/ T. viride* at 50g/m^2 along with neem/pongamia cake at 200 g/m^2 before sowing.

6.17.3. Treatment of Coco Peat or any other Substrate

The coco peat or any other substrate used for producing the nursery seedlings in portrays is treated with *P. fluorescens/T. harzianum/T. viride* at 1 kg/tonne before filling with the substrate.

6.17.4. Treatment of Soil Mixture

The soil mixture used for producing the nursery seedlings in polythene bags in a conventional manner is treated with *P. fluorescens/T. harzianum/T. viride* at 2 kg/tonne along with neem/ pongamia cake at 50 kg/tonne before sowing.

6.18. TRADITIONAL PEST CONTROL

Product	Preparation	Pests controlled
Neem leaf extract	Chop 6 kg fresh neem leaves and make it into a paste; soak the paste in 6 litres of water overnight; on the following morning boil the suspension and filter by pouring over a coarse cloth; add 150 g of detergent soap powder to the filtrate; make up the volume to 60 litres; this is ready to spray.	Most of the leaf feeding caterpillar pests.
Vitex leaf extract (Nochi)	Grind 6 kg fresh leaves into a paste; soak the paste overnight in 3 litres of cow's urine; on the following morning, stir the solution and boil for 3 hours; allow it to cool; pour it over a coarse cloth and squeeze; to the filtrate, add 150 g of detergent soap powder; the filtrate is then made up to 60 litres by adding water; this is ready to spray.	Castor semi-looper.
Calotropis leaf extract	Chop1 kg calotropis leaves and make it into a fine paste; mix the paste with 50 litres water and filter the suspension and spray.	Caterpillar pests.
Jatropha leaf extract	Grind 5 kg of leaves and stem of Jatropa into fine paste; soak the paste in 5 litres of cow's urine overnight; on the following morning, add 2 litres of water and boil for 3 hours; the solution is then concentrated to 5 litres and filtered; this is ready to spray.	All pests.
Tobacco leaf extract	Cut 5 kg of tobacco waste into pieces and soak in 5 litres of cow's urine for 1 week; filter the suspension and mix 150 g of detergent soap powder and 250 ml of water; spray at 50 ml/l water.	Aphids, thrips, beetles, lepidopteran caterpillars.
Agave flesh extract	Grind chopped agave leaves and make it to a paste by thoroughly mixing it; soak it in water for 24 hr; mix the suspension and filter it with a coarse cloth; add 10 g of detergent soap powder and 500 ml of cow's urine to every 100 ml of extract and spray.	Leaf folder.
Garlic clove extract	Chop garlic cloves and make it into a fine paste; soak it in equal volume of kerosene over night; mix and filter it; add 150 g of detergent soap powder.	Brown leaf hopper, green leaf hopper, thrips, diamond back moth.

Table. (Contd.)

Product	Preparation	Pests controlled
Mahua oil spray	Mix 30 ml of mahua oil and 5 g of detergent soap powder with 1litre of water and spray.	
Nerium seed extract	Seeds of nerium are dried, powdered, sieved and soaked in water over night; filter with coarse cloth and spray the filtrate.	Rhinocerous beetle.
Cow dung extract	Mix 100 g cow dung in 100 ml water; filter it and spray.	
Ipomea leaf extract	Clip 5 kg of fresh Ipomea leaves and grind to fine paste; mix paste with equal volume of water and boil; add a little of cow's urine to the boiling mixture; when the mixture is ¼ its volume stop boiling and allow it to cool overnight; filter the solution; repeat the washing with cow's urine to make a volume of 4.5 litres; add 250 g of detergent soap powder in 500 ml water; mix the extract with the soap solution; use 1.25 litres of solution in 25 litres water.	All pests.

7 SOURCES OF CRITICAL INPUTS FOR THE MANAGEMENT OF PESTS AND DISEASES

7.1. INSECTICIDES

Table 7.1. Common and trade names and sources of availability of insecticides

Group	Common Name	Trade Name/s (Source)
Chlorinated Hydrocarbons	Endosulfan	Agrosulfan (16), Dawn (42), Devigor (9), Endocel (14), Endocin (23), Endogaurd (15), Endodhan (13), Endotaf (37), Endonil (11), Endoin (20), Endosul (46), Endosol (28), Hexasulfan (32), Hildan (17), Hysulfan (18), Lusulfan (26), Nagusulphan (29), Parrysulfan (18), Sedostar (41), Speed (31), Thiodan (4), Tagsulfan (50), Thiokill (52), Topsulfan (49).
	Gamma BHC	Lindane
Organophos-phates	Acephate	Acatin (23), Acehero (39), Acekill (5), Acemil 75 WP (28), Acet (19), Acetop (49), Acifete (26), Aimthene (1), Artin (32), Asataf (37), Dhanraj (11), Hilphate (17), Hythene 75 SP (18), Lancer (52), Laser (20), Orthane (9), Pace (31), Specthane (43), Starthane (41), Tamaron Gold (4), Tarpedo 75 SP (40), Tremor (55), Volace (54).
	Chlorfenvinphos	Chlorfenvinphos
	Chlorpyriphos	Agro Chlore (16), Blaze (21), Chorobon (52), Chlorgaurd (15), Classic (26), Coroban (7), Corocin (23), Cyphos (19), Dhanwan 20 (11), Doomer (5), Durmet 20 EC (8), Force (31), Hexaban (32), Hilban (17), Hyban (18), Nuklor 20 EC (12), Pyrivol (54), Radar (40), Robust (39), Ruban (9), Scout (1), Specphos20 (42), Starban (41), Strike (55), Sulban (46), Tafaban (37), Tagban 20 EC (50), Tarkash (3).
	Diazinon	Basudin (47), Tik-20 (37).

Table. 7.1 (Contd.)

Group	Common Name	Trade Name/s (Source)
	Dichlorvos (DDVP)	Agro-DDVP (16), Amidos (1), Dicotop (49), Divap 100 (35), Doom (52), Hilfol (17), Luvan (26), Marvex Super (32), Nuvan (47), Specvos (43), Suchlor (45), Vantaf (37), Vapona (3).
	Dimethoate	Hygro (18), Rogar (37), Rogrin (20), Rogar (24), Tara-909.
	Ethion	Acarin (18), Crush (20), Dhanumit (11), Fosmite (35), Indothion (21), Laser (26), MIT-505 (41), Phosmil (21), Tafethion (37), Volthion (54).
	Fenitrothion	Folithion (4), Hexafen (32), Sumithion (37).
	Fenthion	Baytex (4), Lebaycid (4).
	Formathion	Formathion
	Malathion	Agromala (16), Hilmala (17), Luthion (26), Malamar (32), Malataf (37), Cythion (8), Malatop (49), Mlamal (20), Maltox (1), Specmal (43), Sulmalathion (46).
	Methyl parathion	Devithion (32), Dhanumar (11), Luthion (26), Metacid (4), Milion (28), Prataf (49), Paratox (1), Profit (20), Tagpar (16).
	Monocrotophos	Agro Monark (16), Atom (21), Azodrin (8), Balwan (37), Bilphos (4), Corophos (7), Croton (24), Gaurdian (15), Hilcron (17), Hycrophos (18), Kadett (35), Luphos (26), Macrophos (50), Milphos (20), Moncar (5), Monochem (30), Monocil (9), Monocin (23), Monocron (31), Monodhan (11), Monophos (32), Monosect (2), Monostar (41), Monovol (54), Nuvacron (47), Parryphos (13), Phoskill (52), Specron (42), Sufos (45), Topcil (49).
	Oxy-demeton methyl	Dhanussystox (11), Hexasystox (32), Metasystox (4), Mode (2).
	Phenthoate	Aimsan (1), Amaze (55), Delsan (7), Elsan (52), Phendal (32), Penthomet (11), Tagsan (50).
	Phorate	Agro-phorate (16), Dhan 100 (11), Foratox (35), Forcin (23), Granutox (1), Helmet (50), Hilphorate (17), Luphate (26), Milate (28), Specphor (43), Starphor (41), Thimet (8), Umet (52), Volphor (53), Warrant (40).

Table. 7.1 (Contd.)

Group	Common Name	Trade Name/s (Source)
	Phoslane	Voltas Phosalone (54), Zolone (38).
	Profenophos	Ajanta (33), Banjo (22), Bolero (24), Carina (35), Curacron (47), Proven (27), Prabal (37).
	Quinalphos	Agroquin (16), Basquin (5), Bayruil (4), Chemlux (30), Dhanulux (11), Ekalux (47), Ekatop (49), Flash (21), Hilquin (17), Hyguin (18), Kinalux (52), Knock (42), Krush (55), Miluax (20), Quick (31), Quinal (46), Quinaltaf (37), Quinatox (1), Quingaurd (15), Shakti (13), Smash (40), Starlux (41), Suquin (45), Tagquin (50), Vazara (26), Volquin (54).
	Triazophos	Hostothion (4), Josh (52), Titan (20), Trelka, Truzo (50), Trihero (39).
Carbamates	Carbaryl	Agroryl (16), Dhanuvin (11), Hexavin (32), Kildiryl (25), Killex Cabaryl (34), Parry Win 50 WP (13), Sevidol (38), Sevin 50 WP (38), Sevin D (38), Sevin Flo 42 (38).
	Carbofuran	Carbocil (9), Caardon (52), Diafuran (33), Furadon (37), Furacarb (1), Furatox (1), Fury (31), Hexafuran (32), Sumo (20).
	Carbosulfan	Marshall (37).
	Indoxacarb	Avant (12), Dash (21), Daksha (37), Dhava (11), Fego (52).
	Methomyl	Astra (13), Dunnet (11), Lannata (12).
	Thiodicarb	Larvin (2).
Pyrethroides	Alphacypermethrin	Forward 10 EC (24), Sherpa Alpha 10 EC (4).
	B-Cyfluthrin	Bulldock (4).
	Cypermethrin	Agro-Cyper (16), Angel (18), Ankush (3), Auzar (55), Bilcyp (4), Challenger (50), Cilcord (9), Colt-25 (35), Cybil 25 EC (4), Cymbush (19), Cyper Top (49), Cypercin (23), Cypergaurd (15), Cypermar (32), Cypermil (28), Cypersul (46), Cyproid (1), Cyrux 25 (52), Hilcyperin (17), Hi-Power (46), Hycyper (18), Indothrin (21), Jawa (12), Laser 10 EC (24), Piothrin (5), Polytrin (47), Ralothrin (37), Ripcord (8), Shakti (26), Sherpa Cyper 25 EC (4), Simper (19), Spec (43), Spec-Cyperin (42), Starcip (41), Superkiller (11), Suraksha (31), Trofy 25 EC (24), Trofy 25 EC (40), Ustad (52), Volcyper (54), White Gold (30).

Table. 7.1 (Contd.)

Group	Common Name	Trade Name/s (Source)
	Decamethrin	Decakill (18), Decasyn (1), Decis (2), Easytab 25 EC (2), Faulad Rukrin (8).
	Ethofenfrox	Punakaso.
	Fenvalerate	Agrofen (16), Anchor (19), Bhaskarfen (5), Bhasma (55), Bilfen (4), Fen Fen (32), Fenchem (30), Fencron (47), Fenkill (52), Fennock 20 (9), Fenny (31), Fenval (40), Flame (20), Hilfen (17), Lufen (26), Milfen (28), Newfen (15), Starfen (41), Sumitox (1), Trfen (50), Tatafen (37), Topfen (49), Trumpcard (11), Viper (43).
	Lambda-Cyhalothrin	Agent (21), Bravo (20), Cliva (6), Karate (19), Reeva 5 (37).
	Permethrin	Ambush (19), Hilthrin (17), Permasect (32).
	Pyrethrum	
Insect growth regulators	Diflubenzuron	Dimilin 18 EC, 25 WP (15).
	Diafenthuran	Polo (4), Pegasus (47).
	Flufenoxuron	Cascade 10 EC (3).
	Lufenuron	Match 5 EC (47).
	Novaluron	Rimon 10 EC (21), Rim.
	Tiflubenzuron	Nomolt (9).
Combi products	B-Cyfluthrin 12.5 g +Chlorpyriphos 250 g	Bulldock 262.5 EC (4).
	Imidachloprid 50 g +B-Cyfluthrin 50 g	Confider Ultra 100 EC (4).
	Alphamethrin 1% + Chlorpyriphos 16%	Ducord 17 EC (4), Alfatoon, Lethal Plus (20).
	Fenvalerate 3% + Acephate 25%	Koranda 28 EC (29).
	Cypermethrin 5% + Ethion 40%	Nagata 45 EC (37), Rimjim (20).
	Cypermethrin 5% + Chlorpyriphos 50%	Noorani 55 EC (37), Nurello D-505, Lrthal Super 50 (20), Synergy 505 (9).
	Cypermethrin 4% + Profenphos 40%	Polytrin 44 EC (36).
	Cypermethrin 3% + Quinalphos 20%	Prachanda 23 EC, Viraat 23 EC (52).
	Deltamethrin 8% + Endosulfan 32%	Sevit 40 EC.

Table. 7.1 (Contd.)

Group	Common Name	Trade Name/s (Source)
	Deltamethrin 1% + Trizophos 35%	Spark 36 EC (4).
New molecules	Acetamaprid	Lift (21), Mnik (37), Pride (3), Sharp (20).
	Chlorfenapyr	Intrepid (3).
	Emamectin benzoate	Proclaim (47), Prabhav (37).
	Fipronil	Reagent (4).
	Imidachloprid	Admit (24), Atom (21), Confidar (4), Chemida (6), Imida Gold (52), Sabmida (39), Tata Mida (48), Victor.
	Spinosad	Tracer (9), Spintar (4).
	Thiomethoxam	Actra (4), Ananth 25 WP (37), Click (21), Crusier (47), Renova (52), Reno 70 WP.
Acaricides	Abamectin	Vertimec (47).
	Diafenthiuron	Polo, Pegasus.
	Dicofol	Kelthane (9), Colonel-S (21).
	Ethion	Acarin (18), Crush (20), Dhanumit (11), Fosmite (35), Indothion (21), Laser (26), MIT-505 (41), Phosmil (21), Tafethion (37), Volthion (54).
	Fenazaquin	Magister (12).
	Fenpropathrin + Pyriproxifen	Prempt 20 EC.
	Sulphur	Dhanusulf (11), Farmsulf (41), Hexasul (32), Insulf (52), Microsulf (32), Sulcol, Wet-Sulf (14), Sulfin M-20, (16), Sulphosan (1), Sultaf (37), Thiovit (47).
	Propagate	Omite (11).

7.2. FUNGICIDES

Table 7.2. Common and trade names and sources of availability of fungicides

Group	Common Name	Trade Name/s (Source)
Copper fungicides	Copper hydroxide	Kocide 101 (12).
	Copper oxychloride	Blitox (37), Blue Copper (47), Captox (1), Cupramar (32), Cuprastar (41), Devicopper (10), Dhanucop (11), Fytolan (19), Hexamar Copper (32), Hil Copper (17), Blue Vit (25), Trucop (21).

Table. 7.2 (Contd.)

Group	Common Name	Trade Name/s (Source)
	Cuprous oxide	Cuprous oxide.
Di-carboxymides	Iprodione	Rovral (38).
	Vinclozolin	
Heterocyclic nitrogenous fungicides	Captafol	Foltaf.
	Captan	Captra (21), Caftap (37), Deltan (7), Dhanutan (11), Hexacap (32).
	Folpet	
Mercury fungicides	Methoxy ethyl mercuric chloride	Bagalol (52).
	Phenyl mercury acetate	
Sulphur fungicides (Inorganic)	Wettable Sulphur	Dhanusulf (11), Farmsulf (41), Hexasul (32), Insulf (52), Microsulf (32), Sulcol, Wet-Sulf (14), Sulfin M-20 (16), Sulphosan (1), Sultaf (37), Thiovit (47).
Sulphur fungicides (Organic–Carbamates)	Ferbam	Hexferb (32).
	Maneb/Mancozeb	Abie (47), Agromanco (16), Aimcozeb (1), Dhanuka (11), Dithane M-45 (4), Hilthane (17), Indofil (21), Luzen (26), Leo M-45 (20), Zebthane M-45, Flowin-HT, Luzim (26), Saviour (9), Shield (35), Sparsh (55), Spic Mancozeb (44), Uthane M-45 (52), Zeb (31), Zebthane (37).
	Nabam	
	Thiram	Thiride (20), Hexathir (32), Vegfru Thiram (35).
	Vapam	
	Zineb	Discon-Z (1), Devizeb (10), Hexathane (32), Indofil Z-78 (21).
	Ziram	Cumin L (47), Hexazir (32), Tagziron (50), Ziride (20), Vegfru Zitox (35), Dhanuka Z-27 (11).
Quinine fungicides	Chloronil	
	Dichlone	
Systemic fungicides Benzimidazoles	Benomyl	Benlate (32), Benofit (13).

Table. 7.2 (Contd.)

Group	Common Name	Trade Name/s (Source)
	Carbendazim	Agni (21), Agrozim (16), Aimocozin (1), Arrest (24), Bavistin (3), Benfin (21), Bengard (9), Calzin (26), Benzin (5), Carzim (26), Derosal (2), Dhamusten (11), Diafuran (33), Hycarb (18), Nirmool (41), Sahara (20), Stare (32), Volzim (54), Zen (31), Zoom (52).
	Thiabendazole	
Thiophanates	Thiophanate	Thiophanate.
	Thiophanate Methyl	Cercobin, Roko (55), Prism (20).
Oxathin compounds	Carboxin	Carboxin.
	Oxy-Carboxin	Morestein.
Phenylamides	Oxadixyl	
	Metalaxyl	Apron (24), Matco, Himilgard.
Triazoles	Bitertanol (Biloxazol)	Baycor (4).
	Dichlobutrazol	Vigil.
	Difenconazole	Score (47).
	Flusilazol	
	Flutriafol	
	Hexaconazole	Avone (20), Contaf (37), Conquer (52), Danzole(6), Manage (24), Sitara (21).
	Myclobutanil	Systhane (9), Boon (21).
	Penconazole	Topas (47).
	Propiconazole	Radar (37), Tilt (47), Swing, Sona (20), Dhan (21).
	Triadimefon	Bayleton (4).
	Triadimenol	
	Tricyclazole	Beam (21), Force-11 (20).
Pyrimidines	Fenarimol	
Morpholines	Dodemorph	Dodemorph.
	Tridemorph	Calixin (3).
Organophosphorus compounds	Fosetyl –AL	Aliette (2).
	Kitazine & Kitazine-P	
	Pyrazophos	Apagon.
Acetamides	Cymoxynil	
Carbamates	Propamocarb	
	Propineb	Antracol (4).

Table. 7.2 (Contd.)

Group	Common Name	Trade Name/s (Source)
Di-caroxymides	Procymidone	
Imidazoles	Fenapanil	
Piperazine	Triforine	
Phenol derivatives	Chloroneb	
Miscellaneous systemic fungicides	Antibiotics	Aureofungin, Cycloheximide, Griseofulvin, Kasugamycin, Validamycin, Streptocycline.
	Etridiazol	
	Pyroxychlor	
Miscellaneous fungicides	Chinomethionat	
	Clorothalanil	Bravo, Kavach (47).
	Clorothaniz	
	Dinocap	Karathane (4).
	Dodine	Sylil (4), Syllet (28), Noor (21).
	Edifenphos	Hinosan (4).
	Glyodin	
	Morocide	Morocide
	Prochloraz	
	Quintozine	
Combi products	Metalaxyl 8% + Mancozeb 64%	Ridomil MZ (47), Master (48), Himil (20), Matco-8-64 (21).
	Iprodione 25% + Carbendazim 25%	Quintal (4).
	Carbendazim 12% + Mancozeb 63%	Care (20), Companion (21), Top-20 (18), Sixer 75 WP (11), Saaf (52).
	Cymoxanil 8% + Mancozeb 64%	Curzate 72 WP (12).
	Propioconazole + Difenoconazole 30%	Armure 30 EC.

7.3. RODENTICIDES

Table 7.3. Common and trade names and sources of availability of rodenticides

Common name	Trade name/s (Source)
Aluminium phosphide	Synfume, Phosfume, Celphos (14), Quickphos (52), Phostoxin, Delacea (32), Alphos (1), Fumino (52).
Bromadiolone	Lainirat (47), Moosh-Moosh (37).
Coumatetralyl	Racumin (4).
Magnesium phosphide	Magnophos.

Table. 7.3 (Contd.)

Common name	Trade name/s (Source)
Warfarin	Rodafarin, Ratafin, Fumarin, Warfotax (1).
Zinc phosphide	Synphos, Rasol (52), Zincfox (1), Rasox.
D-D 1: 1 mixture (Dichloro propene – Dichloro propane)	Telone, Vidden-D.
EDCT 3: 1 mixture (Ethylene dichloride carban tetrachloride)	Kiloptera, Fumex, Dichromek.

Source of Pesticides

1. AIMCO
2. Aventis
3. BASF
4. Bayer
5. Bhaskar Agro Chemicals
6. Cheminova
7. Coromandal Indag
8. Cynamid Agro
9. De Nocil
10. Devi Dayal
11. Dhanuka
12. Du Pont
13. EID Parry
14. Excel
15. Gardha Chemicals
16. Gujarat Agro
17. Hindusthan Insecticides
18. Hyderabad Chemicals
19. ICI Zeneca
20. IIL
21. Indofil
22. Insecticides India
23. IOCL
24. IS Agro
25. Kilpest
26. Lupin
27. Meghamani Organics Ltd.
28. Montari
29. Nagarjuna Chemicals
30. New Chemi
31. NFCL
32. Parry Chemicals
33. Parry India Ltd.
34. Paushak
35. Pesticides India
36. Polychem
37. Rallis India
38. Rhone-Poulenc
39. Sabero
40. Searle India
41. Shaw Wallace
42. Southern Pesticides
43. SPEC
44. Spic
45. Sudarshan Chemicals
46. Sulphur Mills
47. Syngenta
48. Tata
49. Thakar Chemicals
50. Tropical Agro
51. T. Stanes
52. United Phosphorus Ltd.
53. Volrho
54. Voltas
55. Wockardt

7.4. NEMATICIDES

Table 7.4. Globally important nematicides and their manufacturers

Active substance	Chemical group	Trade name	Manufacturer
1, 3-Dichloropropene	Halogenated hydrocarbon	Telone	Bayer Crop Science
Dazomet	Methyl isothiocyanate liberator	Basamid	BASF Corporation
Metham sodium	Methyl isothiocyanate liberator	Vapam	Amvac Chemical Corpn.
Ethoprophos	Organophosphorus	Mocap	Bayer Crop Science
Phenamiphos	Organophosphorus	Nemacur	Bayer Crop Science
Fosthiazate	Organophosphorus	Nemathorin	Syngenta
Cadusafos	Organophosphorus	Rugby	FMC Corporation
Aldicarb	Oxime carbamate	Temik	Bayer Crop Science
Oxamyl	Oxime carbamate	Vydate	Du Pont
Carbofuran	Carbamate	Furadan	FMC Corporation

7.5. PREDATORS

Table 7.5. Sources of availability of predators

Predators	State	Sources of availability
Amblyscius sp.	Tamil Nadu	Rajendra Foundn for Agri Res, Chittar
Apertochrysa sp.	Karnataka	Project Directorate of Biol Control, Bangalore
Blastostethus pallescens	Karnataka	Project Directorate of Biol Control, Bangalore
Brumoides suturalis	Karnataka	Project Directorate of Biol Control, Bangalore; Central IPM Centre, Bangalore
	New Delhi	Entomol Divn, Indian Agri Res Inst, New Delhi
Cardiastethus exiguous	Karnataka	Project Directorate of Biol Control, Bangalore
Cheilomenes sexmaculata	Gujarat	Entomol Dept, Anand Agri Univ, Anand
	Karnataka	Project Directorate of Biol Control, Bangalore; Central IPM Centre, Bangalore
	New Delhi	Entomol Divn, Indian Agri Res Inst, New Delhi
Chilocorus nigrita	Himachal Pradesh	Central IPM Centre, Solan; Ent Dept, Y S Paramar Univ of Hort & Forestry, Solan
	Jammu & Kashmir	Central IPM Centre, Jammu; Central IPM Centre, Srinagar; Entomol Dept, SK Univ Agri Sci & Technol, Srinagar
	Karnataka	Project Directorate of Biol Control, Bangalore; Bio Control Res Lab, PCI Pvt Ltd., Bangalore
	Tamil Nadu	Rajendra Foundn for Agri Res, Chittar

Table. 7.5 (Contd.)

Predators	State	Sources of availability
	West Bengal	Central IPM Centre, Burdwan
Chrysoperla carnea	Andhra Pradesh	Central Tobacco Res Inst, Rajahmundry; Ent Dept, Acharya N G Ranga Agri Univ, Hyderabad; Pest Control (India) Pvt Ltd, Hyderabad; SOM-IPM Systems (I) Ltd, Hyderabad
	Assam	Pest Control (India) Pvt Ltd, Guwahati
	Bihar & Jarkhand	Pest Control (India) Pvt Ltd, Jamshedpur
	Gujarat	Entomol Dept, Anand Agri Univ, Anand; Pest Control (India) Pvt Ltd, Vadodara
	Haryana	Pest Control (India) Pvt Ltd, Chandigarh
	Karnataka	Project Directorate of Biol Control, Bangalore; Central IPM Centre, Bangalore; Univ of Agri Sci, Entomol Dept, Dharwad; Kadur Agro Pvt. Ltd., Bangalore; Pest Control (India) Pvt Ltd, Bangalore
	Kerala	Pest Control (India) Pvt Ltd, Cochin
	Madhya Pradesh	Indore Biotech Inputs & Res (P) Ltd., Indore; Pest Control (India) Pvt Ltd, Bhopal
	Maharashtra	Entomol Dept, Mahatma Phule Agri Univ, Rahuri; Nathkrupa Biocontrol Lab, Nagpur; Pest Control (India) Pvt Ltd Mumbai-400 063; Pest Control (India) Pvt Ltd, Mumbai-400 059; Pest Control (India) Pvt Ltd, Thane-400 601; Pest Control (India) Pvt Ltd, New Mumbai; Pest Control (India) Pvt Ltd, Nashik-422 009; Pest Control (India) Pvt Ltd, Pune-411 033; Pest Control (India) Pvt Ltd, Pune-411 030
	New Delhi	Biotech Intern Ltd., New Delhi; Pest Control (India) Pvt Ltd, New Delhi
	Orissa	Pest Control (India) Pvt Ltd, Bhubaneswar
	Punjab	Pest Control (India) Pvt Ltd, Chandigarh
	Rajasthan	Pest Control (India) Pvt Ltd, Jaipur
	Tamil Nadu	Entomol Dept, TNAU, Madurai; Basarass Biocontrol Res, Vallavar; Greentech Agro Prod P Ltd, Coimbatore; Pest Control (India) Pvt Ltd, Chennai; Rajendra Foundn for Agri Res, Chittar; TN Co-op Sugar Fedn Ltd, Chengalpattu

Table. 7.5 (Contd.)

Predators	State	Sources of availability
	Uttarakhand	Pest Control (India) Pvt Ltd, Dehradun
	Uttar Pradesh	Pest Control (India) Pvt Ltd, Bareilly; Pest Control (India) Pvt Ltd, Lucknow; Pest Control (India) Pvt Ltd, Kanpur
	West Bengal	Pest Control (India) Pvt Ltd, Kolkata
Coccinella septumpunctata	Tamil Nadu	Jeypee Bio Techs, Virudhunagar
Cryptolaemus montrouzieri	Andhra Pradesh	Acharya N G Ranga Agri Univ, Hyderabad; Pest Control (India) Pvt Ltd, Hyderabad
	Assam	Pest Control (India) Pvt Ltd, Guwahati
	Bihar & Jarkhand	Pest Control (India) Pvt Ltd, Jamshedpur
	Gujarat	Pest Control (India) Pvt Ltd, Vadodara
	Haryana	Pest Control (India) Pvt Ltd, Chandigarh
	Karnataka	Project Directorate of Biol Control, Bangalore; Agri Technol Infmn Centre, IIHR, B'lore; Kadur Agro Pvt. Ltd, Bangalore; Pest Control (India) Pvt Ltd, Bangalore
	Kerala	Pest Control (India) Pvt Ltd, Cochin
	Madhya Pradesh	Central IPM Centre, Indore; Indore Biotech Inputs & Res (P) Ltd, Indore; Pest Control (India) Pvt Ltd, Bhopal
	Maharashtra	Pest Control (India) Pvt Ltd, Mumbai-400 063; Pest Control (India) Pvt Ltd, Mumbai-400 005; Pest Control (India) Pvt Ltd, Thane; Pest Control (India) Pvt Ltd, New Mumbai; Pest Control (India) Pvt Ltd, Nashik; Pest Control (India) Pvt Ltd, Pune-411 033; Pest Control (India) Pvt Ltd, Pune-411 030
	New Delhi	Entomol Divn, Indian Agri Res Inst, New Delhi; Pest Control (India) Pvt Ltd, New Delhi
	Orissa	Pest Control (India) Pvt Ltd, Bhubaneswar
	Punjab	Pest Control (India) Pvt Ltd, Chandigarh
	Tamil Nadu	Basarass Biocontrol Res, Vallavar; Greentech Agro Prod P Ltd, Coimbatore; Pest Control (India) Pvt Ltd; Rajendra Foundn for Agri Res, Chittar; Shri Durga Agro Services, Coimbatore
	Uttarakhand	Pest Control (India) Pvt Ltd, Dehradun
	Uttar Pradesh	Pest Control (India) Pvt Ltd, Bareilly; Pest Control (India) Pvt Ltd, Lucknow; Pest Control (India) Pvt Ltd, Kanpur
	West Bengal	Pest Control (India) Pvt Ltd, Kolkata

Table. 7.5 (Contd.)

Predators	State	Sources of availability
Curinus coeruleus	Karnataka	Project Directorate of Biol Control, Bangalore
Mallada spp.	Gujarat	Entomol Dept, Anand Agri Univ, Anand
	Karnataka	Project Directorate of Biol Control, Bangalore
	New Delhi	Entomol Divn, Indian Agri Res Inst, New Delhi
	Punjab	Entomol Dept, Punjab Agri Univ, Ludhiana
M. astur	Karnataka	Project Directorate of Biol Control, Bangalore
M. boninensis	Karnataka	Project Directorate of Biol Control, Bangalore
Orius tantillus	Karnataka	Project Directorate of Biol Control, Bangalore
Parena nigrolineata	Karnataka	Project Directorate of Biol Control, Bangalore
Pharoscymnus horni	Karnataka	Project Directorate of Biol Control, Bangalore
Scymnus coccivora	Karnataka	Project Directorate of Biol Control, Bangalore

7.6. PARASITOIDS

Table 7.6. Sources of availability of parasitoids

Parasitoids	State	Sources of availability
Apanteles angaleti	Karnataka	Project Directorate of Biol Control, Bangalore
	Maharashtra	Entomol Dept, Agri College, Pune
	New Delhi	Entomol Divn, Indian Agri Res Inst, New Delhi
Botryoideclava bharatiya	Karnataka	Project Directorate of Biol Control, Bangalore
Brachymeria nosatoi	Andhra Pradesh	Ent Dept, Acharya N G Ranga Agri Univ, Hyderabad
	Karnataka	Central IPM Centre, Bangalore
	Kerala	CPCRI Regional Station, Kayangulam
B. nephantidis	Andhra Pradesh	Ent Dept, Acharya N G Ranga Agri Univ, Hyderabad
	Karnataka	Project Directorate of Biol Control, Bangalore Central IPM Centre, Bangalore
	Kerala	CPCRI Regional Station, Kayangulam
Bracon brevicornis	Karnataka	Project Directorate of Biol Control, Bangalore
	New Delhi	Entomol Divn, Indian Agri Res Inst, New Delhi Pest Control (India) Pvt Ltd, New Delhi
	Tamil Nadu	Central IPM Centre, Tiruchirapalli Basarass Biocontrol Res, Vallavar Rajendra Foundn for Agri Res, Chittar TN Coconut Parasite Breeding Centre, Coimbatore TN Co-op Sugar Fedn Ltd, Chengalpattu

Table. 7.6 (Contd.)

Parasitoids	State	Sources of availability
B. hebetor	Andhra Pradesh	Ent Dept, Acharya N G Ranga Agri Univ, Hyderabad
	Karnataka	Project Directorate of Biol Control, Bangalore Central IPM Centre, Bangalore
	Kerala	CPCRI Regional Station, Kayangulam
	Maharashtra	Entomol Dept, Agri College, Pune Nathkrupa Biocontrol Lab, Nagpur
	New Delhi	Entomol Divn, Indian Agri Res Inst, New Delhi
	Tamil Nadu	Jeypee Bio Techs, Virudhunagar Rajendra Foundn for Agri Res, Chittar Sakthi Biocontrol & Biofert Centre, Kallipatti Gobi
B. kirkpatricki	Karnataka	Project Directorate of Biol Control, Bangalore
	Kerala	CPCRI Regional Station, Kayangulam
Campoletis chloridae	Karnataka	Project Directorate of Biol Control, Bangalore
Cephalonomia stephanoderis	Gujarat	Entomol Dept, Anand Agri Univ, Anand
	Karnataka	Coffee Research Sub Station, Chethalli
	Kerala	Entomol Dept, Kerala Agri Univ, Trichur
	Maharashtra	Central Inst of Cotton Res, Nagpur
	Tamil Nadu	Regional Coffee Res Stn, Thandigudi Centre for Pl Prot Studies, TNAU, Coimbatore
Chelonus blackburni	Karnataka	Project Directorate of Biol Control, Bangalore
Coccidoxenoides peregrinus	Karnataka	Project Directorate of Biol Control, Bangalore
Copidosoma koehleri	Bihar & Jarkhand	Central IPM Centre, Patna
	Karnataka	Project Directorate of Biol Control, Bangalore
	Maharashtra	Entomol Dept, Mahtma Phule Agri Univ, Rahuri
	New Delhi	Entomol Divn, Indian Agri Res Inst, New Delhi
Cotesia flaviceps	Karnataka	Project Directorate of Biol Control, Bangalore
	Orissa	Central IPM Centre, Bhubaneswar
	Punjab	Entomol Dept, Punjab Agri Univ, Ludhiana
	Tamil Nadu	Divn of Pl Prot, SBI, Coimbatore
	Uttar Pradesh	Entomol Divn, IISR, Lucknow
C. plutellae	Karnataka	Project Directorate of Biol Control, Bangalore
C. africanus	Andhra Pradesh	Central Tobacco Res Inst, Rajahmundry
	Karnataka	Project Directorate of Biol Control, Bangalore
	Maharashtra	Central IPM Centre, Nagpur
Dipha aphidivora	Karnataka	Agri Res Stn, UAS, Bheemarayanagudi

Table. 7.6 (Contd.)

Parasitoids	State	Sources of availability
Elasmus nephantidis	Andhra Pradesh	Ent Dept, Acharya N G Ranga Agri Univ, Hyderabad
	Karnataka	Project Directorate of Biol Control, Bangalore
		Agri Technol Infmn Centre, IIHR, Bangalore
	Kerala	CPCRI Regional Station, Kayangulam
	Tamil Nadu	Basarass Biocontrol Res, Vallavar
Encarsia spp.	Himachal Pradesh	Central IPM Centre, Solan
	Jammu & Kashmir	Entomol Dept, SK Univ Agri & Technol, Srinagar
	Karnataka	Central IPM Centre, Bangalore
Encarsia guadeloupae	Karnataka	Project Directorate of Biol Control, Bangalore
		Agri Technol Infmn Centre, IIHR, Bangalore
Epiricania melanoleuca	Maharashtra	Vasanthdada Sugar Institute, Pune
	Tamil Nadu	Central IPM Centre, Tiruchirapalli
		Rajendra Foundn for Agri Res, Chittar
	Uttar Pradesh	Entomol Divn, IISR, Lucknow
		Central IPM Centre, Gorakhpur
Eriborus argenteopilosus	Karnataka	Project Directorate of Biol Control, Bangalore
Goniozus nephantidis	Andhra Pradesh	Ent Dept, Acharya N G Ranga Agri Univ, Hyderabad
	Karnataka	Project Directorate of Biol Control, Bangalore
		Pest Control (India) Pvt Ltd, Bangalore
	Kerala	CPCRI Regional Station, Kayangulam
	Tamil Nadu	Central IPM Centre, Tiruchirapalli
		Greentech Agro Prod P Ltd, Coimbatore
		Rajendra Foundn for Agri Res, Chittar
		TN Coconut Parasite Breeding Centre, Coimbatore
Leptomastix dactylopii	Karnataka	Project Directorate of Biol Control, Bangalore
		Agri Technol Infmn Centre, IIHR, Bangalore
		Entomol Divn, Central Coffee Res Inst, Chikmagalur
	Kerala	Entomol Dept, Kerala Agri Univ, Trichur
	Tamil Nadu	Greentech Agro Prod P Ltd, Coimbatore
Micromus igarotus	Karnataka	Regional Res Stn, UAS, Raichur
Oomyzus sokolowskii	Karnataka	Project Directorate of Biol Control, Bangalore
Sturmiopsis inferens	Karnataka	Agri Technol Infmn Centre, IIHR, Bangalore
Telenomus spp.	Karnataka	Project Directorate of Biol Control, Bangalore
T. remus	Andhra Pradesh	Central Tobacco Res Inst, Rajahmundry

Table. 7.6 (Contd.)

Parasitoids	State	Sources of availability
	Karnataka	Project Directorate of Biol Control, Bangalore
	Madhya Pradesh	Central IPM Centre, Indore
Tetrastichus howardi	Karnataka	Project Directorate of Biol Control, Bangalore
T. israeli	Gujarat	Entomol Dept, Anand Agri Univ, Anand
	Karnataka	Project Directorate of Biol Control, Bangalore Central IPM Centre, Bangalore
	Kerala	CPCRI Regional Station, Kayangulam
	Tamil Nadu	Jeypee Bio Techs, Virudhunagar Rajendra Foundn for Agri Res, Chittar TN Coconut Parasite Breeding Centre, Coimbatore
Trichogramma achaeae	Karnataka	Project Directorate of Biol Control, Bangalore Agri Technol Infmn Centre, IIHR, Bangalore Central IPM Centre, Bangalore
	Punjab	Entomol Dept, Punjab Agri Univ, Ludhiana
	Tamil Nadu	Rajendra Foundn for Agri Res, Chittar Shri Durga Agro Services, Coimbatore Sun Agro Ind India, Chennai
T. brasiliensis	Karnataka	Project Directorate of Biol Control, Bangalore Agri Technol Infmn Centre, IIHR, Bangalore
	Tamil Nadu	Khatau Agro Tech, Mumbai National Biocontrol Laboratory, Nagpur
T. chilonis *T. japonicum*	Andhra Pradesh	Central Tobacco Res Inst, Rajahmundry Ent Dept, Acharya N G Ranga Agri Univ, Hyderabad Pest Control (India) Pvt Ltd, SOM-IPM Systems (I) Ltd., Hyderabad
	Assam	State Biocontrol Laboratory, Guwahati Pest Control (India) Pvt Ltd, Guwahati
	Bihar & Jarkhand	Pest Control (India) Pvt Ltd, Jamshedpur
	Goa	Central IPM Centre, Madgaon
	Gujarat	Pest Control (India) Pvt Ltd, Vadodara
	Haryana	Entomol Dept, CCS Haryana Agri Univ, Hisar Regional Agri Centre, CCSHAU, Karnal State Biocontrol Laboratory, Sirsa Biocontrol Laboratory, Yamuna Nagar
	Karnataka	Project Directorate of Biol Control, Bangalore Agri Technol Infmn Centre, IIHR, Bangalore Central IPM Centre, Bangalore Entomol. Dept, Univ of Agri Sci, Dharwad Inst of Pulses & Oilseeds Res, Gulbarga

Table. 7.6 (Contd.)

Parasitoids	State	Sources of availability
		Regional Res Stn, UAS, Raichur
		AAO, Parasite Laboratory, Bailhongal
		AO, Parasite Laboratory, Gangavathi
		AAO, Parasite Laboratory, Mandya
		Bio Solujan, Bangalore
		Hospet Sugar Mills, Hospet
		Kadur Agro Pvt. Ltd., Bangalore
		Karnataka Agri Res Foundn, Sameerwadi
		Pest Control (India) Pvt Ltd, Bangalore
	Kerala	Pest Control (India) Pvt Ltd, Cochin
	Madhya Pradesh	Indore Biotech Inputs & Res (Pvt.) Ltd., Indore
		Pest Control (India) Pvt Ltd, Bhopal
	Maharashtra	Central Inst of Cotton Res, Nagpur
		Entomol Dept, Mahtma Phule Agri Univ, Rahuri
		Entomol Dept, Marathwada Agri Univ, Parbhani
		Entomol Dept, Konkan Agri Univ, Dapoli
		KNS Biotech, Nanded
		Nathkrupa Biocontrol Lab, Nagpur
		Pest Control (India) Pvt Ltd, Mumbai-400 063
		Pest Control (India) Pvt Ltd, Mumbai-400 059
		Pest Control (India) Pvt Ltd, Thane-400 601
		Pest Control (India) Pvt Ltd, New Mumbai
		Pest Control (India) Pvt Ltd, Nashik-422 009
		Pest Control (India) Pvt Ltd, Pune-411 030
		Pest Control (India) Pvt Ltd, Pune-411 033
		R.S. Kherdekar, Talpathri
		Sandoz (I) Ltd., Mumbai
	New Delhi	Entomol Divn, Indian Agri Res Inst, New Delhi
		Biotech Intern Ltd., New Delhi
		Navdanya (NGO), New Delhi
		Pest Control (India) Pvt Ltd, New Delhi
	Orissa	Pest Control (India) Pvt Ltd, Bhubaneswar
	Punjab	Pest Control (India) Pvt Ltd, Chandigarh
	Rajasthan	Biocontrol Lab, ARS, RAU, Sri Ganganagar
		JDA (Pl Path), Biocontrol Lab, Durgapura
		DDA (Agro), IPM Lab, Ajmeer
		DDA (Agro), IPM Lab, Malikpur
		DDA (Agro), IPM Lab, Rampura

Table. 7.6 (Contd.)

Parasitoids	State	Sources of availability
		DDA (Agro), IPM Lab, Cjjatrapura
		DDA (Agro), IPM Lab, Chittorgarh
		DDA (Agro), IPM Lab, Hanumangarh
		DDA (Pl Path), IPM Lab, Banaswara
		Adarsh Biocontrol Laboratories, Alwar
		Pest Control (India) Pvt Ltd, Jaipur
	Tamil Nadu	Divn of Pl Prot, SBI, Coimbatore
		Regional Coffee Res Stn, Thandigudi
		Bio-control Lab., Papparapatty
		Bio-control Lab., Coimbatore
		Bio-control Lab., Salem
		Bio-control Lab, Villupuram
		Bio-control Lab., Trichy
		Bio-control Lab., Panjupettai
		Bio-control Lab., Kuruppanaichenpalyam
		Bio-control Lab., Vinayagapuram
		Bio-control Lab., Tirunelveli
		Bio-control Lab., Tanjavur
		Bio-control Lab., Namakkal
		Agro Res Lab, Coimbatore
		Basarass Biocontrol Res, Vallavar
		Biocontrol Res Centre, Kootham Poondi
		Greentech Agro Prod P Ltd, Coimbatore
		Jeypee Bio Techs, Virudhunagar
		Main Biocontrol Res Lab, Chengalpattu
		Pest Control (India) Pvt Ltd, Chennai
		Rajendra Foundn for Agri Res, Chittar
		Sakthi Biocontrol & Biofert Centre, Kallipatti, Gobi
		Shri Durga Agro Services, Coimbatore
		Sinagro Industries India, Chennai
		Sun Agro Ind India, Chennai
		Vigneswara Biocontrol, Pollachi
	Uttarakhand	Pest Control (India) Pvt Ltd, Dehradun
	Uttar Pradesh	Entomol Divn, IISR, Lucknow
		Biocontrol Lab, SBBP Agri Univ, Meerut
		Bioved Res Soc, Allahabad
		Crop Health Bioprod Res Centre, Ghaziabad
		Pest Control (India) Pvt Ltd, Bareilly

Table. 7.6 (Contd.)

Parasitoids	State	Sources of availability
		Pest Control (India) Pvt Ltd, Lucknow
		Pest Control (India) Pvt Ltd, Kanpur
	West Bengal	Pest Control (India) Pvt Ltd, Kolkata
T. dendrolimi	Karnataka	Project Directorate of Biol Control, Bangalore
		Agri Technol Infmn Centre, IIHR, Bangalore
	Tamil Nadu	Rajendra Foundn for Agri Res, Chittar
T. embryophagum	Himachal Pradesh	Central IPM Centre, Solan
		Ent Dept, Y S Paramar Univ of Hort & Forestry, Solan
	Jammu & Kashmir	Central IPM Centre, Jammu
		Central IPM Centre, Srinagar
		Entomol Dept, SK Univ Agri Sci & Technol, Srinagar
	Karnataka	Project Directorate of Biol Control, Bangalore
T. evanescens	Himachal Pradesh	Central IPM Centre, Solan
		Y S Paramar Univ of Hort & Forestry, Solan
	Jammu & Kashmir	Central IPM Centre, Jammu
		Central IPM Centre, Srinagar
		Entomol Dept, SK Univ Agri & Technol, Srinagar
	Karnataka	Project Directorate of Biol Control, Bangalore
T. maidis	Karnataka	Project Directorate of Biol Control, Bangalore
T. pretiosum	Himachal Pradesh	Central IPM Centre, Solan
		Ent Dept, Y S Paramar Univ of Hort & Forestry, Solan
	Jammu & Kashmir	Central IPM Centre, Jammu
		Central IPM Centre, Srinagar
		Entomol Dept, SK Univ Agri Sci & Technol, Srinagar
	Karnataka	Project Directorate of Biol Control, Bangalore
Trichogrammatoidea armigera	Karnataka	Project Directorate of Biol Control, Bangalore
T. bactrae	Karnataka	Project Directorate of Biol Control, Bangalore
Trichospilus pupivorus	Karnataka	Central IPM Centre, Bangalore
	Kerala	CPCRI Regional Station, Kayangulam
Xanthopimpla punctata	Karnataka	Project Directorate of Biol Control, Bangalore
		Central IPM Centre, Bangalore
	Kerala	CPCRI Regional Station, Kayangulam

7.7. PATHOGENS

7.7.1. Arbuscular Mycorrhizal Fungi

Table 7.7. Trade names and source of availability of Arbuscular Mycorrhizal Fungi

Arbuscular	Trade Name	State	Source of Availability
Glomus mosseae	Symbion	Karnataka	Agri Technol Infmn Centre, IIHR, B'lore
	VAM	Maharashtra	Bharatiya Agro Industries Foundation, Pune
		Tamil Nadu	T Stanes & Co, Coimbatore
G. fasciculatum	Symbion	Karnataka	Agri Technol Infmn Centre, IIHR, B'lore
	VAM	Maharashtra	Bharatiya Agro Industries Foundation, Pune
		Tamil Nadu	T Stanes & Co, Coimbatore
G. aggregatum		Maharashtra	Bharatiya Agro Industries Foundation, Pune
Gigaspora margarita		Maharashtra	Bharatiya Agro Industries Foundation, Pune
Arbuscular mycorrhizal fungi	Multiplex Trishul	Karnataka	Karnataka Agro Chemicals, B'lore

7.7.2. Antagonistic Fungi Against Plant Diseases and Nematodes

Table 7.8. Trade names and sources of availability of antagonistic fungi against Plant Diseases and nematodes

Antagonistic Fungi/ Trade Name	State	Source of Availability
Ampelomyces quisqualis (Bio-Dewcon)	Tamil Nadu	T. Stanes & Co, Coimbatore
Arthrobotrys cladodes var. *macroides*	Karnataka	Project Directorate of Biol Control, Bangalore
Arthrobotrys oligospora	Karnataka	Project Directorate of Biol Control, Bangalore
Aspergillus nizer (Kalisena SD, Kalisena SL, Sanjeevni)	Gujarat	Cadila Pharmaceuticals Ltd., Ahmedabad
	New Delhi	Plant Path Divn, Indian Agri Res Inst, New Delhi
Dactylella brochophaga	Karnataka	Project Directorate of Biol Control, Bangalore
Fusarium oxysporum	Karnataka	Project Directorate of Biol Control, Bangalore
Fusarium sporotrichoides	Karnataka	Project Directorate of Biol Control, Bangalore
Gliocladium catenulatum	Karnataka	Project Directorate of Biol Control, Bangalore
G. deliquescens	Karnataka	Project Directorate of Biol Control, Bangalore
G. roseum	Karnataka	Project Directorate of Biol Control, Bangalore
Phoma glomerata	Karnataka	Project Directorate of Biol Control, Bangalore

Table. 7.8 (Contd.)

Antagonistic Fungi/ Trade Name	State	Source of Availability
Pochonia chlamydosporia (Bionema)	Karnataka	Project Directorate of Biol Control, Bangalore Agri Technol Infmn Centre, IIHR, Bangalore Central Sericultural Res &Trng Inst, Mysore
Paecilomyces lilacinus Bio-Nemator, Biocon, Niyantran, Multiplex Sparsha, Abtec Paecilomyces)	Karnataka	Project Directorate of Biol Control, Bangalore Agri Technol Infmn Centre, IIHR, Bangalore
T. aureoviride	Karnataka	Project Directorate of Biol Control, Bangalore
T. hamatum	Karnataka	Project Directorate of Biol Control, Bangalore
T. harzianum (Binab-T, Ecoderma, F-Stop, Supravit, Tricodex, Trichoderma)	Andhra Pradesh	ADA (BCL), P.M. Palem ADA (BCL), Kakinada ADA (BCL), Nidadavole ADA (BCL), Ibrahimpatnam ADA (BCL), Ongole ADA (BCL), Nellore DDA (FTC), Nandyal ADA (BCL), Anantapur DDA (FTC), Rajendranagar ADA (Oilseeds) (BCL), Mahabubanagar ADA (BCL), Nalagonda ADA (BCL), Mulugu Road ADA (BCL), Karimnagar ADA (BCL), Adilabad Pest Control (India) Pvt Ltd, Hyderabad
	Assam	State Biocontrol Laboratory, Guwahati Pest Control (India) Pvt Ltd, Guwahati
	Bihar & Jarkhand	Pest Control (India) Pvt Ltd, Jamshedpur
	Gujarat	Pest Control (India) Pvt Ltd, Vadodara
	Haryana	State Biocontrol Laboratory, Sirsa Pest Control (India) Pvt Ltd, Chandigarh
	Karnataka	Project Directorate of Biol Control, Bangalore Agri Technol Infmn Centre, IIHR, Bangalore Central IPM Centre, Bangalore Central Sericultural Res & Trng Inst, Mysore Coffee Research Sub Station, Chethalli Central Coffee Res Inst, Entomol Divn, Chikmagalur Univ of Agri Sci, Entomol Dept, Dharwad Inst of Pulses & Oilseeds Res, Gulbarga

Table. 7.8 (Contd.)

Antagonistic Fungi/ Trade Name	State	Source of Availability
		Regional Res Stn, UAS, Raichur
		Agri Res Stn, UAS, Bijapur
		Kadur Agro Pvt. Ltd., Bangalore
		Pest Control (India) Pvt Ltd, Bangalore
		P.J. Margo Pvt. Ltd., Bangalore
	Kerala	Travancore Organic Fertilizers, Kottayam
		VRM Biotech Products, Changanacherry
	Madhya Pradesh	Pest Control (India) Pvt Ltd, Bhopal
	Maharashtra	Bio Era Technologies, Nagpur
		Hoechst Schering Agr. Envt, Mumbai
		KNS Biotech, Nanded
		Nathkrupa Biocontrol Lab, Nagpur
		Om Agro Organics, Yavatmal
		Pest Control (India) Pvt Ltd, Mumbai-400 063
		Pest Control (India) Pvt Ltd, Mumbai-400 059
		Pest Control (India) Pvt Ltd, Thane
		Pest Control (India) Pvt Ltd, New Mumbai
		Pest Control (India) Pvt Ltd, Nashik
		Pest Control (India) Pvt Ltd, Pune-411 033
		Pest Control (India) Pvt Ltd, Pune-411 030
		Sanvardhini Agro Pvt. Ltd., Satara
		Soman Biofertilizers, Pune
		Vidyas Biotech Lab, Nagpur
		West Coast Herbochem P Ltd, Mumbai
	New Delhi	Biotech Intern Ltd., New Delhi
		Pest Control (India) Pvt Ltd, New Delhi
	Orissa	Pest Control (India) Pvt Ltd, Bhubaneswar
	Punjab	Pest Control (India) Pvt Ltd, Chandigarh
	Rajasthan	Biocontrol Lab, ARS, RAU, Sri Ganganagar
		JDA (Pl Path), Biocontrol Lab, Durgapura
		DDA (Agro), IPM Lab, Ajmeer
		DDA (Agro), IPM Lab, Malikpur
		DDA (Agro), IPM Lab, Rampura
		DDA (Agro), IPM Lab, Cjjatrapura
		DDA (Agro), l IPM Lab, Chittorgarh
		DDA (Agro), IPM Lab, Hanumangarh
		DDA (Pl Path), IPM Lab, Banaswara
		Pest Control (India) Pvt Ltd, Jaipur
	Tamil Nadu	Centre for Pl Prot Studies, TNAU, Coimbatore
		Bio-control Lab, Papparapatty

Table. 7.8 (Contd.)

Antagonistic Fungi/ Trade Name	State	Source of Availability
		Bio-control Lab, Coimbatore
		Bio-control Lab, Salem
		Bio-control Lab, Villupuram
		Bio-control Lab, Trichy
		Bio-control Lab, Panjupettai
		Bio-control Lab, Kuruppanaichenpalyam
		Bio-control Lab, Vinayagapuram
		Bio-control Lab, Tirunelveli
		Bio-control Lab, Tanjavur
		Bio-control Lab, Namakkal
		Basarass Biocontrol Res, Vallavar
		Greentech Agro Prod P Ltd, Coimbatore
		Jeypee Bio Techs, Virudhunagar
		Pest Control (India) Pvt Ltd, Chennai
		Rajendra Foundn for Agri Res, Chittar
		Shri Durga Agro Services, Coimbatore
		Shrishti Bioprod P Ltd, Coimbatore
		Sun Agro Ind India, Chennai
		TN Co-op Sugar Fedn Ltd, Chengalpattu
		TAP Industries, Chennai
	Uttarakhand	Pl Path Dept, GB Pant Uni Agri & Technol, Pantnagar
		Pest Control (India) Pvt Ltd, Dehradun
	Uttar Pradesh	Crop Health Bioprod Res Centre, Ghaziabad
		Pest Control (India) Pvt Ltd, Bareilly
		Pest Control (India) Pvt Ltd, Lucknow
		Pest Control (India) Pvt Ltd, Kanpur
	West Bengal	Pest Control (India) Pvt Ltd, Kolkata
T. koningii	Karnataka	Project Directorate of Biol Control, Bangalore
T. longibactriatum	Karnataka	Project Directorate of Biol Control, Bangalore
T. polysporum	Karnataka	Project Directorate of Biol Control, Bangalore
T. pseudokoningii	Karnataka	Project Directorate of Biol Control, Bangalore
(Nursery Guard)		Central Sericultural Res &Trng Inst, Mysore
T. virens (Gligard, WRC-GV)	Karnataka	Project Directorate of Biol Control, Bangalore
T. viride (Bio-Cure, Bio-Cure F, Agroderma, Antogon TV, Bioderma, Dermapak, Ecofit, Monitor, Multiplex Nisarga,	Andhra Pradesh	ADA (BCL), P.M. Palem
		ADA (BCL), Kakinada
		ADA (BCL), Nidadavole
		ADA (BCL), Ibrahimpatnam
		ADA (BCL), Ongole

Table. 7.8 (Contd.)

Antagonistic Fungi/ Trade Name	State	Source of Availability
Rakshak, Trichosan, Trichoft, Trichoderm, Trichonik, Tricho-X, Trichodermin)		ADA (BCL), Nellore DDA (FTC), NandyalADA (BCL), Anantapur DDA (FTC), Rajendranagar ADA (Oilseeds) (BCL), Mahabubanagar ADA (BCL), Nalagonda ADA (BCL), Mulugu Road ADA (BCL), Karimnagar ADA (BCL), Adilabad Pest Control (India) Pvt Ltd, Hyderabad
	Assam	State Biocontrol Laboratory, Guwahati Pest Control (India) Pvt Ltd, Guwahati
	Bihar & Jarkhand	Pest Control (India) Pvt Ltd, Jamshedpur
	Gujarat	Pest Control (India) Pvt Ltd, Vadodara
	Haryana	State Biocontrol Laboratory, Sirsa Pest Control (India) Pvt Ltd, Chandigarh
	Karnataka	Project Directorate of Biol Control, Bangalore Coffee Research Sub Station, Chethalli Central Coffee Res Inst, Entomol Divn, Chikmagalur Kadur Agro Pvt. Ltd., Bangalore Pest Control (India) Pvt Ltd, Bangalore P.J.Margo Pvt. Ltd., Bangalore Travancore Org Fert Co P Ltd., Bangalore
	Kerala	Travancore Organic Fertilizers, Kottayam VRM Biotech Products, Changanacherry
	Madhya Pradesh	Pest Control (India) Pvt Ltd, Bhopal
	Maharashtra	Bio Era Technologies, Nagpur Hoechst Schering Agr. Envt, Mumbai KNS Biotech, Nanded Nathkrupa Biocontrol Lab, Nagpur Om Agro Organics, Yavatmal Pest Control (India) Pvt Ltd, Mumbai-400 063 Pest Control (India) Pvt Ltd, Mumbai-400 059 Pest Control (India) Pvt Ltd, Thane Pest Control (India) Pvt Ltd, New Mumbai Pest Control (India) Pvt Ltd, Nashik Pest Control (India) Pvt Ltd, Pune-411 033 Pest Control (India) Pvt Ltd, Pune-411 030 Sanvardhini Agro Pvt. Ltd., Satara Soman Biofertilizers, Pune

Table. 7.8 (Contd.)

Antagonistic Fungi/ Trade Name	State	Source of Availability
		Vidyas Biotech Lab, Nagpur
		West Coast Herbochem P Ltd, Mumbai
	New Delhi	Biotech Intern Ltd., New Delhi
		Pest Control (India) Pvt Ltd, New Delhi
	Orissa	Pest Control (India) Pvt Ltd, Bhubaneswar
	Punjab	Pest Control (India) Pvt Ltd, Chandigarh
	Rajasthan	Biocontrol Lab, ARS, RAU, Sri Ganganagar
		JDA (Pl Path), Biocontrol Lab, Durgapura
		DDA (Agro), IPM Lab, Ajmeer
		DDA (Agro), IPM Lab, Malikpur
		DDA (Agro), IPM Lab, Rampura
		DDA (Agro), IPM Lab, Cjjatrapura
		DDA (Agro), 1 IPM Lab, Chittorgarh
		DDA (Agro), IPM Lab, Hanumangarh
		DDA (Pl Path), IPM Lab, Banaswara
		Pest Control (India) Pvt Ltd, Jaipur
	Tamil Nadu	Centre for Pl Prot Studies, TNAU, Coimbatore
		Bio-control Lab, Papparapatty
		Bio-control Lab, Coimbatore
		Bio-control Lab, Salem
		Bio-control Lab, Villupuram
		Bio-control Lab, Trichy
		Bio-control Lab, Panjupettai
		Bio-control Lab, Kuruppanaichenpalyam
		Bio-control Lab, Vinayagapuram
		Bio-control Lab, Tirunelveli
		Bio-control Lab, Tanjavur
		Bio-control Lab, Namakkal
		Basarass Biocontrol Res, Vallavar
		Greentech Agro Prod P Ltd, Coimbatore
		Jeypee Bio Techs, Virudhunagar
		Pest Control (India) Pvt Ltd, Chennai
		Rajendra Foundn for Agri Res, Chittar
		Shri Durga Agro Services, Coimbatore
		Shrishti Bioprod P Ltd, Coimbatore
		Sun Agro Ind India, Chennai
		TN Co-op Sugar Fedn Ltd, Chengalpattu
		TAP Industries, Chennai
		T. Stanes & Co, Coimbatore

Table. 7.8 (Contd.)

Antagonistic Fungi/ Trade Name	State	Source of Availability
	Uttarakhand	Pl Path Dept, GB Pant Univ Agri & Technol, Pantnagar
		Pest Control (India) Pvt Ltd, Dehradun
	Uttar Pradesh	Crop Health Bioprod Res Centre, Ghaziabad
		Pest Control (India) Pvt Ltd, Bareilly
		Pest Control (India) Pvt Ltd, Lucknow
		Pest Control (India) Pvt Ltd, Kanpur
	West Bengal	Pest Control (India) Pvt Ltd, Kolkata
Verticillium lecanii	Karnataka	Project Directorate of Biol Control, Bangalore
		Pest Control (India) Pvt Ltd, Bangalore
	Maharashtra	Sio Agro Research Lab, Mumbai
	Tamil Nadu	Sakthi Biocontrol & Biofert Centre, Kallipatti Gobi
V. suchlosporium	Karnataka	Project Directorate of Biol Control, Bangalore
Arthrobotrys sp. + *Plilacinus* + *Verticillium* sp. (Nemastin)	Maharashtra	Kumar Krishimitra Bioproducts, Pune
T. harzianum + *T. viride* (Ecoderma, NIPROT)	Karnataka	Bio Control Res Lab, PCI Pvt. Ltd., Bangalore
		P.J. Margo Pvt. Ltd., Bangalore
T. harzianum + *T. viride* + *T. virens* (Combat)	Maharashtra	Kumar Krishimitra Bioproducts, Pune

7.7.3. Entomopathogenic Fungi

Table 7.9. Trade names and sources of availability of entomopathogenic fungi against insect pests

Entomopathogenic Fungi	State	Source of Availability
Beauveria bassiana (Bio-Power, Tracobea, Dispel, Larvocel, Don Muscardin, BABA, Abtech Beauveria)	Karnataka	Project Directorate of Biol Control, Bangalore
		Entomol Dept, Univ of Agri Sci, Dharwad
		Travancore Org Fert Co P Ltd., Bangalore
	Maharashtra	Vidyas Biotech Lab, Nagpur
	New Delhi	Defence Inst of Physiol, New Delhi
		Navdanya (NGO), New Delhi
	Rajasthan	Pesticides India Ltd., Udaipur
	Tamil Nadu	Divn of Pl Prot, SBI, Coimbatore

Table. 7.9 (Contd.)

Entomopathogenic Fungi	State	Source of Availability
		Basarass Biocontrol Res, Vallavar
		Greentech Agro Prod P Ltd, Coimbatore
		Jeypee Bio Techs, Virudhunagar
		Rajshree Sugars & Chem Ltd, Varadaraj Nagar
		Shri Durga Agro Services, Coimbatore
		T. Stanes & Co, Coimbatore
Botryodiplodia theobromae	Karnataka	Project Directorate of Biol Control, Bangalore
Fusarium cocophilum	Karnataka	Project Directorate of Biol Control, Bangalore
F. moniliformae	Karnataka	Project Directorate of Biol Control, Bangalore
F. roseum	Karnataka	Project Directorate of Biol Control, Bangalore
Metarhizium anisopliae (Bio-Magic)	Gujarat	Entomol Dept, Anand Agri Univ, Anand
	Karnataka	Project Directorate of Biol Control, Bangalore
		Entomol Dept, Univ of Agri Sci, Dharwad
	Kerala	CPCRI Regional Station, Kayangulam
	Maharashtra	Bio Era Technologies, Nagpur
		National Biocontrol Laboratory, Nagpur
	New Delhi	Biotech Intern Ltd., New Delhi
		Navdanya (NGO), New Delhi
	Tamil Nadu	Divn of Pl Prot, SBI, Coimbatore
		Basarass Biocontrol Res, Vallavar
		Biocontrol Res Centre, Kootham Poondi
		Rajendra Foundn for Agri Res, Chittar
		Shri Durga Agro Services, Coimbatore
Nomuraea rileyi	Andhra Pradesh	Directorate of Oil Seeds Res, Hyderabad
	Karnataka	Project Directorate of Biol Control, Bangalore
		Agri Technol Infmn Centre, IIHR, Bangalore
		Entomol Dept, Univ of Agri Sci, Dharwad
Paecilomyces fumosoroseus (Priority)	Tamil Nadu	Jeypee Bio Techs, Virudhunagar
		T. Stanes & Co, Coimbatore
Scopulariopsis brevicaulis	Karnataka	Project Directorate of Biol Control, Bangalore
Verticillium lecanii (Bio-Catch, Tracovert, Verticel, Varsha, Abtech Verticillium)	Karnataka	Project Directorate of Biol Control, Bangalore
		Entomol Dept, Univ of Agri Sci, Dharwad
		Agri Res Stn, UAS, Bijapur
		Travancore Org Fert Co P Ltd., Bangalore
	Maharashtra	Wockhardt Ltd., Mumbai
	Tamil Nadu	T. Stanes & Co, Coimbatore

7.7.4. Bacterial Biopesticides

Table 7.10. Trade names and sources of availability of bacterial biopesticides

Bacterial Biopesticides/ Trade Name/s	State	Source of Availability
Botryodiplodia theobromae	Karnataka	Project Directorate of Biol Control, Bangalore
B. sphaericus	New Delhi	Biotech Intern Ltd, New Delhi
	Tamil Nadu	Sinagro Industries India, Chennai
B. subtilis (Kill Dew DP Epic, Kodiac)	Maharashtra	Kumar Krishimitra Bioproducts, Pune
B. thuringiensis var. *kurstaki* (Biolep, Biosap, Delfin, Dipel, Bibot, Spicturin, Halt)	Andhra Pradesh	Indian Inst of Chem Technol, Hyderabad
	Madhya Pradesh	Indore Biotech Inputs & Res (Pvt.) Ltd., Indore
	Maharashtra	Lupin Laboratories (I) P Ltd., Mumbai Sandoz (I) Ltd., Mumbai Wockhardt Ltd., Mumbai
	New Delhi	Biotech Intern Ltd., New Delhi Hindustan Insecticides, New Delhi
	Tamil Nadu	Tuticorn Alkali Chem & Fert Ltd, Chennai
Pseudomonas fluorescens (Bio-Cure B, Bioshield, Tracomonas, Sudocel, Dagger-G)	Assam	State Biocontrol Laboratory, Guwahati
	Haryana	State Biocontrol Laboratory, Sirsa
	Karnataka	Agri Technol Infmn Centre, IIHR, Bangalore Travancore Org Fert Co P Ltd., Bangalore
	Kerala	Plantrich Chem & Fert Ltd, Kottayam
	New Delhi	Biotech Intern Ltd., New Delhi
	Tamil Nadu	Bio-control Lab, Papparapatty Bio-control Lab, Coimbatore Bio-control Lab, Salem Bio-control Lab, Villupuram Bio-control Lab, Trichy Bio-control Lab, Panjupettai Bio-control Lab, Kuruppanaichenpalyam Bio-control Lab, Vinayagapuram

Table. 7.10 (Contd.)

Bacterial Biopesticides/ Trade Name/s	State	Source of Availability
		Bio-control Lab, Tirunelveli
		Bio-control Lab, Tanjavur
		Bio-control Lab, Namakkal
		Basarass Biocontrol Res, Vallavar
		Greentech Agro Prod P Ltd, Coimbatore
		Jeypee Bio Techs, Virudhunagar
		Rajendra Foundn for Agri Res, Chittar
		Sun Agro Ind India, Chennai
		T. Stanes & Co, Coimbatore

7.7.5. Combination Products of Biocontrol Agents

Table 7.11. Source of availability of combination products of biocontrol agents

Combination Products of Biocontrol Agents/ Trade Name/s	State	Source of Availability
B. pumilis + *P. chlamydosporia* (Bacillus Plus-P)	Karnataka	Agri Technol Infmn Centre, IIHR, Bangalore
B. subtilis + *P. chlamydosporia* (Bacillus Plus-S)	Karnataka	Agri Technol Infmn Centre, IIHR, Bangalore
P. fluorescens + *P. chlamydosporia* (Pseudomonas Plus)	Karnataka	Agri Technol Infmn Centre, IIHR, Bangalore
Arthrobotrys sp. + *Plilacinus* + *Verticillium* sp. (Nemastin)	Maharashtra	Kumar Krishimitra Bioproducts, Pune
Aspergillus niger + *T. viride* + *P. lilacinus* (Multiplex Shock)	Karnataka	Karnataka Agro Chemicals, B'lore
P. chlamydosporia + *B. pumilis* (Bacillus Plus-P)	Karnataka	Agri Technol Infmn Centre, IIHR, Bangalore
P. chlamydosporia + *B. subtilis* (Bacillus Plus-S)	Karnataka	Agri Technol Infmn Centre, IIHR, Bangalore

Table. 7.11 (Contd.)

Combination Products of Biocontrol Agents/ Trade Name/s	State	Source of Availability
P. chlamydosporia + *P.lilacinus* (Biovert Plus)	Karnataka	Agri Technol Infmn Centre, IIHR, Bangalore
P. chlamydosporia + *P.fluorescens* (Pseudomonas Plus)	Karnataka	Agri Technol Infmn Centre, IIHR, Bangalore
P. chlamydosporia + *T. harzianum* (Trichorich)	Karnataka	Agri Technol Infmn Centre, IIHR, Bangalore
P. chlamydosporia + *T. viride* (Trichovert)	Karnataka	Agri Technol Infmn Centre, IIHR, Bangalore
T. harzianum + *T. viride* (Ecoderma, NIPROT)	Karnataka	Bio Control Res Lab, PCI Pvt Ltd, Bangalore Margo Pvt Ltd, Bangalore
T. harzianum + *T. viride* + *T. virens* (Combat)	Maharashtra	Kumar Krishimitra Bioproducts, Pune

7.7.6. Entomopathogenic Viruses

Table 7.12. Trade names and sources of availability of entomopathogenic viruses

Entomopathogenic Viruses/Trade Name/s	State	Source of Availability
Ha NPV of *Helicoverpa armigera* (Helivax, Heliokill, Helicide, Helinash, Heliocel)	Andhra Pradesh	ADA (BCL), P.M. Palem ADA (BCL), Kakinada ADA (BCL), Nidadavole ADA (BCL), Ibrahimpatnam ADA (BCL), Ongole ADA (BCL), Nellore DDA (FTC), Nandyal ADA (BCL), Anantapur DDA (FTC), Rajendranagar ADA (Oilseeds) (BCL), Mahabubanagar ADA (BCL), Nalagonda ADA (BCL), Mulugu Road ADA (BCL), Karimnagar ADA (BCL), Adilabad Kamal Herbal & Pheromone Exp (P) Ltd., Hyderabad

Table. 7.12 (Contd.)

Entomopathogenic Viruses/Trade Name/s	State	Source of Availability
		Pest Control (India) Pvt Ltd, Hyderabad
		Sneha Biotech, Vijayawada
	Assam	State Biocontrol Laboratory, Guwahati
		Pest Control (India) Pvt Ltd, Guwahati
	Bihar & Jarkhand	Pest Control (India) Pvt Ltd, Jamshedpur
	Gujarat	Pest Control (India) Pvt Ltd, Vadodara
	Karnataka	Project Directorate of Biol Control, Bangalore
		Agri Technol Infmn Centre, IIHR, Bangalore
		Entomol Dept, Univ of Agri Sci, Dharwad
		Inst of Pulses & Oilseeds Res, Gulbarga
		Regional Res Stn, UAS, Raichur
		Agri Res Stn, UAS, Bheemarayanagudi
		Bio-Pest Management P Ltd, Bangalore
		Cee Kay Associates, Bangalore
		Innovative Pest Control Lab, Bangalore
		Kadur Agro Pvt. Ltd., Bangalore
		Pest Control (India) Pvt Ltd, Bangalore
		P.J.Margo Pvt. Ltd., Bangalore
	Kerala	Pest Control (India) Pvt Ltd, Cochin
	Madhya Pradesh	Indore Biotech Inputs & Res (Pvt.) Ltd., Indore
		Pest Control (India) Pvt Ltd, Bhopal
		Samriddi Bioculture Pvt. Ltd., Indore
	Maharashtra	Central Inst of Cotton Res, Nagpur
		Entomol Dept, Mahtma Phule Agri Univ, Rahuri
		Entomol Dept, Marathwada Agri Univ, Parbhani
		Entomol Dept, Punjabrao Desmukh Agri Univ, Akola
		Entomol Dept, Konkan Agri Univ, Dapoli
		Ajay Biotech Lab (P) Ltd, Pune
		Anand Biocontrol Lab, Jalgaon
		Ankur Seeds Pvt. Ltd., Ratnagiri
		Arag Biotech, Nagpur
		Bio Era Technologies, Nagpur
		Chintawar Biocontrol Lab, Yeotmal
		Harit Biocontrol Laboratories, Yeotmal
		JR Biocontrol Lab, Yeotmal
		KNS Biotech, Nanded
		Nathkrupa Biocontrol Lab, Nagpur
		National Biocontrol Laboratory, Nagpur
		Pest Control (India) Pvt Ltd, Mumbai-400 063

Table. 7.12 (Contd.)

Entomopathogenic Viruses/Trade Name/s	State	Source of Availability
		Pest Control (India) Pvt Ltd, Mumbai-400 059
		Pest Control (India) Pvt Ltd, Thane
		Pest Control (India) Pvt Ltd, New Mumbai
		Pest Control (India) Pvt Ltd, Nashik
		Pest Control (India) Pvt Ltd, Pune-411 033
		Pest Control (India) Pvt Ltd, Pune-411 030
		Sanvardhini Agro Pvt. Ltd., Satara
		Tirupati Agro Inputs, Latur
		Universal Biotech, Nagpur
	New Delhi	Biotech Intern Ltd., New Delhi
		JK Traders, Delhi
		Pest Control (India) Pvt Ltd, New Delhi
	Orissa	Pest Control (India) Pvt Ltd, Chandigarh
	Punjab	Pest Control (India) Pvt Ltd, Chandigarh
	Rajasthan	Biocontrol Lab, ARS, RAU, Sri Ganganagar
		Adarsh Biocontrol Laboratories, Alwar
		Pest Control (India) Pvt Ltd, Jaipur
	Tamil Nadu	Centre for Pl Prot Studies, TNAU, Coimbatore
		Entomol Dept, TNAU, Madurai
		Bio-control Lab, Papparapatty
		Bio-control Lab, Coimbatore
		Bio-control Lab, Salem
		Bio-control Lab, Villupuram
		Bio-control Lab, Trichy
		Bio-control Lab, Panjupettai
		Bio-control Lab, Kuruppanaichenpalyam
		Bio-control Lab, Vinayagapuram
		Bio-control Lab, Tirunelveli
		Bio-control Lab, Tanjavur
		Bio-control Lab, Namakkal
		Basarass Biocontrol Res, Vallavar
		Greentech Agro Prod P Ltd, Coimbatore
		Jeypee Bio Techs, Virudhunagar
		Pest Control (India) Pvt Ltd, Chennai
		Rajendra Foundn for Agri Res, Chittar
		Shri Durga Agro Services, Coimbatore
		Spinco Biotech Pvt. Ltd, Chennai
		Sun Agro Ind India, Chennai
		T. Stanes & Co, Coimbatore
	Uttarakhand	Pest Control (India) Pvt Ltd, Dehradun

Table. 7.12 (Contd.)

Entomopathogenic Viruses/Trade Name/s	State	Source of Availability
	Uttar Pradesh	Biocontrol Lab, SBBP Agri Univ, Meerut
		Crop Health Bioprod Res Centre, Ghaziabad
		Pest Control (India) Pvt Ltd, Bareilly
		Pest Control (India) Pvt Ltd, Lucknow
		Pest Control (India) Pvt Ltd, Kanpur
	West Bengal	Pest Control (India) Pvt Ltd, Kolkata
Sl NPV of *Spodoptera litura* (Spodovax, Spodokill, Spodonash)	Andhra Pradesh	Central Tobacco Res Inst, Rajahmundry
		ADA (BCL), P.M. Palem
		ADA (BCL), Kakinada
		ADA (BCL), Nidadavole
		ADA (BCL), Ibrahimpatnam
		ADA (BCL), Ongole
		ADA (BCL), Nellore
		DDA (FTC), Nandyal
		ADA (BCL), Anantapur
		DDA (FTC), Rajendranagar
		ADA (Oilseeds) (BCL), Mahabubanagar
		ADA (BCL), Nalagonda
		ADA (BCL), Mulugu Road
		ADA (BCL), Karimnagar
		ADA (BCL), Adilabad
		Kamal Herbal & Pheromone Exp (P) Ltd., Hyderabad
		Pest Control (India) Pvt Ltd, Hyderabad
	Assam	Pest Control (India) Pvt Ltd, Guwahati
	Bihar & Jarkhand	Pest Control (India) Pvt Ltd, Jamshedpur
	Gujarat	Pest Control (India) Pvt Ltd, Vadodara
	Karnataka	Project Directorate of Biol Control, Bangalore
		Agri Technol Infmn Centre, IIHR, Bangalore
		Entomol Dept, Univ of Agri Sci, Dharwad
		Regional Res Stn, UAS, Raichur
		Bio-Pest Management Pvt. Ltd., Bangalore
		Pest Control (India) Pvt Ltd, Bangalore
	Kerala	Pest Control (India) Pvt Ltd, Cochin
	Madhya Pradesh	Pest Control (India) Pvt Ltd, Bhopal
	Maharashtra	National Biocontrol Laboratory, Nagpur
		Pest Control (India) Pvt Ltd, Mumbai-400 063
		Pest Control (India) Pvt Ltd, Mumbai-400 059
		Pest Control (India) Pvt Ltd, Thane-400 601
		Pest Control (India) Pvt Ltd, New Mumbai

Table. 7.12 (Contd.)

Entomopathogenic Viruses/Trade Name/s	State	Source of Availability
		Pest Control (India) Pvt Ltd, Nashik-422 009
		Pest Control (India) Pvt Ltd, Pune-411 030
		Pest Control (India) Pvt Ltd, Pune-411 033
	New Delhi	Biotech Intern Ltd., New Delhi
		Pest Control (India) Pvt Ltd, New Delhi
	Orissa	Pest Control (India) Pvt Ltd, Chandigarh
	Punjab	Pest Control (India) Pvt Ltd, Chandigarh
	Rajasthan	Biocontrol Lab, ARS, RAU, Sri Ganganagar
		Adarsh Biocontrol Laboratories, Alwar
		Pest Control (India) Pvt Ltd, Jaipur
	Tamil Nadu	Centre for Pl Prot Studies, TNAU, Coimbatore
		Entomol Dept, TNAU, Madurai
		Basarass Biocontrol Res, Vallavar
		Jeypee Bio Techs, Virudhunagar
		Pest Control (India) Pvt Ltd, Chennai
		Shri Durga Agro Services, Coimbatore
		Sun Agro Ind India, Chennai
	Uttarakhand	Pest Control (India) Pvt Ltd, Dehradun
	Uttar Pradesh	Pest Control (India) Pvt Ltd, Bareilly
		Pest Control (India) Pvt Ltd, Lucknow
		Pest Control (India) Pvt Ltd, Kanpur
	West Bengal	Pest Control (India) Pvt Ltd, Kolkata

7.7.7. Entomopathogenic Nematodes

Table 7.13. Trade names and sources of availability of entomopathogenic nematodes

Entomopathogenic Nematodes/Trade Name/s	State	Source of Availability
Heterorhabditis spp. (Green Cammandos)	Karnataka	Project Directorate of Biol Control, Bangalore
	Maharashtra	Bio Sense Crop Protection, Mumbai
Steinernema spp. (Soil Commandos)	Karnataka	Project Directorate of Biol Control, Bangalore
	Maharashtra	Bio Sense Crop Protection, Mumbai

7.8. PHEROMONES

Table 7.14. Source of availability of pheromones of fruit crop pests

Pheromone	State	Source of Availability
Oriental fruit moth	Maharashtra	Agri Sense BCS, Mumbai Bio Sense Crop Protection, Mumbai
Gypsy moth	Maharashtra	Bio Sense Crop Protection, Mumbai
Codling moth	Maharashtra	Agri Sense BCS, Mumbai Bio Sense Crop Protection, Mumbai
San Jose scale	Maharashtra	Agri Sense BCS, Mumbai
Apple leaf roller	Maharashtra	Agri Sense BCS, Mumbai
Apple fruit miner	Maharashtra	Agri Sense BCS, Mumbai
Red banded leaf roller	Maharashtra	Agri Sense BCS, Mumbai
Pear leaf roller	Maharashtra	Agri Sense BCS, Mumbai
Peach tree borer	Maharashtra	Agri Sense BCS, Mumbai
Peach twig borer	Maharashtra	Agri Sense BCS, Mumbai
Grape torteix	Maharashtra	Agri Sense BCS, Mumbai
Mediterranean fruit fly	Maharashtra	Agri Sense BCS, Mumbai
California red scale	Maharashtra	Agri Sense BCS, Mumbai

7.9. NEEM-BASED AND OTHER BOTANICALS

Table 7.15. Source of availability of neem–based and other botanicals

Neem-Based & other Botanicals	State	Source of Availability
Achook	Maharashtra	Bahar Agrochemicals, Ratnagiri Mahagrapes, Pune
Aphidin	Maharashtra	Eco-Max Agrosystems Ltd, Mumbai
Avana	Tamil Nadu	EID Parry (I) Ltd., Bioprod Divn, Chennai
Aza	Andhra Pradesh	SOM-IPM Systems (I) Ltd., Hyderabad
Azadit	Rajasthan	Pesticides India Ltd., Udaipur
Biogold	Tamil Nadu	SPIC, Bio-Tech Divn, Chennai
Bioneem	Maharashtra	Ajay Biotech Lab (P) Ltd, Pune
Bionema	Uttar Pradesh	Bioved Res Soc, Allahabad
Biopest	Uttar Pradesh	Bioved Res Soc, Allahabad
Biosol	Tamil Nadu	A.V. Thomas Pvt. Ltd., Chennai
Eco-Neem	Karnataka	Murkumbi Bio Agro P Ltd., Belgaum Pest Control (India) Pvt Ltd, Bangalore
Eco-Neem Plus	Karnataka	Murkumbi Bio Agro P Ltd., Belgaum Pest Control (India) Pvt Ltd, Bangalore
Ecotin	Karnataka	Monarch Biofert & Res Centre, Bangalore
Field Marshal	Gujarat	Ganesh Biocontrol System, Gondal

Table. 7.15 (Contd.)

Neem-Based & other Botanicals	State	Source of Availability
Fortuna Aza	Andhra Pradesh	Fortune Bio-tech Limited, Secunderabad
Jai Neem	Haryana	Jai Chemicals, Faridabad
Jawan	Maharashtra	Mc DA Agro Pvt. Ltd., Mumbai
Jeevan Crop Protector	Maharashtra	Mc DA Agro Pvt. Ltd., Mumbai
Juerken	Tamil Nadu	Madurai Chem & Agro Ind Ltd., Madurai
Limonool	Karnataka	Bio-Multi Tech Pvt. Ltd., Bangalore
Margocide-CK	Karnataka	Monofix Agro Products Ltd., Hubli
Margosan-O	Karnataka	Monofix Agro Products Ltd., Hubli
Neemactin	Maharashtra	Wockhardt Ltd., Mumbai
Neemak	Karnataka	West Coast Herbochem P Ltd, Bangalore
	Maharashtra	West Coast Herbochem P Ltd, Mumbai
Neemarin	New Delhi	Biotech Intern Ltd., New Delhi
Neemark	Karnataka	West Coast Herbochem P Ltd, Bangalore
Neemasol	Tamil Nadu	EID Parry (I) Ltd., Pesticides Divn, Chennai
Neemax	Maharashtra	Eco-Max Agrosystems Ltd, Mumbai
Neemazal-F	Tamil Nadu	EID Parry (I) Ltd., Bioprod Divn, Chennai
Neemazal-T/S	Tamil Nadu	EID Parry (I) Ltd., Bioprod Divn, Chennai
Neemocide	West Bengal	Vinayak Fats & Proteins P Ltd, Kolkata
Neemol	Andhra Pradesh	Ramson Laboratories, Vijayawada
	Tamil Nadu	Agro Links, Trichy
Neemolin	Maharashtra	Khatau Agro Tech, Mumbai
Neemosan	Tamil Nadu	Agronule Industries, Trichy
Neemox	Maharashtra	Prakash Farm Chemicals, Mumbai
Neempourn	Tamil Nadu	Prabhakar Oil Mills, Trichy
Neemta	Gujarat	A.J. Chemicals, Ahmedabad
Neethrin	Uttar Pradesh	Amitul Agro Chem Pvt. Ltd., Gorakhpur
Neem Base	New Delhi	Agrochem Divn, Indian Agri Res Inst, New Delhi
Neem Based pesticides	Maharashtra	Bahar Agrochemicals, Ratnagiri
		KNS Biotech, Nanded
		Om Agro Organics, Yavatmal
	Tamil Nadu	Ecobiocides & Botanicals Pvt. Ltd., Theni
		Jeypee Bio Techs, Virudhunagar
Neem Cake	Uttarakhand	Sri Ram Solvent Extraction Plant, Jaspur
Neem Gold	Tamil Nadu	EID Parry (I) Ltd., Bioprod Divn, Chennai
		SPIC, Bio-Tech Divn, Chennai
Neem Guard	Maharashtra	Gharda Chemicals Pvt. Ltd., Mumbai
Neem Hit	Maharashtra	Skylark Agro Chemicals, Thane
Neem Oil	Karnataka	Agro Extracts Ltd., Bangalore
	Maharashtra	Mc DA Agro Pvt. Ltd., Mumbai

Table. 7.15 (Contd.)

Neem-Based & other Botanicals	State	Source of Availability
		National Organic Chem Ind Ltd., Mumbai
		Sio Agro Research Lab, Mumbai
	Uttarakhand	Sri Ram Solvent Extraction Plant, Jaspur
Neem Plus	Maharashtra	BD Keathen & Co., Mumbai
Neem Rich	Maharashtra	National Chem Lab, Pune
Neem seed	Maharashtra	Cheminova Agro-Tech, Nandubar
Neem Soap	Karnataka	Agri Technol Infmn Centre, IIHR, Bangalore
Neem Top	Tamil Nadu	Sri Krishna Co, Coimbatore
Nim-76	New Delhi	Defence Inst of Physiol, New Delhi
Nimba	New Delhi	Entomol Divn, Indian Agri Res Inst, New Delhi
Nimbasal	New Delhi	Nimba Foods & Agrochem, New Delhi
		Pesto-Chem India Ltd, New Delhi
Nimbicidine	Tamil Nadu	T. Stanes & Co, Coimbatore
Nimbitor	Maharashtra	Zandu Pharmaceutical Works, Mumbai
Nimbosol	Tamil Nadu	Victoria Laboratories, Salem
Nimitox	Karnataka	Rallis India Ltd., Bangalore
Nimlin	Maharashtra	Sunline Agro Chem P Ltd, Dhulia
Nimbo Bas	Tamil Nadu	Basarass Biocontrol Res, Vallavar
Ozoneem	Haryana	Ozone Biotech, Faridabad
Phytowin	Tamil Nadu	Phyto Products Ltd., Thiruthuripoondi
Pongamia oil-based biopesticide	Tamil Nadu	Jeypee Bio Techs, Virudhunagar
Pongamia Soap	Karnataka	Agri Technol Infmn Centre, IIHR, Bangalore
Rakshak	Karnataka	Murkumbi Bio Agro P Ltd., Belgaum
		P.J. Margo Pvt. Ltd., Bangalore
RD-9 Repellin	Andhra Pradesh	ITC Ltd., Hyderabad
Replin 555	Tamil Nadu	Micro Chemicals Ltd., Madurai
Shaktiman	Gujarat	Khetiwadi Corner, Vadodara
		Krishna Biotech Pvt. Ltd., Veravel
		Oceon Agro India Ltd., Baroda
Sukrina	Tamil Nadu	Canster Chemicals Ltd., Chennai
Suneem	Maharashtra	Sunida Exports Ltd., Mumbai
Swaticure	Maharashtra	Swathi Industries Ltd., Mumbai
Tric	Uttar Pradesh	Amitul Agro Chem Pvt. Ltd., Gorakhpur
Tricure	Karnataka	Murkumbi Bio Agro P Ltd., Belgaum
		P.J. Margo Pvt. Ltd., Bangalore
Vapacide	Andhra Pradesh	Indian Inst of Chem Technol, Hyderabad

REFERENCES

Abraham, A. and Vatsala, P. 1981. *Introduction to Orchids*. Tropical Botanic Garden and Research Institute, Tiruvananthapuram.

Ahuja, S. 1983. Studies on the variability in infection of root-knot nematode *Meloidogyne incognita* on potato. *Third Nematol. Symp.*, Y.S. Paramar Uni. Hort. and Forestry, Solan, Himachal Pradesh, p. 4.

Alam, M. and Husain, A. 1976. Leaf blight of Java citronella (*Cymbopogon winterianus* Jowiff) caused by *Curvularia andropogonis* (Zimm.) Boedigin. *New Botanist* **3**: 54–56.

Alam, M., Khan, A.M. and Husain, A. 1986. Leaf blight and leaf spot diseases of Java citronella caused by *Curvularia andropogonis*. *Indian Phytopath.* **35**: 480–83.

Amalraj, S.F.A. and Rao, G.R. 1992. Potato cyst nematode pathogens in the southern hills of India. *Indian J. Nematol.* **22**: 82–85.

Anita, B. and Vadivelu, S. 1997. Management of root-knot nematode, *Meloidogyne hapla* on scented geranium, *Pelargonium graveolans*. *Indian J. Nematol.* **27**: 123–25.

Anon, 1985. *Kufri Swarna, a High Yielding Late Blight and Cyst Nematode Resistant Potato Cultivar Suitable for Cultivation in Tamil Nadu Hills*. No. 21(E), Central Potato Research Institute, Shimla, 4 pp.

Anon. 1990. *Annual Report of the Central Tuber Crops Research Institute*, Tiruvananthapuram.

Anusuya, K. and Vadivelu, S. 2002. Mutualistic symbiosis of VAM fungi on plant growth promoting rhizobacteria (PGPR) on micro propagated carnation. *J. Microbiol. World* **4**: 105–07.

Atu, U.G., Odurukwe, S.O. and Ogbuji, R.O. 1983. Root-knot nematode damage to *Dioscorea rotundata*. *Plant Dis.* **67**: 814–15.

Ayala, A. and Acosta, N. 1971. Observations on yam (*Dioscorea alata*) nematodes. *Nematropica* **1**: 39–40.

Bandare, J.M.R.S. 1981. Puccinia rust of citronella and lemon grass in Sri Lanka. *Plant Dis.* **65**: 164.

Barua, A. and Bordoloi, D.N. 1983. Record of a new disease of lemon grass (*Cymbopogon flexuosus* Stapf.) caused by *Curvularia verruciformis* Agarwal and Sahni. *Curr. Sci.* **52**: 640–41.

Bergeson, G.B. 1963. Influence of *Pratylenchus penetrans* alone and in combination with *Verticillium albo-utrum* on growth of peppermint. *Phytopathology* **53**: 1164–66.

Bergeson, G.B. and Green, Jr. R.J. 1979. Damage to cultivars of pepper mint *Mentha piperita* by the lesion nematodes, *Pratylenchus penetrans* in India. *J. Nematol.* **7**: 54.

Bhattacharyya, S.K., Shekhawat, G.S. and Singh, B.P. 1990. *Potato Late Blight*. Tech. Bull. No. 27, Central Potato Research Institute, Shimla, India, 40 pp.

Butani, D.K. 1982. Insect pests of Thulasi (*Ocimum sanctum* L.) and their control. *Pesticides* **16**: 11–12.

Calabretta, C. and Nucifora, A. 1985. Considerations on the possibility of biological and integrated control against *Liriomyza trifolii* (Burgess) (Diptera: Agromyzidae) on gerbera grown under glass in Sicily. *Academia Nazionale Italiano di Entomologia*, pp. 807–13.

Cappel, E.L.1892. *Report of Department of Land Records in Agriculture, Bombay,* p. 16.

Caveness, F.E. 1981. Root-knot nematodes on cassava. *In: Proceedings of the Third Research Planning Conference on Root-knot Nematodes,* Meloidogyne spp., *Region IV and V,* Ibadan, Nigeria.

Conijin, C.G.M. and Groen, N.P.A. 1990. Physical, biological and chemical control of *Taeniothrips simplex* in gladiolus. *Acta Hort.* **266**: 539–46.

Darling, H.M. 1959. *Plant Dis. Reptr.* **23**: 239–42.

David, P.M.M., Hanifa, A.M. and Natarajan, S. 1992. Evaluation of some insecticides and neem oil against jasmine blossom midge. *Madras Agric. J.* **79**: 15–17.

David, P.M.M. and Natarajan, S. 1998. Effect of insecticides on blossom midge-caused discolouration and drying of flower buds in *Jasminum sambac cv.* Gundumalli. *Indian Perfumer* **32**: 292–94.

Decker, H., Casamayor, G.R. and Bosch, D. 1967. Observaciones sobre la presencia del nematode *Scutellonema bradys* en el tuberculo die name en la provincea de Oriente (Cuba). *Centro, Boletin de Ciencias y Tecnologia, Universidad Central de Las Villas* **2**: 67–75.

Doharey, K.L. and Singh, R.P. 1989. Evaluation of neem (*Azadirachta indica* A. Juss.) seed kernel extracts against chaffer beetles. *Indian J. Ent.* **51**: 217–20.

Faulkner, L.R. and Skotland, C.B. 1965. Interaction of *Verticillium dahliae* and *Pratylenchus miniyus* in *Verticillium* wilt of peppermint. *Phytopathology* **55**: 583–86.

Fassuliotis, G. and Sparrow, A.H. 1955. Preliminary report of x-ray studies on the golden nematode. *Plant Dis. Reptr.* **39**: 572.

Ferraz, S. and Lear, B. 1976. Interaction of four plant parasitic nematodes and *Fusarium oxysporum* f. sp. *dianthi* on carnation. *Experientiae* **22**: 272–77.

Freitas, O.M.B.L. De and Moura, R.M. 1986. Response of cultivars of cassava (*Manihot esculenta* Crantz.) to *Meloidogyne incognita* and *M. javanica* (Nematoda: Heteroderidae) and comparison with hydrogen cyanide content. *Nematol. Medit.* **10**: 109–31.

French, E.R. 1986. *ACIAR Bacterial Wilt Newsletter No.* **1**: 5-6.

Gadewar, A.V., Trivedi, T.P. and Shekhawat, G.S. 1991. *Potato in Karnataka*. Tech. Bull. No. 17, Central Potato Research Institute, Shimla, India, 72 pp.

Gardner, D.E. 1985. Lemon grass rust caused by *Puccinia makanishikii* in Hawaii. *Plant Dis.* **69**: 1100.

Gaspin, R.M. and Valdez, R.B. 1979. Pathogenicity of *Meloidogyne* spp. and *Rotylenchulus reniformis* on sweet potato. *Ann. Trop. Res.* **1**: 20–25.

Gill, J.S. 1981. Fighting the foliar nematode of chrysanthemum. *Indian Hort.* **25**: 21–22.

Gill, J.S. and Sharma, N.K. 1976. Additional hosts of the foliar nematode, *Aphelenchoides ritzemabosi* from India. *Indian J. Nematol.* **6**: 169–71.

Gill, J.S. and Walia, R.K. 1980. Control of chrysanthemum foliar nematode, *Aphelenchoides ritzemabosi* on *Zinnia elegans. Indian J. Nematol.* **10**: 86–88.

Gupta, M.L., Janardhan, K.K. and Husaini, A. 1977. Curvularia leaf blotch—A new disease of palmarosa in India. *Indian Phytopath.* **29**: 312–13.

Gupta, M.L. and Husaini, A. 1977. *Ellesiella* leaf blight of palmarosa. *Sci. Cult.* **43**: 412.

Gurfan, S.M. and Chakravarti, B.P. 1960. *Proc. Bihar Acad. Agri. Sci.* **8 & 9**: 126–27.

Hague, N.G.M. 1972. Nematode diseases of flower bulbs, glasshouse crops and ornamentals. *In: Economic Nematology* (J.M. Webster, *ed.*), pp. 409–34. Academic Press, London.

Hara, A.H., Mau, R.F.F.L., Sata, D. and Mmandbushe, B.C. 1987. Effects of seasons and insecticides on orchid thrips injuring anthuriums in Hawaii. *Hort. Sci.* **22**: 77–79.

Haseeb, A. 1992. Nematode pests of medicinal and aromatic plants. *In "Nematode Pests of Crops"* (D. S. Bhatti and R.K. Walia, *eds.*), pp. 239–49. CBS Publishers & Distributors, Delhi, 381 pp.

Haseeb, A. and Butool, F. 1991. Evaluation of different pesticides and oilcake on the control of root-knot nematode (*Meloidogyne incognita*) infesting davana (*Artemisia pallens*). *Curr. Nematol.* **2**: 201–204.

Haseeb, A., Butool, F. and Pandey, R. 1988. Influence of different initial population densities of *Meloidogyne incognita* on the growth and oil yield of *Ocimum basilicum cv.* French. *Indian J. Mycol. Pl. Path.* **6**: 176–79.

Haseeb, A. and Pandey, R. 1987. Incidence of root-knot nematodes in medicinal and aromatic plants-new host records. *Nematropica* **17**: 209–12.

Haseeb, A. and Pandey, R. 1989a. Root-knot disease of Henbane, *Hyscyamus niger*—a new disease record. *Trop. Pest Mangmt.* **35**: 212–13.

Haseeb, A. and Pandey, R. 1989b. Observations on *Meloidogyne* spp. affecting Japanese mint—new disease records. *Nematropica* **19**: 93-97.

Haseeb, A. and Pandey, R. 1990. Root-knot nematodes—a constraint to cultivation of Davana, *Artemisia pallens*. *Trop. Pest Mangmt.* **36**: 317–19.

Hassan, A. 1988. Interaction between *Pratylenchus coffeae* and *Pythium aphanidermatum* and/or *Rhizoctonia solani* on chrysanthemum. *Zeist Phytopath.* **123**: 227–32.

Hata, T.Y. and Hara, A.H. 1992. Anthurium whitefly, *Aleurotulus anthuricola* Nakahara biology and control in Hawaii. *Trop. Pest Mangmt.* **38**: 152–54.

Hide, G.A. and Corbett, D.C.M. 1974. Field experiments in the control of *Verticillium dahliae* and *Heterodera rostochiensis* on potatoes. *Ann. Appl. Biol.* **78**: 295-307.

Ingham, R.E., Newcomb, G.B. and Morris, M.A. 1988. Spring nematicide application for control of root lesion and pin nematodes of peppermint, 1987. *Fungicide and Nematicide Tests* **43**: 170.

Inserra, S. and Calbretta, C. 1985. Attack by *Spodoptera littoralis* (Boisd.) on gerbera grown in greenhouses and results of treatment with *Bacillus thuringiensis* Berliner and control techniques. *Difesa delle Piante* **8**: 255–62.

Janardhan, K.K., Gupta, M.L. and Husaini, A. 1980. Effect of *Curvularia* leaf blotch disease on essential oil contents of palmarosa. *Indian J. Expt. Biol.* **18**: 439–40.

Jatala, P. 1988. Nematodes in tuber and root crops and their management—Improvement of sweet potato (*Ipomea batatas*) in East Africa with some references of other tuber and root crops. *Report of the Workshop on Sweet Potato Improvement in Africa.* ILRAD, Nairobi, Kenya, UNDP Project CIAT–CIP–IITA, pp. 91–100.

Jatala, P., Martin, C., Mendoza, H.A., Hidalgo, O.A. and Rincon, R.H. (*eds.*). 1990. Role of nematodes in the expression of *Pseudomonas solanacearum* and strategies for screening and breeding for combined resistance. *Advances en el Mejoramiento Genetico de la Papa en los Paises del cono sur.*, pp.187–89.

Jayaraman, V., Rajendran, G. and Muthukrishnan, T.S.1975. Occurrence of root-knot nematodes in *Polianthes tuberosa* L. in Tamil Nadu. *Indian J. Nematol.* **5**: 101–102.

Jesitor, L.J. 1976. The injurious effects of the two spotted spider mite (*Tetranychus urticae* Koch) on greenhouse carnations. *Ekologia Polska* **24**: 413–19.

Jhansi Rani, B. and Jagan Mohan, N. 1997. Pest management in ornamental crops. *In*: *Progresssive Floriculture* (I.S. Yadav and M.L. Choudhary, *eds.*), pp. 169–81. House of Sarpan, Bangalore.

Jhansi Rani, B. and Jagan Mohan, N. 1998. Cascade, a potential acaricide for management of two spotted mite on rose. *Insect Envt.* **4** (1): 12.

Joginder Singh and Sohi, A.S. 1990. *Helicoverpa* (= *Heliothis*) *armigera* (Hub.) a pest of ornamentals in Punjab, India. *23rd Int. Hort. Cong.*, Firenze, Italy, p. 541.

Johnson, N. 2000. *Studies on Nematode Pests of Cut Flowers with Special Reference to Carnation, Cystera, Gladiolus and Asiatic Lily*. Ph. D. thesis, Tamil Nadu Agri. Univ., Coimbatore, 160 pp.

Jones, F.C.E. 1961. The potato root eelworm *Heterodera rostochiensis* Woll. in India. *Curr. Sci.* **30**: 187.

Jones, F.C.E. 1969. Integrated control of the potato cyst nematode. *Proc. Fifth British Insecticide Fungicide Conf.* **3**: 646–56.

Jones, V.P. and Parella, M.P. 1986. Biological control of leafminers in greenhouse chrysanthemums. *Calif. Agri.* **40**: 10–12.

Jothi, G., Ramakrishnan, S. and Sundara Babu, R. 2004. Integrated management of *Pratylenchus delattrei* in crossandra. *National Symposium on Paradigms in Nematological Research for Biodynamic Farming,* Univ. of Agri. Sci., Bangalore, p. 97.

Karuppuchamy, N., Dandapani, N., Sundarababu, P. and Mohanasundaram, M.C. 1986. Efficacy of certain acaricides for control of carmine mite spider mite, *Tetranychus cinnabarinus* (Boisd.) on jasmine. *Madras Agric. J.* **73**: 236–39.

Kerenyi Nemestothy, K. and Tusandi, C.K. 1985. Plant protection: A dangerous gerbera pest. *Kerteszetes Szolesze* **34** (13): 6.

Khan, R.M. and Parvatha Reddy, P. 1989. *Jasminum pubescens* hitherto unrecorded host of *Radopholus similis* (Cobb, 1893) Thorne, 1949 from India. *Int. Nematol. Network Newsl.* **6**: 19–20.

Khan, R.M. and Parvatha Reddy, P. 1992. Nematode pests of ornamental crops. *In: Nematode Pests of Crops* (D.S. Bhatti and R.K. Walia, *eds.*), pp. 250-257. CBS Publishers & Distributors, Delhi, 381 pp.

Khan, R.M. and Parvatha Reddy, P. 1994. Nematode problems of ornamental crops and their management. *In: Floriculture-Technology, Trades and Trends* (J. Prakash and K.R. Bhandary, *eds.*), pp. 468–72. Oxford & IBH Publishing Co. Pvt. Ltd., New Delhi, 667 pp.

Kolodge, C., Radewald, J.D. and Shibaya, F. 1987. Revised host range and studies on life cycle of *Longidorus africanus*. *J. Nematol.* **19**: 77–81.

Koshy, P.K. and Jasy, T. 1990. Effect of the burrowing nematode, *Radopholus similis* on sweet potato (*Ipomea batatas* L.) (Abstr.). *National Symposium on Recent Advances in the Production and Utilization of Tropical Tuber Crops,* ISRC, Central Tuber Crops Research Institute, Thiruvanathapuram, India, p. 47.

Krishnaiah, K., Prasad, V.G. and Jagan Mohan, N. 1975. New records of alternate hosts of tobacco caterpillar, *Spodoptera litura* (Fab.) (Noctuidae: Leoidoptera). *Curr. Sci.* **44**: 138.

Labanowski, G. and Mynett, M. 1990. Control of the gladiolus thrips, *Taeniothrips simplex* (Morison) on gladiolus crop. *Acta Hort.* **266**: 547–51.

Lindquest, R.K., Casey, M.I. and Krueger, H.R. 1986. Evaluation of neem extract, a natural plant product, for leafminer control on chrysanthemum. *Ohio Florists' Assoc. Bull. No.* **680**: 7–8.

Machmud, 1986. *In: Bacterial Wilt Disease in Asia and the South Pacific* (G.J. Persley, *ed.*), pp. 30–34. ACIAR Proc. No. 13.

Mai, W.F., Brodie, B.B., Harrison, M.B. and Jatala, P. 1981. Nematodes. *In: Compendium of Potato Diseases* (W.J. Hooker, *ed.*), pp. 93–101. American Phytopathological Society.

Mallikarjunappa, S., Misra, R.K., Mithyantha, M.S. and Sivasankaran, K. 1991. Bio-efficacy of pesticides against the eryophid mite infesting jasmine. *Curr. Res.* **20**: 242–43.

Mohandas, C. 2001. Nematode diseases of tuber crops and their management. *Natl. Cong. on Centenary of Nematol. in India—Appraisal & Future Plans*, Indian Agri. Research Institute, New Delhi, pp. 35–36.

Mohandas, C. and Palaniswami, M.S. 1990a. Resistance in sweet potato (*Ipomea batatas* L.) to *Meloidogyne incognita* (Kofoid and White) Chitwood in India. *J. Root Crops* **16**: 148–49.

Mohandas, C. and Palaniswami, M.S. 1990b. Effect of Meloidogyne incognita on three popular cultivars of Colocasia esculenta (Abstr.). National Symposium on Recent Advances in the Production and Utilization of Tropical Tuber Crops, ISRC, Central Tuber Crops Research Institute, Thiruvanathapuram, India.

Mohandas, C., Palaniswami, M.S. and Potty, V.P. 1990. Survey, identification and pathogenicity of nematodes in tuber crops. *Annual Report of Central Tuber Crops Research Institute (1989-90)*, Thiruvanthapuram, pp. 72–73.

Mustafa, U. and Khan, M.R. 2004. Management of disease complex caused by *Fusarium oxysporum* f. sp. *gladioli* and *Meloidogyne incognita* on gladiolus. *National Symposium on Paradigms in Nematological Research for Biodynamic Farming*, Univ. of Agri. Sci., Bangalore, p. 91–92.

Muthukrishnan, T.S., Chandrasekaran, J., Lakshmanan, P.L. and Chinnarajan, A.M. 1975. Occurrence of the nematode *Hemicycliophora labiata* Colbran 1960 on roses in Tamil Nadu. *South Indian Hort.* **23**: 75–76.

Muthukrishnan, T.S., Vadivelu, S. and Rajendran, G. 1977. *Longidorus africanus* and its pathogenic effect on *Crossandra undulaefolia*. *Indian J. Nematol.* **7**: 69–72.

Nadakal, A.M. and Thomas, A.N. 1967. Observations on nematodes associated with dry rot of *Dioscorea alata* L. *Sci. Cult.* **33**: 142–43.

Nagesh, M., Hussaini, S.S., Singh, S.P. and Biswas, S.R. 2003. Management of root-knot nematode, *Meloidogyne incognita* (Kofoid & White) Chitwood in chrysanthemum using *Paecilomyces lilacinus* (Thom.) Samson in combination with neem cake. *J. Biol. Control* **17**: 125–31.

Nagesh, M. and Janakiram, T. 2004. Root-knot nematode problem in polyhouse roses and its management using dazomet, neem cake and *Pochonia chlamydosporia* (*Verticillium chlamydosporium*). *J. Ornamental Hort.* **7**: 147–52.

Nagesh, M., Janakiram, T., Rao, T.M. and Parvatha Reddy, P. 1995. Evaluation of China aster varieties/pure lines for resistance to *Meloidogyne incognita* (Kofoid & White, 1919) Chitwood, 1949. *Pest Mangmt. Hortil. Ecosystems* **1**: 101–04.

Nagesh, M., Meenakshi, S., Murthy, N. and Parvatha Reddy, P. 1995. Screening of some tuberose cultivars/hybrids for resistance against *Meloidogyne incognita* race-1. *Indian Perfumer* **39**: 138–40.

Nagesh, M. and Parvatha Reddy, P. 1995. Comparative efficacy of *Paecilomyces lilacinus* (Thom.) Samson and *Verticillium lecanii* (A. Zimmerman) Viegas in combination with botanicals against *Meloidogyne incognita* (Kofoid & White, 1919) Chitwood, 1949 infecting *Crossandra undulaefolia* L. *J. Biol. Control* **9**: 109–12.

Nagesh, M. and Parvatha Reddy, P. 1996a. Management of *Meloidogyne incognita* on carnation and gerbera in commercial polyhouses. *Second Int. Crop Sci. Cong.*, New Delhi, p. 349.

Nagesh, M. and Parvatha Reddy, P. 1996b. Eco-friendly management of *Meloidogyne incognita* on *Crossandra undulaefolia* using vesicular arbuscular mycorrhiza, *Glomus mosseae* and oil cakes. *Mycorrhiza News* **9** (1): 12–14.

Nagesh, M. and Parvatha Reddy, P. 2000. Crop loss estimation in carnation and gerbera due to the root-knot nematode, *Meloidogyne incognita* (Kofoid & White) Chitwood. *Pest Managmt. Hortil. Ecosystems* **6**: 158–59.

Nagesh, M. and Parvatha Reddy, P. 2005. Management of carnation and gerbera to control the root-knot nematode, *Meloidogyne incognita*, in commercial polyhouses. *Nematol. Medit.* **33**: 157–62.

Nagesh, M., Parvatha Reddy, P., Janakiram, T. and Rao, T.M. 1999. Sequential biochemical changes in roots of *Callistephus chinensis* lines resistant and susceptible to *Meloidogyne incognita* race 1. *Nematol. Medit.* **27**: 39–42.

Nagesh, M., Parvatha Reddy, P. and Ramachandran, N. 1998. Integrated management of *Meloidogyne incognita* and *Fusarium oxysporum* f sp. *gladioli* in gladiolus using antagonistic fungi and neem cake. *In: Nematology—Challenges & Opportunities in 21st Century* (Usha K. Mehta, *ed.*), pp. 263–66. Sugarcane Breeding Inst., Coimbatore.

Nagesh, M., Parvatha Reddy, P. and Rao, M.S. 1995. Management of *Meloidogyne incognita* on *Crossandra undulaefolia* using *Paecilomyces lilacinus* and *Pasteuria penetrans* formulations. *National Symposium on Perspectives in Eco-friendly Approaches to Plant Protection*. Indian Agri. Research Institute, New Delhi, pp. 47–48.

Nagesh, M., Parvatha Reddy, P. and Rao, M.S. 1997. Integrated management of *Meloidogyne incognita* on tuberose using *Paecilomyces lilacinus* in combination with plant extracts. *Nematol. Medit.* **25**: 3–7.

Nagesh, M., Parvatha Reddy, P. and Rao, M.S. 1998. Comparative efficacy of VAM fungi in combination with neem cake against *Meloidogyne incognita* on *Crossandra undulaefolia. Mycorrhiza News* **11**: 11–13.

Nagesh, M., Parvatha Reddy, P., Vijaya Kumar, M.V. and Nagaraju, B.M. 1998. Management of *Meloidogyne incognita* on *Polianthes tuberosa* using parasitic fungus, *Paecilomyces lilacinus* in combination with oil cakes. *In "Advances in IPM for Horticultural Crops"* (P. Parvatha Reddy, N. K. Krishna Kumar and A. Verghese, *eds.*), pp. 339–44. Association for Advancement of Pest Management in Hortil. Ecosystems, Bangalore.

Nair, V.V., Reghunath, P. and Visalakshy, A. 1991. Control of thrips, *Scirtothrips dorsalis* Hood on rose. *Entomon* **16**: 327–29.

Nath, V.R., Vaheduddin, S. and Reddy, D.P. 1958. *Sci. Cult.* **23**: 605–06.

Nelson, S.J., Venugopal, M.S., Janarthan, R. and Natarajan, S. 1993. Efficacy of certain plant products against jasmine bud worm, *Hendicasis duplifasialis* Hmp. (Pyraustidae: Lepidoptera). *Indian Perfumer* **27**: 236–39.

Nirula, K.K. 1959. Root-knot nematode on *Colocasia*. *Curr. Sci.* **28**: 125–26.

Nirula, K.K. and Bassi, K.K. 1965. Thermotherapy for root-knot nematode, *Meloidogyne incognita* in potato tubers. *Indian Potato J.* **7**: 9–11.

Nisha, M.S. and Sheela, M.S. 2004. Rhizome treatment—A low cost technology for the management of nematodes in Kacholam, an important medicinal plant. *National Symposium on Paradigms in Nematological Research for Biodynamic Farming,* Univ. of Agri. Sci., Bangalore, pp. 62–63.

Nisha, M.S and Sheela, M.S. 2006. Bio-management of *Meloidogyne incognita* on Coleus, *Solenostemon rotundifolius* by integrating solarization, *Paecilomyces lilacinus, Bacillus macerans* and neem cake. *Indian J. Nematol.* **36**: 136–38.

Nisha, M.S., Sheela, M.S. and Beena, R. 2001. Effect of mulching for the management of plant parasitic nematodes associated with kacholam, *Kaempferia galangal* L. *Proceedings of the Second National Symposium on Integrated Pest Management in Horticultural Crops: New Molecules, Biopesticides and Environment*, Bangalore, pp. 145–47.

Nucifora, A. and Calabretta, C.A. 1986. Advances in integrated control of gerbera protected crops. *Acta Hort.* **176**: 191–97.

Ohkawa, K. and Saigusa, T. 1981. Resistance of rose rootstocks to *Meloidogyne hapla, Pratylenchus penetrans* and *P. vulnus*. *Hort. Sci.* **16**: 559–60.

Padwick, G.W. 1946. Notes on Indian fungi IV. *Mycol. Pap. Imp. Mycol. Inst.* **17**, 12 pp.

Paharia, K.D. 1963. *Proc. Second All India Potato Workers Conf.*, Central Potato Research Institute, Shimla, India, pp. 198–202.

Pandaolfo, F.M. 1984. Chemical control test against the red spider (*Tetranychus urticae* Koch) on carnation in greenhouse. *Informatore Fitopatologico* **34**: 45–46.

Pandey, R. 1980. *Studies on root-knot nematode,* Meloidogyne incognita *(Kofoid and White) Chitwood, infesting Japanese mint (*Mentha arvensis L. *sub sp.* Haplocalyx briquette *var.* piperascens *Holmes) and its control.* Ph. D. thesis, Kanpur Univ., Kanpur (UP), India, 163 pp.

Pandey, R. 1990. Pathogenicity and reproduction of *Meloidogyne incognita* on *Hyoscyamnus albus* and rhizosphere association of other phytoparasitic nematodes. *Int. Nematol. Network Newsl.* **7**: 34–37.

Pandey, R. 1994a. Bionomics of phytonematodes in relation to medicinal and aromatic plants. *Indian J. Pl. Path.* **12**: 16–23.

Pandey, R. 1994b. Comparative performance of oil seed cakes and pesticides in management of root-knot nematode disease of davana. *Nematol. Medit.* **22**: 17–19.

Pandey, R. 2001. Influence of *Meloidogyne incognita, Pratylenchus thornei* and *Tylenchorhynchus vulgaris* on growth and oil yield of menthol mint. *Indian J. Nematol.* **31**: 111–14.

Pandey, R. 2002. Application of botanicals for management of root-knot nematode disease of *Ammi majus* L. *Indian J. Nematol.* **32**: 198–200.

Pandey, R. 2003. Mint nematology—Current status and future needs. *In: Advances in Nematology* (P.C. Trivedi, *ed.*), pp. 155–66. Scientific Publishers (India), Jodhpur.

Pandey, R. 2005a. Field application of bio-organics in the management of *Meloidogyne incognita* in *Mentha arvensis. Nematol. Medit.* **33**: 51–54.

Pandey, R. 2005b. Management of *Meloidogyne incognita* in *Artemisia pallens* with bioagents. *Phytoparasitica* **33**: 304–08.

Pandey, R., Gupta, M.L., Singh, H.N. and Kumar, S. 1999. The influence of vesicular arbuscular mycorrhizal fungi alone and IPM combination with *Meloidogyne incognita* on *Hyoscyamnus niger*. *Bioresource Technol.* **69**: 275–78.

Pandey, R., Haseeb, A. and Husain, A. 1992. Distribution, pathogenicity and management of *Meloidogyne incognita* on *Mentha arvensis cv.* MAS-1. *Afro-Asian J. Nematol.* **2**: 27–34.

Pandey, R. and Kalra, A. 2005a. Chemical activators: A novel and sustainable management approach for *Meloidogyne incognita* (Kofoid & White) Chitwood in *Chamomile recutita* L. *Archives of Phytopathology and Plant Protection* **38**: 107–11.

Pandey, R. and Kalra, A. 2005b. Root-knot disease of ashwagandha *(Withania somnifera)* and its eco-friendly cost effective management. *J. Mycol. Pl. Path.* **33**: 240–45.

Pandey, R., Kalra, A.,Gupta, M.L and Sharma, P. 2003. Phytonematodes: Major pest of MAPs. In *Proceedings of First National Interactive Meet on Medicinal and Aromatic Plants* (A.K. Mathur, S. Dwivedi, D.D. Patra, G.D. Bagchi, N.S. Sangwan, A. Sharma and S.P.S. Khanuja, *eds.*), pp. 188–97. CIMAP, Lucknow, India.

Pandey, R., Kalra, A., Gupta, M.L and Sharma, S.C. 2002. *Bacopa monnieri* damaged by root-knot nematode and remedial measures. *J. Trop. Medicinal Plants.*

Pandey, R., Kalra, A., Singh, H.N., Gupta, M.L., Singh, H.B. and Kumar, S. 2000. Suppressive response of bioagent, organic material and pesticide on population development of *Meloidogyne incognita* on *Matricartia chamomilla. J. Spices and Aromatic Plant Sci.* **22**: 649–51.

Pandey, R. and Patra, N.K. 2001. Screening mint (*Mentha* spp.) accessions against root-knot nematode infection. *J. Spices and Aromatic Crops* **10**: 55–56.

Pandey, R., Singh, H.B. and Gupta, M.L. 1997. Antagonistic impact of vesicular arbuscular mycorrhizae (VAM) on *Meloidogyne incognita* population development in Japanese mint. *Int. J. Trop. Pl. Dis.* **15**: 237–45.

Pandey, R., Singh, H.B., Gupta, M.L., Singh, H.N., Sharma, S. and Kumar, S. 1998. Efficacy of VAM fungi, green manure, organic materials and pesticides in the management of root-knot nematode in menthol mint. *Proc. of National Symp. on Rational Approaches in Nematode Mangmt. for Sustainable Agri.* (S.C. Dhawan and K.K. Kaushal, *eds.*), pp. 47–50. Gujarat Agri. Univ., Anand.

Pandey, R., Singh, H.B. and Kumar, S. 1999. Pathogenicity of *Meloidogyne incognita* (Kofoid & White) Chitwood on *Matricaria chamomilla* L. *J. Spices and Aromatic Crops* **8**: 201–203.

Parella, M.P., Christie, G.D. and Robb, K.L. 1983. Compatibility of insect growth regulators and *Chrysocharis parksi* (Homoptera: Eulopidae) for control of *Liriomyza trifolii* (Diptera: Agromyzidae). *J. Econ. Ent.* **76**: 949–51.

Parvatha Reddy, P. and Nagesh, M. 1995. Nematode problems of ornamental plants in India. *In: Nematode Pest Management—An Appraisal of Eco-friendly Approaches* (G. Swarup, D.R. Dasgupta and J.S. Gill, *eds.*), pp. 244–51. Nematol. Soc. of India, New Delhi.

Parvatha Reddy, P., Singh, D.B. and Rao, V.R. 1979. The root-knot nematode on Gladiolus from India. *Curr. Sci.* **48**: 82.

Patel, J.R.1980. *The Bionomics and Control Measures of the Potato Tuber Moth,* Phthorimaea (Gnorimoschema) operculella *Zeller (Gelechidae: Lepidoptera).* M. Sc. Thesis, Gujarat Agri. Univ., Anand Campus, Anand, 76 pp.

Patwardhan, P.G. 1965. *Ellisiella caudata* (P.K.) Sacc. A new record from India. *Curr. Sci.* **34**: 88–89.

Pillai, K.S. 1976. Nematicidal control of root-knot nematode on *Coleus parviflorus. J. Root Crops* **2**: 60–63.

Pirone, P.P. 1978. *Diseases and Pests of Ornamental Plants*. John Wiley & Sons, New York, 566 pp.

Pokharkar, D.S., Jadhav, D.B., Pokharkar, R.N. and Ajri, D.S. 1986. Comparative efficacy of different pesticides against two spotted red mite (*Tetranychus urticae* Koch) on rose. *Curr. Res. Reptr. Mahatma Phule Agri. Univ., Rahuri* **2**: 174–78.

Prasad, D., Singh, K.M., Singh, R.N. and Mehto, D.N. 1983. A new predator of new pest of jasmine in Delhi. *Bull. Ent.* **24**: 140–41.

Prasad, P.R.K 1978. *Studies on the Root-knot Nematodes Infecting Patchouli* (Pogostemon cablin *Bench*). M. Sc. (Agri.) thesis, Univ. of Agri. Sci., Bangalore.

Prasad, P.R.K. and Reddy, D.D.R. 1984. Pathogenicity and analysis of crop losses on patchouli due to *Meloidogyne incognita. Indian J. Nematol.* **14**: 36–38.

Prasad, K.S.K. 1989. Nematological problems and progress of research on potato in India. *Fourth Group Meeting on Nematological Problems of Plantation Crops*, Univ. Agri. Sci., Bangalore.

Prasad, K.S.K. 1990. *Helicotylenchus dihystera*, a potential pest of potato. *J. Indian Potato Assn.* **17**: 204–05.

Prasad, K.S.K., Raj, D. and Sharma, R.K. 1983. Effect of planting time on *Meloidogyne incognita* infesting potato at Shimla Hills. *Third Nematol. Symp*, Y.S Parmar Univ. Hort. And Forestry, Solan, Himachal Pradesh, pp. 5–6.

Prasad, K.S.K. and Sharma, R.K. 1985. Parasitic nematodes associated with potato and the pathogenicity of *Quinisulcius capitatus* on variety Kufri Jyoti. *J. Indian Potato Assn.* **12**: 56–62.

Prasad, S.K. and Dasgupta, D.R. 1964. Plant parasitic nematodes associated with roses in Delhi. *Indian J. Ent.* **26**: 12–122.

Prem Sagar. 1992. Efficacy of insecticides against the aphid, *Aphis gossypii* Glover on isabgol (*Plantago ovata* Forsk.) in the Punjab. *Indian J. Ent.* **54**: 399–401.

Prem Sagar and Ramji. 1991. Life history of cabbage semilooper, *Thysanoplusia orichalcea* (Fab.) on Japanese mint in the Punjab. *Indian Perfumer* **38**: 23–25.

Prem Sagar and Singh, D.P. 1991. Chemical control of the aphid, *Aphis offinis* Del (Homoptera: Aphididae), a pest of Japanese mint, *Mentha arvensis* Linn. in Punjab. *Entomon* **6**: 73–79.

Puranik, S.B. and Suryanarayana, D. 1967. A new record of *Gloeocercospora* on khas (*Vetiveria zizanoides* Stap.). *Indian Phytopath.* **19**: 394.

Raghunath, T. 1965. A new smut on lemon grass from India. *Pl. Dis. Reptr.* **49**: 916.

Raj, B.T. and Nirula, K.K. 1969. Trials on the eradication of root-knot nematodes from potato fields with DD (Abstr.). *All India Nematol. Symp.*, Indian Agri. Research Institute, New Delhi, India.

Rajani, T.S., Sheela, M.S. and Sivaprasad, P. 1998. Management of nematodes associated with kacholam, *Kaempferia galanga* L. *In: Advances in IPM of Horticultural Crops* (P. Parvatha Reddy, N.K. Krishna Kumar and A. Verghese, *eds.*), pp. 326–27. Association for Advancement of Pest Management in Horticultural Ecosystems, Bangalore.

Rajendran, B. and Rajendran, G. 1979. Report of *Meloidogyne incognita* in *Jasminum flexile. South Indian Hort.* **27**: 70.

Rajendran, G., Vadivelu, S. and Muthukrishnan, T.S. 1976. Pathogenicity of *Meloidogyne incognita* on crossandra. *Indian J. Nematol.* **6**: 115–16.

Rajagopal, D. and Kadire Gowda. 1994. Efficacy of "Limonool" (neem based pesticide) on rose aphid, *Macrosiphum rosae* Linn. (Hemiptera: Aphididae). *The Lalbaugh* **36**: 22–24.

Rai, P.S. 1997. The shining leaf chaffer – A destructive pest of rose in Karnataka. *The Lalbaugh* **39**: 46–47.

Ramana, K.V. and Mohandas, C. 1988. Occurrence of potato cyst nematode, *Globodera pallida* (Stone, 1973) in Kerala. *Indian J. Nematol.* **18**: 141.

Rashid, A. and Khan, A.M. 1975. Studies on the lesion nematode, *Pratylenchus coffeae* associated with chrysanthemum. *Indian Phytopath.* **28**: 22–28.

Ray, S., Sahoo, N.K. and Mohanty, K. 1990. Plant parasitic nematodes associated with tuber crops in Orissa (Abstr.). *National Symposium on Recent Advances in the Production and Utilization of Tropical Tuber Crops,* ISRC, Central Tuber Crops Research Institute, Thiruvanthapuram.

Reddy, D.N.R., Puttaswamy, N. and Hegde, N.S. 1981. Pests infesting *Catharanthus roseus*—a medicinal plant and their control. *The Lalbaugh* **26**: 49–51.

Regupathy, A., Vadivel, E. and Kulasekaran, M. 1979. Occurrence of leaf webber (*Pachyzancla aegrotalis* Zell.) on patchouli (*Pogostemon patchouli* Bentt) in lower pulneys (Tamil Nadu). *Indian Perfumer* **23**: 131–32.

Rhoades, H.L. 1984. Control of *Pratylenchus scribneri* on spearmint, *Mentha spicata* with nonfumigant nematicides. *Nematropica* **14**: 85–89.

Rhoades, H.L. 1988. Effect of several phytoparasitic nematodes on the growth of basil, *Ocimum basilicum*. *Ann. Appl. Nematol.* **2**: 22–24.

Rowe, R.C., Davis, J.R., Powelson, M.L. and Rouse, D.I. 1987. Potato early dying: causal agents and management strategies. *Plant Dis.* **71**: 482–89.

Sandhu, G.S. and Shukla, G.K. 1984. Chemical control of Jasminum leaf web-worm. *Pestology* **8**: 17–19.

Sangwan, N.K., Verma, K.K., Verma, B.S., Malik, M.S. and Dhindsa, K.S. 1985. Nematicidal activity of essential oils of *Symbopogon* grasses. *Nematologica* **13**: 93–99.

Santhosh Kumar, T. and Sheela, M.S. 2004. Status of nematode pests of Chethikoduveli, *Plumbago rosea* L. *National Symposium on Paradigms in Nematological Research for Biodynamic Farming,* Univ. of Agri. Sci., Bangalore, p. 28–29.

Saradamma, K. and Dale, D. 1983. Hemiptera as a pest of lemon grass. *Agric. Res. J. Kerala* **21**: 86.

Sarwar, M., Parameswaran, T.N., Shanmugam, C. and Narayan, M.R. 1982. Appraisal of Patchouli in India. Root-knot nematode problem and control. *Indian Perfumer* **26** (2-4): 117–21.

Sarwar, M. and Srinath, K.V. 1963. Occurrence of *Curvularia trifolii* (Kauf.) Boed. on khas - *Vetiveria zizanoides* Stapf. *Curr. Sci.* **34**: 668.

Sasser, J.N. 1989. *The Plant-parasitic Nematodes, The Farmers Hidden Enemy.* Dept. of Plant Path., North Carolina State Univ. & Consortium for Intnl. Crop Prot., Raleigh, North Carolina, 115 pp.

Sasser, J.N. and Freckman, D.W. 1987. A world perspective on Nematology: The role of the Society. *In: Vistas on Nematology* (J.A. Veech and D.W. Dickson, *eds.*), pp. 7–14. Society of Nematologists, Hyattsville, Maryland.

Scopes, N.E.A. 1969. The potential of *Chrysopa carnea* as a biological control agent of *Myzus persicae* on glasshouse chrysanthemums. *Ann. Appl. Biol.* **64**: 433–39.

Senthamarai, M., Poornima, K. and Subramanian, S. 2006a. Nematode-fungal disease complex involving *Meloidogyne incognita* and *Macrophomina phaseolina* on *Coleus forskohlii* Briq. *Indian J. Nematol.* **36**: 181–84.

Senthamarai, M., Poornima, K. and Subramanian, S. 2006b. Bio-management of root-knot nematode, *Meloidogyne incognita* on *Coleus forskohlii* Briq. *Indian J. Nematol.* **36**: 206–08.

Senthamarai, M., Poornima, K. and Subramanian, S. 2006c. Assessment of avoidable yield loss on *Coleus forskohlii* due to *Meloidogyne incognita*. *Indian J. Nematol.* **36**: 296–97.

Sharma, H.K. and Gill, J.S. 1992. Prevalence of races in root-knot nematode, *Meloidogyne incognita* in India. *Indian J. Nematol.* **22**: 43–49.

Sheela, M.S., Hebsybai and Anitha, N. 1996. Nematodes associated with medicinal and aromatic plants in Kerala. *Eigth Kerala Science Congress*, pp. 450–52.

Sheela, M.S., Nisha, M.S. and Mohandas, C. 2004. Eco-friendly management of nematodes associated with Chinese potato (Coleus), *Solanostemon rotundifolius* (Poir) Morton. *National Symposium on Paradigms in Nematological Research for Biodynamic Farming,* Univ. of Agri. Sci., Bangalore, p. 90.

Sheela, M.S. and Rajani, T.S. 1998. Status of phytonematodes as a part of medicinal plants in Kerala. *In: Nematology – Challenges and Opportunities in 21st Century* (Usha K. Mehta, *ed.*), pp. 2–3. Sugarcane Breeding Inst., Coimbatore.

Shetty, K.D. and Reddy, D.D.R. 1984. Pathogenicity of *Meloidogyne incognita* on *Solanum khasianum*. *Indian J. Nematol.* **14**: 185–86.

Shetty, D.K and Reddy, D.D.R. 1985. Control of root-knot nematodes on *Solanum khasianum* by chemical bare-root dips. *Indian J. Nematol.* **15**: 227–29.

Shivaramu, K. and Jhansi Rani, B. 1993. Evaluation of insecticides for the control of eriophyid mite *Aceria jasmini* (Acarina: Eriophyidae) on jasmine (Abstr.). *Golden Jubilee Symp. of Hort. Soc. India on Hort. Res.—Changing Scenario*, Bangalore, p. 262.

Shukla, P.K. and Haseeb, A. 1996. Effectiveness of some nematicides and oil cakes in the management of *Pratylenchus thornei* on *Mentha citrata, M. piperita* and *M. spicata*. *BITE* **57**: 307–09.

Shylaja, M. 2004. Studies on the development of bio-intensive management strategies of nematode induced wilt disease complex in tuberose (*Polyanthes tuberosa*) and carnations (*Dianthus caryophyllus*) using biological control agents. *National Symposium on Paradigms in Nematological Research for Biodynamic Farming,* Univ. of Agri. Sci., Bangalore, p. 136–37.

Shylaja, M., Rao, M.S. and Rahiman, A.B. 2004. *Studies on the Development of Bio-intensive Disease Management Strategies in Tuberose (*Polyanthes tuberosa *L.*) and Carnations (Dianthus caryophyllus *L.) Using Biological Control Agents*. Ph. D. thesis, Kuvempu Univ., Shimoga.

Sitaramaiah, K. and Naidu, H. 2003. Plant parasitic nematodes-management opportunities and challenges. *In: Nematode Management in Plants* (P.C. Trivedi, *ed.*), pp. 57–80. Scientific Publishers (India), Jodhpur.

Sreeja, P., Nageswari, S. and Mohandas, C. 1998. Reaction of cassava cultivars to root-knot nematode, *Meloidogyne incognita* (Kofoid & White, 1919) Chitwood, 1949. *In: Nematology: Challenges and Opportunities in 21st Century* (Usha K. Mehta, *ed.*), pp. 137-139. Sugarcane Breeding Institute, Coimbatore.

Srinivasan, A. 1974. *Studies on Root Lesion Nematode*, Pratylenchus delattrei. M. Sc. (Agri.) thesis, Tamil Nadu Agricultural University, Coimbatore.

Srinivasan, A. and Muthurksihnan, T.S. 1975. A note on nematode fungal complex in crossandra (*Crossandra undulaefolia*) in Coimbatore. *Curr. Sci.* **44**: 743–44.

Srinivasan, A. and Muthukrishnan, T.S. 1976. Control of root lesion nematode *Pratylenchus delattrei* Luc 1958 on crossandra. *Indian J. Nematol.* **6**: 111–14.

Srivastava, A.S., Verma, R.S. and Pandey, R.C. 1971. Comparative efficacy of different nematicides tested against root-knot nematode infesting *Colocasia antiquorum*. *Labdev J. Sci. Technol.* **9B**: 142–44.

Sumathi, M., Parthiban, S. and Samuel, S.D. 2006. Effect of graded doses of nitrogen and VAM on herbage, oil yield and nematode population on patchouli (*Pogostemon patchouli* Pellet). *Indian J. Nematol.* **36**: 209–13.

Sundarababu, P.C. 1969. Note on the possible use of *Bacillus thuringiensis* Berliner for the control of certain lepidopterous pests. *Madras Agric. J.* **56**: 824–26.

Sundarababu, R. 1992. Nematodes in jasmine. *Tamil Nadu Agricultural University Newsletter* **21** (11): 2.

Sundarababu, R. and Vadivelu, S. 1988a. Studies on the nematodes associated with rose. *Indian J. Nematol.* **18**: 139.

Sundarababu, R. and Vadivelu, S. 1988b. Pathogenicity of *Meloidogyne* species to tuberose (*Polyanthes tuberosa* L.). *Indian J. Nematol.* **18**: 146–48.

Thachuck, V.K., Mitrofanov, V.I. and Soboleva, L.E. 1983. Effectiveness of pyrithroid preparations for greenhouse whitefly control on gerberas. *Bull. Gosudarstvennogo Nikitskogo Botanicheskogo Sada* **51**: 82–87.

Thakur, R.N. and Husain, A. 1975. A new leaf spot disease of lemon grass. *Indian Phytopath.* **28**: 100–102.

Thangaraju, D. 1983. Distribution of potato cyst nematodes in Kodaikanal Hills, Madurai district, Tamil Nadu. *Indian J. Nematol.* **13**: 222–23.

Thirumalachar, M.J. 1951. Root-knot nematode on potato tubers in Shimla. *Curr. Sci.* **20**: 20, 104.

Thomas, J. 1987. Status paper on essential oil yielding plants with special reference to Kerala. Aromatic and Medicinal Plants Res. Stn., Odakali, Kerala.

Thorne, G. 1961. *Principles of Nematology*. McGraw-Hill, New York and London, 553 pp.

Timmerman, D. and Van De, M. 1982. Thrips infestation and control in orchids. *Vakbald voorde Bloemisterij* **37** (27): 38–39.

Tusandi, C.K. 1984. Leaf aphids on gerberas. *Kerteszet Szoleszet* **33** (24): 5.

Udayagiri, S. 1985. Insect pests of China aster—first record From India. *Indian J. Ent.* **47**: 78–82.

Upadhyaya, D.N. and Bordoloi, D.N. 1975. New records of diseases of cultivated essential oil bearing plants from North-East India. *Indian Phytopath.* **28**: 532–34.

Vadivelu, R.S. and Muthukrishnan, T.S. 1979. Control of root lesion nematode, *Pratylenchus delattrei* on crossandra. *Indian J. Nematol.* **9**: 86.

Vadivelu, S., David, B.V. and Rajamohan, N. 1976. Pests of *Crossandra infundibuliformis* in South India. *Pesticides* **10**: 39–40.

Vyas, H.N. 1994. Biology of castor semilooper, *Achea janata* L. on rose, *Rosa indica*. *Crop Res., HAU, Hisar* 7: 109–11.

Wallace, H.R. 1960. Observations on the behaviour of *Aphelenchoides ritzemabosi* in chrysanthemum leaves. *Nematologica* **5**: 315–21.

Wang, C.I. and Lin, R.T. 1984. Study on the quarantine treatments of insect pests on chrysanthemum cut flowers–the application of protection bags and the improved dipping methods. *J. Agric. Res. China* **33**: 325–30.

SUBJECT INDEX

D

T

U

V

W

X

Y